Crop Improvement

Crop Improvement

Biotechnological Advances

Edited by

PANKAJ KUMAR AND AJAY KUMAR THAKUR

CRC Press
Taylor & Francis Group
Boca Raton London New York

CRC Press is an imprint of the
Taylor & Francis Group, an **informa** business

First edition published 2021
by CRC Press
6000 Broken Sound Parkway NW, Suite 300, Boca Raton, FL 33487-2742
and by CRC Press
2 Park Square, Milton Park, Abingdon, Oxon, OX14 4RN

CRC Press is an imprint of Taylor & Francis Group, LLC

Library of Congress Cataloging-in-Publication Data

Names: Kumar, Pankaj (Biotechnology scientist), editor. | Kumar Thakur, Ajay, editor.
Title: Crop improvement : biotechnological advances / edited by Pankaj Kumar and Ajay Thakur.
Description: First edition. | Boca Raton : CRC Press, 2021. | Includes bibliographical references and index.
Identifiers: LCCN 2020058435 | ISBN 9780367567095 (hardback) | ISBN 9781003099079 (ebook)
Subjects: LCSH: Crop improvement. | Crops–Genetic engineering.
Classification: LCC SB106.I47 C74 2021 | DDC 631.5/233–dc23
LC record available at https://lccn.loc.gov/2020058435

ISBN: 9780367567095 (hbk)
ISBN: 9780367567101 (pbk)
ISBN: 9781003099079 (ebk)

Typeset in Times LT Std
by KnowledgeWorks Global Ltd.

Dedicated

To

Our Most Respected Parents

&

Ph.D. Supervisor

Contents

Preface

The development of a recombinant DNA molecule in 1972, along with the genomics revolution in biotechnology, initiated the whole genome sequencing of the first plant species, i.e., *Arabidopsis thaliana*, in 2000. In real terms, it expedited the pace of research in agricultural biotechnology. This led to the development of an array of transgenic crops conferring resistance to various biotic and abiotic stresses, quality improvement and yield enhancement. On the other hand, the congregation of information technology with biological sciences led to the development of a new branch of science called Bioinformatics/ Genoinformatics, which equipped scientists with tools and techniques to analyse large whole genome sequence data sets, comparative genomics, gene annotation, transcriptomics and metabolomics studies discovering many trait-related genes, which can further be targeted by genetic engineering to develop better crops.

Advances in understanding plant biology, novel genetic resources, genome modification and omics technologies generated new solutions for food security and strengthened the farming community to grow more food on less land using cultivation practices that are environmentally sustainable through biotechnology. Then came the era of genome editing, which involved modifying the inherent genes of the plant species to yield better crops. The advent of nanotechnology increased the efficiency of agricultural inputs and thus enhanced crop yields. Trait (genes or quantitative loci) mapping and further introgression of the desired traits into various cultivars have been made much more precise, less cumbersome and less time-consuming with the aid of various molecular markers including polymer chain reaction (PCR)-based and non-PCR-based markers and now the genome sequence–based molecular markers (single nucleotide polymorphisms). Marker-assisted breeding has now become breeders' first choice to develop cultivars with the desired trait.

In the food processing industry, microbes have been found to play a pivotal role in catalysing various kinds of bioconversions and biotransformations. Plant growth-promoting rhizobacteria sprays are being given to crops for enhanced growth. Microbes are also employed as bioremediation agents in contaminated soils or polluted water. The genetic engineering technology is being used to develop robust superbugs to clear the pollutants from water resources.

In a nutshell, the agri-biotechnology sector is going through a rapid technological change. In this book, the authors have tried to address the various latest technological interventions in the crop-biotechnology sector. We feel that this book will provide a ready reckoner to various researchers and scholars engaged in agri-biotechnology research.

Editors

Dr. Pankaj Kumar presently works as Assistant Professor, Department of Biotechnology, at Dr. Yashwant Singh Parmar University of Horticulture and Forestry, Solan, Himachal Pradesh, India. Formerly, he worked as a SERB-National Post Doctoral Fellow (DST Young Scientist Scheme) at the Council of Scientific and Industrial Research–Institute of Himalayan Bioresource Technology, Palampur, Himachal Pradesh, India. Dr. Kumar currently works on plant secondary metabolite enhancement through cell and tissue engineering approaches, identifying molecular cues linked with enhanced metabolites using a comparative transcriptomics approach. He has also worked on the hydro-aero cultivation of medicinal plants for industrial importance. During his doctoral studies, he (and his research team) generated technology at Dr. Yashwant Singh Parmar University of Horticulture and Forestry Nauni, Solan, for developing insect pest-resistant transgenic plants in economically important vegetable crops such as cauliflower cv. Pusa Snowball, cabbage cv. Pride of India and broccoli cv. Solan Green Head (with *cry-IAa* gene). He has been awarded the DST INSPIRE Junior/Senior Research Fellowship (JRF/SRF) Fellowship, Department of Science and Technology, Ministry of Science and Technology, Government of India for a Ph.D. full doctoral program. He has qualified for Indian Council of Agricultural Research, All India Competitive Examination – Senior Research Fellowship (ICAR AICE-SRF), Indian Council of Agricultural Research, Agricultural Scientists Recruitment Board (ICAR ASRB) National Eligibility Test (NET). He has published 32 research/review papers in various journals of international and national repute and contributed 13 book chapters and 5 popular articles. Dr. Kumar has received many awards from various societies and scientific organizations for his scientific contribution, i.e., the 2018 Young Scientist Award Biotechnology by the Society for Plant Research; 2019 Young Scientist Award by the Society of Tropical Agriculture, New Delhi, India and the 2019 Excellence in Research Award by the Agro Environmental Development Society of India.

Dr. Ajay Kumar Thakur presently works as a Senior Scientist (Biotechnology) at ICAR-Directorate of Rapeseed-Mustard Research, Bharatpur, Rajasthan. He obtained his graduation, postgraduate and doctoral degrees from Dr. Yashwant Singh Parmar University of Horticulture and Forestry, Solan, Himachal Pradesh, India. He was induced into Agricultural Research Services (ARS) in 2008. He has published 40 research/review papers in various journals of international and national repute, authored one book titled *Agricultural Biotechnology at a Glance* and contributed 8 book chapters and 22 popular articles. He has developed high-efficiency plant regeneration and genetic transformation protocols in a number of crops including *Populus ciliata*, *P. deltoides*, *Punica granatum*, *Capsicum annum* and *Cucumis sativus*. Dr. Thakur is associated with a *Brassica juncea* improvement programme using biotechnological interventions for the last 11 years. He has developed a core set of microsatellite markers for *B. juncea* genomics and is presently working on germplasm characterization and association mapping of various agronomically important traits in oilseed crops. He has also filed one patent associated with the development of a high-yield Indian mustard variety, Giriraj; a white rust-resistant Indian mustard genetic stock DRMR MJA 35, which is a *Moricandia* system-based cytoplasmic male sterile line of *B. juncea* and a multiple disease-resistant (*Alternaria* blight, white rust and powdery mildew) Indian mustard genetic stock, DRMRIJ 12-48. Dr. Thakur has received many awards from various societies and scientific organizations for his scientific contributions. He is also an elected Member of the Plant Tissue Culture Association of India.

Contributors

Sangita Bansal
ICAR-National Bureau of Plant Genetic Resources
New Delhi, India

Priya Bhargava
Department of Plant Pathology
CSKHPKV
Palampur, India

Pritom Biswas
Plant Tissue Culture Lab
Department of Biotechnology
Central University of South Bihar
Bihar, India

R.S. Dadarwal
Department of Agronomy
CCSHAU
Hisar, Haryana, India

Devki
Department of Microbiology
JECRC University
Jaipur, Rajasthan, India

Swati Gupta
Manipal University Jaipur
Jaipur, Rajasthan, India

Aradhana L. Hans
Department of Biotechnology
Babhasaheb Bhimrao Ambedkar University
Uttar Pradesh, India

Rohit Jain
Manipal University Jaipur
Rajasthan, India

Arunabh Joshi
Department of MBBT, RCA, MPUAT
Udaipur, Rajasthan, India

Sumita Kachhwaha
University of Rajasthan
Jaipur, Rajasthan, India

Fahad Khan
Noida Institute of Engineering and Technology
Greater Noida, Uttar Pradesh, India

Nisha Khatik
Department of Botany
Maharshi Dayanand Saraswati University
Ajmer, Rajasthan, India

Usha Kiran
Vanercia Education and Research Centre
Gautam Buddh Nagar, Uttar Pradesh, India

S.L. Kothari
Amity University Rajasthan
Jaipur, Rajasthan, India

Deepak Kumar
Department of Zoology
University of Rajasthan
Jaipur, Rajasthan, India

Naveen Kumar
ICAR-Indian Institute of Maize Research
New Delhi, India

Nitish Kumar
Plant Tissue Culture Lab
Department of Biotechnology
Central University of South Bihar
Bihar, India

Rahul Kumar
Faculty of Agricultural Sciences
DAV University
Sarmastpur, Jalandhar, India

Satish Kumar
Dr. Yashwant Singh Parmar University of Horticulture and Forestry
Nauni, Solan, Himachal Pradesh, India

V.S. Varun Kumar
School of Agriculture
Suresh Gyan Vihar University
Jaipur, Rajasthan, India

Vikas Kumar
Department of Food Science and Technology
Punjab Agricultural University
Ludhiana, Punjab, India

Dalpat Lal
Department of Agriculture
Jagannath University
Jaipur, Rajasthan, India

Era Vaidya Malhotra
ICAR-National Bureau of Plant Genetic Resources
New Delhi, India

S.R. Maloo
Pacific University
Udaipur, India

Manas Mathur
School of Agriculture
Suresh Gyan Vihar University
Jaipur, Rajasthan, India

Ritesh Mishra
Institute of Plant Sciences and Genetics in Agriculture
The Robert H. Smith Faculty of Agriculture, Food and Environment
The Hebrew University of Jerusalem
Rehovot, Israel

Deepesh Neelam
Department of Microbiology
JECRC University
Jaipur, Rajasthan, India

Neelam Prabha Negi
University Institute of Biotechnology
Chandigarh University
Mohali, Punjab, India

Taru Negi
Department of Food Science and Technology
Govind Ballabh Pant University of Agriculture and Technology
Pantnagar, Uttarakhand, India

Saurabh Pandey
National Institute of Plant Genome Research
Aruna Asaf Ali Marg
New Delhi, India

Rakesh Kumar Prajapat
School of Agriculture
Suresh Gyan Vihar University
Jaipur, Rajasthan, India

Manoj Prasad
National Institute of Plant Genome Research
Aruna Asaf Ali Marg
New Delhi, India

Rashmi
Department of Genetics and Plant Breeding
College of Agriculture
Govind Ballabh Pant University of Agriculture and Technology
Pantnagar, Uttarakhand, India

Rajni Saini
Food Technology and Nutrition
School of Agriculture
Lovely Professional University
Phagwara, Punjab, India

Sangeeta Saxena
Department of Biotechnology
Babhasaheb Bhimrao Ambedkar University
Lucknow, Uttar Pradesh, India

Aditi Sharma
Department of Plant Pathology
Dr. Yashwant Singh Parmar University of Horticulture and Forestry
Nauni, Solan, Himachal Pradesh, India

Ashutosh Sharma
Faculty of Agricultural Sciences, DAV University
Sarmastpur, Jalandhar, India

Deepak Sharma
School of Agriculture
JECRC University
Jaipur, Rajasthan, India

Indu Sharma
Department of Botany
Sant Baba Bhag Singh University
Khiala, Jalandhar, India

Manju Sharma
School of Agriculture
Suresh Gyan Vihar University
Jaipur, Rajasthan, India

Shweta Sharma
Department of Vegetable Science and Floriculture
CSKHPKV
Palampur, Himachal Pradesh, India
and
Shoolini University
Solan, Himachal Pradesh, India

Bharti Shree
Department of Agricultural Biotechnology
CSKHPKV
Palampur, Himachal Pradesh, India

Shabnam Sircaik
School of Life Sciences
Jawaharlal Nehru University
New Delhi, India

Priyanka Sood
National Institute of Plant Genome Research
Aruna Asaf Ali Marg
New Delhi, India

Jebi Sudan
Department of Plant Biotechnology
JECRC University
Jaipur, Rajasthan, India

Deepak Kumar Surolia
ICAR-CIAH
Bikaner, Rajasthan, India

Priyanka Suthar
Food Technology and Nutrition
School of Agriculture
Lovely Professional University
Phagwara, Punjab, India

Pitambri Thakur
University Institute of Biotechnology
Chandigarh University
Mohali, Punjab, India

Mahak Tufchi
Department of Biotechnology
GCW Parade
Jammu, India

Tarun Kumar Upadhyay
Parul Institute of Applied Sciences
Parul University
Vadodara, Gujarat, India

Akshay Kumar Vats
Department of Genetics and Plant Breeding
Maharishi Markandeshwar University
Haryana, India

1

Biotechnology in Agriculture: Development, Potentials and Safety

Shabnam Sircaik
Jawaharlal Nehru University

CONTENTS

1.1 Introduction

The history of improving plants for desirable characteristics through breeding is older than 10,000 years and has produced domestically useful plants and animals. However, the advancement of the 20th century has introduced more sophisticated tools to agriculture that contribute to crop development (Kumar et al. 2020). The area of biotechnology comprises the tools that employ the use of different organisms or parts of an organism to improve microorganisms, plants or animals for the benefit of humans. Particularly, agricultural biotechnology is the term generally used for crop and livestock improvement. Some significant contributions of biotechnology to agriculture are the production of the improved high-yielding crop properties, production of low-cost disease-free plant material and development of novel techniques that could be used for prevention, diagnosis and treatment of different kinds of disease associated with agriculturally important plants and animals. The studies in the past have shown that agricultural biotechnology is a powerful tool that significantly contributes to overall agriculture development. This chapter focuses on the key achievements biotechnology has made in the field of agriculture in current years and a brief introduction to the biotechnological tools available for the improvement of agriculture produce.

1.2 Plant Tissue Culture

Plant tissue culture involves an aseptic culture of plant tissue or plant parts like pollen grain, anthers or any other tissue in controlled conditions so that a whole new organism could be obtained from these cells (Thorpe 2007). These controlled conditions are proper supply of essential nutrients, pH, temperature and gaseous and liquid environment. The types of tissue culture methods are named based on the part of the explant used for culture purposes, such as anther culture, embryo culture and so forth. Micropropagation is the technology that has been used for the large-scale production of plant material and has recently gained importance in the commercial plant propagation for production of disease-free plant material and variety improvement (Brown and Thorpe 1995). It has been made possible through micropropagation that a single explant can produce thousands of plants round the year, in a small time span (Idowu, Ibitoye and Ademoyegun 2009). The technology has contributed significantly to the conservation of endangered, threatened and rare plant species. In addition to this, the tissue culture has been used to produce virus-free plant materials by culturing meristem tips on growth-inducing media (García-Gonzáles et al. 2010). One good example is the meristem culture for producing banana varieties free from viruses like brome mosaic virus (BMV) and banana bunchy top virus (BBTV) (El-Shamy 2011). Tissue culture also allows the successful production of genetically homogenous plant material, and variability in plant tissue culture has successfully allowed the production of novel useful genetically stable phenotypes. Embryo rescue is another tissue culture technique that has been used to culture immature embryos in a unique medium to prevent early abortion and support germination (Stewart 1981). A famous example of embryo rescue is the production of variety NERICA – New Rice that was a resultant of the cross between Asian and the African rice (*Oryza sativa* and *Oryza glaberrima*). It is quite pertinent to mention that plant tissue culture plays an important role on the lower end of agricultural biotechnology as the plant tissue's ability to regenerate to full plant could be used as a brick to employ advanced biotechnological approaches in agriculture.

1.3 Genetic Engineering

DNA is responsible for carrying a particular trait from one generation to the next. Genetic engineering (GE) employs the transfer of a DNA fragment of the desired trait to a host genome, generating a modified genome. This gene transformation is followed by tissue culture to obtain a whole new plant that will contain the genetic constituent from the parent cell. The technology is termed *rDNA technology* and the developed organism is named *transgenic organism*. GE for crop improvement has become a powerful tool to replace conventional plant improvement methods. These methods allow access to the novel gene groups that can be used to insert desired characters into the economically significant genetically modified (GM) crops (Datta 2012; Meli et al. 2010; Ghosh et al. 2011; Kamthan et al. 2012). Most of the commercialized GM crops have utilized microbial genes as a source for a particular trait (Kumar et al. 2018).

1.3.1 Introduction of Foreign Genes into a Plant

The gene introduction to a plant could either be accomplished by the use of a vector (indirect) or without using a vector (direct) (Potrykus 2007).

1.3.1.1 Agrobacterium-Mediated Gene Transfer

This method involves the vector-based gene transfer system using *Agrobacterium tumefaciens*. The ability of this bacterium to insert Ti plasmid DNA to the host plant cell was first detected in 1977 (Chilton et al. 1977). Following the discovery, this plasmid has been routinely used as a vector to introduce a gene to the host plant cell to develop the transgenic plant. In 1983, the first transgenic antibiotic-resistant petunia and tobacco were developed using this technology (Fraley et al. 1983; Herrera-Estrella et al. 1983). This event was followed by the successful development of other transgenic crops (potato, maize and cotton) with insect-resistant properties expressing the *Bt* gene from *Bacillus thrungenesis*. In 1994, 'Flavr Savr' GM tomato with a delayed ripening and an extended shelf life was developed by the U.S.-based company Calgene (Monsanto), which is the first transgenic crop approved for commercial cultivation. Other transgenic crops that received approval for commercialization in the United States in the initial years of the technology were bromoxynil herbicide–resistant cotton and glyphosate-resistant soybean (James 1997).

1.3.1.2 Gene Transfer Using Plant Protoplast

Other methods employed for gene transfer involve DNA uptake by plant protoplast using methods like calcium phosphate precipitation, polyethylene glycol (PEG) treatment, electroporation or a combination of all these techniques (Potrykus 2007). Although this method of gene transfer is not the extensively adopted method for transgenics, transgenic rice has been produced by delivery of free DNA to the protoplast followed by successful regeneration (Toriyama et al. 1988).

1.3.1.3 Gene Transfer Using 'Particle Gun' or High-Velocity Microprojectile System

The techniques involving direct transfer of DNA to the intact cell is always a preferred choice over the protoplast-mediated transfer as this method saves the time it takes to make protoplast. A particle gun system is one such system involving DNA transfer through the cell wall into the cytoplasm. In this method, DNA is attached to small metal particle surfaces (0.5–5 μM) followed by high-speed acceleration (hundred meters per second), injecting these particles directly into the cell (Christou 1993). The system has been successfully applied to different crops like castor (Sailaja, Tarakeswari and Sujatha 2008), tobacco (Klein et al. 1988) and soybean (McCabe et al. 1988). Besides these routine gene transfer methods, direct gene transfer also includes gene transfer to pollen, reproductive organs and microinjection into the embryos (Klöti et al. 1993; Saunders and Matthews 1997). There are certain reports supporting gene transfer using these techniques; the range of these techniques needs to be addressed. As per the records, there were about 32 transgenic crops available for commercial purposes until 2019 (ISAAA database 2019) and the number certainly promises to rise in the coming years. Figure 1.1 summarizes the major transgenic crops developed with novel resistant properties.

1.3.2 Marker-Assisted Genetic Analysis/Selection

Molecular markers are the heritable difference between two nucleotide sequences of the two homologous chromosomes of two individuals that follow a simple Mendelian pattern of inheritance. The various types of molecular markers with routine use in research are (1) first-generation or non-polymerase

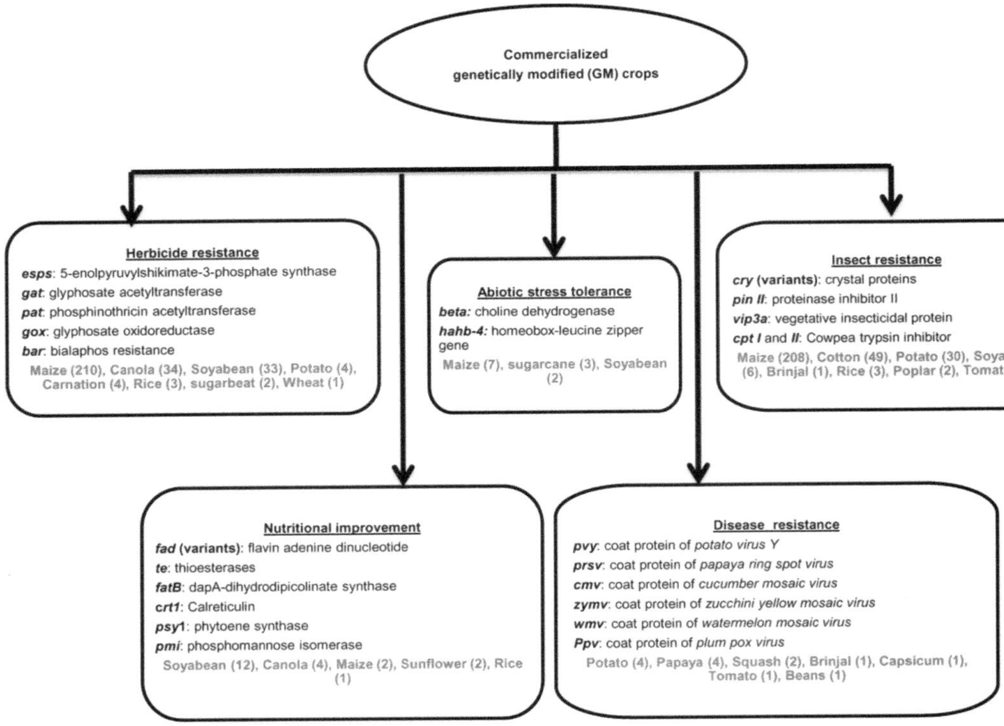

FIGURE 1.1 Diagrammatic illustration of the commercialized transgenic (GM) crops with improved genetic traits like disease and insect resistance, herbicide and abiotic stress tolerance and improved nutritional traits followed by the GM crops with the number of crops commercialized transgenic events in parenthesis.

chain reaction (PCR)-based markers, (2) second-generation or PCR-based markers, (3) new-generation molecular markers and (4) advance markers.

Depending on the need, certain modifications have been employed in these routinely used markers leading to the generation of the more advanced molecular markers. The ideal marker system should have these characteristics: (1) a polymorphic nature; (2) provide adequate information on genetic differences; (3) be simple, quick and cost-effective; (4) have linkage to different phenotypes and (5) be independent of any prior information about the genome. Based on these characteristics, PCR-based markers are the most commonly used tool in agricultural biotechnology. The PCR involves 30–40 cycles of denaturation-annealing-renaturation yielding millions of identical copies of the template DNA. These markers have significantly helped in developing different gene tags, molecular cloning of agronomically important genes, phylogenetic analysis and marker-assisted selection (MAS) of a desirable genotype. The markers have the following different uses in crop genetics:

- Identification and fingerprinting of the genotype.
- MAS.
- Successful assessment of genetic variability during species.
- Assessment of genetic distance between two species, population or different breeds.
- Sequence identification for useful candidate genes.
- Detection of quantitative trait loci (QTL).

Through MAS, biotechnology has significantly reduced the time for crop production in agriculture by reducing crop development time, as the concept of MAS has been used extensively for identifying variability between different clones across species (Song et al. 1995; Ji et al. 2016).

1.4 Molecular Diagnostics

Agricultural crops are always at high risk of damage by different kinds of plant-associated diseases and various pests, resulting in crop damage and production of low-quality fruits and vegetables. The traditional method of disease detection employs identification of the diseased crops by visual examination, limiting the treatment alternatives, and delaying the crop protection from disease. The shortcoming of traditional disease detection necessitates the development of a new method to identify an infection before it becomes visible. The advancements in biotechnology and molecular biology techniques make the early detection of disease easy by identifying a pathogen or the pathogen-associated molecules produced during infection. The routinely used diagnostic kits are based on the detection of either DNA or protein associated with the pathogen and are described herein.

1.4.1 Direct Detection

a. **DNA-based diagnostic kits:** This kind of detection employs PCR using gene-specific primers. Real-time PCR is a modified PCR method that involves the use of a fluorescent dye and is more sensitive and rapid than traditional PCR (Schaad and Frederick 2002). DNA-based detection has been successfully used to detect various diseases like Sigatoka disease of banana, *Phytophthora* disease in potato and *Fusarium* disease in cotton (Trout et al. 1997; Moricca et al. 1998; Johanson and Jeger 1993). DNA microarray is another DNA-based diagnostic tool in which an array of DNA specific to pathogens immobilized on a surface is used for detection. The sample DNA from the infected plant is amplified by PCR followed by labelling with a fluorescent dye and hybridisation to the array. The intensity of fluorescence is directly proportional to the pathogen inoculum and vice versa.
b. **Protein-based detection:** This detects either a pathogen or a host-associated protein produced during infection. Enzyme-linked immunosorbent assay (ELISA) involves antigen detection of infected plants using substrate-linked antibodies that produce colour products on successful detection (Koenig 1981). Fluorescence *in situ* hybridization (FISH) is another detection technique that has also been used to detect bacterial infections in plants (Kempf, Trebesius and Autenrieth 2000). Reports suggest the successful use of FISH to detect viral and fungal plant infections (Hijri 2009; Kliot et al. 2014). Although both DNA and protein-based systems are routinely used for disease detection, PCR-based detection systems are preferred as they are more specific, reliable and rapid.

1.4.2 Indirect Detection

Indirect detection methods are used to detect other abiotic and biotic stresses in plants (Bravo et al. 2011; Moshou et al. 2005; West et al. 2003). This kind of detection includes thermography (Oerke et al. 2006), fluorescence imaging (involving the effect on the photosynthetic apparatus) (Bürling, Hunsche and Noga 2011; Kuckenberg, Tartachnyk and Noga 2009), hyperspectral imaging (for large-scale agricultural disease detection) (Mahlein et al. 2012) and gas chromatography (Fang, Umasankar and Ramasamy 2014). In addition to these detection methods, biosensor-based pathogen detection is a novel system that was developed only recently. Biosensors have been successfully used for the detection of potato virus *Y* (PVY), cucumber mosaic virus (CMV) and tobacco rattle virus (TRV) (Eun and Wong 2000) (Perdikaris et al. 2011; Das and Chaudhuri 2015).

1.5 Vaccine Production

The procedure for vaccine development using the plant as a host was described in the 1990s (Haq et al. 1995; Mason and Arntzen 1995), where the vaccine is produced by expressing a microbial gene in the plant cell. The produced vaccine is either consumed as a plant or can be purified from the plant culture.

Biotechnology has successfully developed vaccines for diseases that do not have any pre-existing treatment alternative. The first reported production of vaccine antigen with plant use was developed in 1990 where a surface antigen A (SpaA) protein of *Streptococcus mutans* was expressed in tobacco (Curtis and Cardineau 1990). Table 1.1 summarizes some vaccines that have been produced using transgenic plants.

1.6 Omics Technology for Plant Improvement

The advancement of technology has contributed towards the use of 'omics' studies in plant sciences. Omics is a term regularly used for overall biological units like genes (genome), RNA (transcript) and proteins (proteome). Modern high-throughput technology has enabled access to this technology at a more economic level by lowering the cost. The advancement of the technology has enabled the researchers to obtain the diversity between gene families of the large group that has been used to define variation in different plant families like *Triticeae* and *Solanaceae* (Jupe et al. 2013; Steuernagel et al. 2016).

To capture the diversity of specific gene families within a large group, genomic DNA samples can be preferentially enriched prior to sequencing. This method has been used to define genetic variation in disease resistance gene repertoires and gluten gene families in bread wheat (Jouanin et al. 2019). Although robotics and other technologies are speeding acquisition of data on crop plants, the high-throughput detection technologies are used to obtain advanced data on DNA, RNA and protein/metabolites. Furthermore, the transcriptomics data can be used to identify different regulatory hubs involved in variable stress responses (Lee et al. 2015; Brooks et al. 2019). On the other hand, proteomics analysis is used to detect protein abundance and alterations in corresponding modifications necessary for different functions (Friso and van Wijk 2015). A recent good example of the role of proteomics in plant advancement is the identification of novel protein complexes and their interacting partners playing a role in different agronomic traits (McWhite et al. 2020). The identification of a particular gene with a large or small effect on plant phenotype is used to direct the manipulation of intricate complex plant characters (Huang et al. 2014). One good example of the application of omics technology for plant improvement is linked to flavour improvement in tomato, as flavour production is governed by several genetic determinants and complex interaction between sugar, acids, metabolites and other volatile aromatic compounds (Zhu et al. 2019).

1.7 Food Safety Associated with GM Crops

GM crops are already commercialized in many parts of the world. However, some consumers and environmental activists are concerned that studies are still too limited to understand the potential dangers associated with transgenic crops. This section focuses on the general concern associated with GM crops. The issues raised could be an outcome of an actual concern or could be due to a limited amount of information or misinformation available regarding the benefits of transgenics. The general issues related to food safety associated with GM crops are related to an effect on human health, environment and society (Mertens 2008; Key, Ma and Drake 2008; Lövei, Bøhn and Hilbeck 2010).

1.7.1 Health-Related Issues

1.7.1.1 Allergens and Toxins

Allergenicity is a concern raised with the consumption of GM foods. Reports have shown that around 2% of people across the world have some sort of allergies towards one or a few allergens in some sort of foods. People with this kind of food allergy are more prone to GM foods as these foods always come with a risk of introducing some sort of toxins or allergens to otherwise safe foods. An example of the allergic reaction was associated to the 'StarLink' variety of maize expressing *Cry9c*, thus banning the approval for this variety for human consumption (Bucchini and Goldman 2002). However, many studies have shown that GM foods like rice, maize and wheat have no negative effects on animal health (Tsatsakis et al. 2017;

TABLE 1.1

List of Commercially Produced Vaccines Using GM Technology

S. No.	Disease	Protein Expressed	Host Plant	Reference
Bacterial Diseases				
1.	Enterotoxigenic *Escherichia coli*	Heat labile enterotoxin B subunit (LT-B)	Tobacco	Haq et al. (1995)
			Potato	Haq et al. (1995)
			Tobacco	Kang et al. (2003)
			Tobacco	Chikwamba et al. (2003)
			Soybean	Moravec and Justine (2007)
2.	*Clostridium tetani*	TetC	Tobacco	Tregoning et al. (2003)
3.	*Staphylococcus aureus*	D2 peptide of fibronectin-binding protein FnBP	Cowpea leaf and tobacco leaf	Brennan et al. (1999)
4.	*Pseudomonas aeruginosa*	Peptides of outer membrane protein F	Cowpea leaf	Gilleland et al. (2000)
			Tobacco leaf	Staczek et al. (2000)
5.	*Mycobacterium tuberculosis*	LT-B and early secretory antigen	*Arabidopsis thaliana*	Rigano et al. (2004)
6.	*Vibrio cholera*	Cholera toxin B-subunit (CT-B)	Potato, tobacco, tomato and rice	Arakawa, Chong and Langridge (1998); Daniell et al. (2001); Jiang et al. (2007); Nochi et al. (2007
Viral Diseases				
7.	HIV type 1	V3 loop of gp120 protein	Tobacco leaf	Yusibov et al. (1997)
		Peptide of V3 loop of gp120 protein	Tobacco leaf	Joelson et al. (1997)
		Nucleocapsid protein p24	Tobacco leaf	Zhang et al. (2002)
		Tat protein	Spinach	Karasev et al. (2005)
		Rotavirus (VP6) protein	Alfalfa	Dong et al. (2005)
8.	Hepatitis B	Surface	Tobacco, potato, carrot, banana and tomato	Mason, Lam and Arntzen (1992); Richter et al. (2000); Joung et al. (2004); Kumar et al. (2006); Li et al. (2011)
9.	Pathogenic avian influenza	H5N1	Tobacco	Shoji et al. (2009)
10.	Human papilloma	Type 16 E7 oncoprotein	Tobacco leaf	Franconi et al. (2002)
11.	Rabies	Glycoprotein	Tomato leaf and fruit	MacGregor et al. (1987)
		Glycoprotein and nucleoprotein	Tobacco and spinach leaf	Yusibov et al. (2002)
Parasite Diseases				
12.	*Plasmodium yoelii*	Merozoite surface protein (PyMSP4/5)	Tobacco	Wang et al. (2009)
13.	*Entamoeba histolytica*	LecA, a surface antigen	Tobacco	Chebolu and Daniell (2007)
14.	*Toxoplasma gondii*	Surface antigen 1 (SAG1)	Tobacco	Clemente et al. (2005)
15.	*Plasmodium falciparum*	Peptides of circumsporozoite protein	Tobacco	Turpen et al. (1995)
Poultry, Swine and Cattle Diseases				
16.	Bronchitis	Viral S1 glycoprotein (IBV)	Tubers fed	Zhou et al. (2017)
17.	Bursal disease	VP2 protein	Arabidopsis, rice	Gómez et al. (2013)
18.	Newcastle disease	Recombinant hemagglutinin-neuraminidase protein	Potato, tobacco and rice	Joensuu, Niklander-Teeri and Brandle (2008)
19.	Porcine transmissible gastroenteritis corona virus (TGEV)	Envelope spike protein		Lamphear et al. (2004)
20.	Post-weaning diarrhoea	Subunit protein of ETEC F4 fimbriae	Leaves of tobacco, alfalfa and seed of barley	Kolotilin et al. (2012)
21.	Foot and- mouth disease virus (FMDV)	Capsid protein VP1 antigenic epitope		Yang et al. (2007)
22.	Herpes virus (BHV)	Protein E2 of bovine herpes virus	Alfalfa	Aguirreburualde et al. (2013)

Vos and Swanenburg 2018), thus doubting this issue. The U.S. Food and Drug Administration (FDA) is a governing body that keeps in check the food standards associated with GM foods.

1.7.1.2 Antibiotic Resistance in Human Flora

The property of antibiotic resistance has been used to select and screen the transgenic crops, helping in ensuring gene transfer. The use of these markers is always a concern as raising an antibiotic resistance in a bacterial strain standing as a major safety challenge (Heritage 2004; Keese 2008), although it has been shown that this kind of gene transfer has a very low occurrence. To address the concern, the FDA has advised using clinically important markers that would add clinical value to GM food (Tuteja and Gill 2013).

1.7.2 Environmental Concern

1.7.2.1 Gene Escape Risks and Generation of 'Superweed'

There has always been a possibility of cross-pollination between transgenic plants and weeds, increasing the chances of generating superweeds that are hard to manage. For instance, the transfer of pollen from a herbicide-resistant crop to a weed may make them herbicide resistant. Following this pattern, around 16 transgenic glyphosate-resistant weed species are reported to be produced from this kind of gene transfer (Heap 2014). *Conyza canadensis* is a glyphosate-resistant weed plant occurring globally; however, the global economic damage was caused by two species of *Amaranthus: Amaranthus palmeri* and *Amaranthus tuberculatus* (Heap and Duke 2018). The probability of this kind of gene transfer is quite low, but it is believed that the risk of escape of the herbicide resistance to the weed into the wild may create an imbalance in the environment. Two certain ways that can minimize the production of the superweeds include:

a. Rotational cultivation of the different herbicide-resistant crops (with an alternative mode of active resistance).
b. Development of novel technologies to prevent gene passage from one species to another; for instance, a strategy could be designed to target the transgene to cytoplasm and not to the pollens.

1.7.2.2 Effect on Non-Target Species

Another concern associated with transgenic crops is an adverse effect on non-target species due to high selection pressure. In 1999, the study has reported that the product from *Bt* maize may kill caterpillars of the Monarch butterfly (*Danaus plexippus*) (Clarke 2001). However, the report is still debatable as follow-up studies in real time have proven that Monarch caterpillars rarely come in contact to the pollens from *Bt* corn, and thus, it does not have any adverse effect on this species (Dively et al. 2004; Sears et al. 2001). Similar events showing a decline in the population of the monarch butterfly have been reported in the United States and Mexico (Brower et al. 2012). As the earlier study has established no particular effect on non-harmful species, studies are underway to clarify the falsehoods associated with GM crops.

1.7.2.3 Development of Resistance in Insects

The development of insecticide resistance in insects is another issue associated with the use of transgenic crops as a crop protection measure. As insects have an inherent potential to adapt to a selective pressure, the fear of the development of resistance in these species always persists. Reports have shown that some major pests have significantly developed resistance to different insecticides (Van den Berg and Campagne 2015). This kind of resistance is reported in *Busseola fusca* (African stem borer) towards *cry1Ab* (Van Rensburg 2007) and *Diabrotica virgifera* (western corn rootworm) to *cry3Bb* (Gassmann 2012) when expressed in corn. Similar pest resistance to *Bt* maize has also been reported for other insect species, such as *Pectinophora gossypiella* (pink bollworm) (Bagla 2010) and *Helicoverpa zea* (cotton bollworm). Though the abovementioned studies support the evolution of insect resistance to GM crops, this issue needs further addressing.

1.7.2.4 Loss to the Local Biodiversity

The loss of natural biodiversity is also a concern to the farmers using transgenic crops. However, biotechnological approaches have taken extensive efforts towards collecting and storing the seeds of the variable varieties of all major crops, minimizing the fear associated with this issue. Furthermore, the use of modern biotechnology not only adds to the limited knowledge to how the genes express themselves, but this has also emphasized the significance of preserving genetic material for maintaining the pool of genetically diverse crops that could be of use in the future. While transgenic crops help ensure a reliable supply of basic foodstuffs, U.S. markets for specialty crop varieties and locally grown produce appear to be expanding rather than diminishing. Thus, the use of GM crops is unlikely to negatively impact biodiversity.

1.7.3 Social Issues

1.7.3.1 Labelling

The GM crops have always been a concern to consumers demanding special labelling of GM food and products. Considering this, the United States has made it compulsory to label these GM foods differently from the conventionally grown food items.

1.7.3.2 'Terminator' Technology

Terminator technology is widely accepted by U.S.-based companies producing GM crops. This technology makes sure that farmers must buy fresh seeds every year as the seeds from the past year are not capable of producing good quality plant products. The technology has been successfully applied to different crops like peppers, tomatoes and corn. The situation is of particular concern for developing countries as the farmers are poor and are not able to purchase seeds every year. Terminator technology simply means that farmers cannot benefit from genetically improved varieties. This is a social concern because it limits the access of common/poor farmers to seed. A few multinational companies (MNCs) own 70% of the transgenic patents including Syngenta, Monsanto, Dupont, Bayer CropScience and Groupe Limagrain (Global Seeds Market Report 2018-2022; In Business wire, 2018).

1.7.3.3 Cost of Commercialization

The high cost for a safety evaluation linked to research and development is also a limiting factor in commercialization of the transgenic crops. Time taken for approving transgenic crops for human consumption is very lengthy and may last 5–7 years (Davison 2010; Miller and Bradford 2010). Furthermore, the cost of global authorization is also excessive and may reach to US$35.01 million for one crop. These issues stand as a bottleneck for small companies and restrict them from adopting transgenic technology.

1.8 Recent Advances in Plant Improvement: Beyond Transgenic

Classical transgenics have some limitations including non-acceptance of GM crops for consumption by a wide range of consumers due to fear of antibiotic resistance. Another concern is the high probability of carrying random insertion of multiple copies of the desired transgene into the host plant leading to random and unpredictable genetic expression, resulting in a difficult and tedious screening of the plant with desired characters (Ow 2005). The shortcomings necessitate an urgent need to develop alternate technology for crop improvement. Along with the conventional crop improvement technology via GE, recent advances use alternative strategies for crop improvement. This section focuses on a few techniques that have been recently used for crop improvements. The most adapted genome improvement technologies are cisgenesis, intragenesis and genome editing. These techniques involve crop improvement without involving foreign gene transfer, thus giving it wide public acceptance.

1.8.1 Recombinase Technology

Site-specific recombination (SSR) involves the precise integration of a transgene onto specific sites with an advantage of a single copy insertion. This technology is quite promising and helps to overcome the technical hurdles that arise during transgenics like the exploitation of multiple genes in a single plant genome (Halpin 2005). The conventionally used SSRs are Cre-lox from *Escherichia coli* bacteriophage P1 (Ow 2007), the short flippase recognition target (FRT) sites by the recombinase flippase (Flp) (FLP-FRT) from *Saccharomyces cerevisiae*, and pSRl recombination system (R-RS) from *Zygosaccharomyces rouxii*. Recently developed methods for plant transformation include the ΦC31-att and λ-att systems. The system ΦC31-att consists of a recombinase protein named ΦC31 (λ integrase) (Int) which catalyzes recombination between attB and attP recombination sites. The commercially developed crop by the Cre-Lox SSR system is corn variety LY038, which consists of a high lysine content developed by Monsanto (Ow 2007).

Co-transformation is another technique involving the transformation of a selectable marker gene (by *Agrobacterium*-mediated transformation) and different T-DNAs in a host parent plant together. The transfer is followed by subsequent segregation in the following generations (Komari et al. 1996; Ebinuma et al. 2001). Non-selectable transformation is yet another similar approach that employs the use of a marker-free plasmid pCAMBIA2300 (Bhatnagar-Mathur, Vadez and Sharma 2008).

1.8.2 Cisgenesis and Intragenesis

Cisgenesis involves the genetic improvement of the crop by introducing a copy of the gene (with intron), with its regulatory components in the sense orientation. The desired gene can be obtained from the host crop itself or from a nearly related/sexually compatible crop species (Schouten, Krens and Jacobsen 2006). The late blight resistance in potato is the first successful example of cisgenesis technology (Haverkort et al. 2009). This event was followed by the incorporation of cisgenesis in crop improvement for scab resistance in apple (Vanblaere et al. 2011) and barley for high phytase activity (Holme et al. 2010). Although studies have demonstrated the use of this technology in the development of different resistant varieties, it is still surprising to know that no cisgenic crop product is commercialized to date.

Intragenesis also involves a similar gene transfer from the same special or sexually compatible species or in an antisense direction with the regulatory elements (e.g., promoter or terminator) from other species (Rommens et al. 2007). The benefit of this technology is that this combination could result in novel and useful gene products. The technique has been employed to reduce the acrylamide levels in potatoes by using the tuber-specific asparagine synthase-1 (*StAst1*) gene (Chawla, Shakya and Rommens 2012). In 2015, J.R. Simplot Company has been credited with the development of the commercial potato variety with multiple developed traits like resistance to discoloration and low acrylamide levels. This was achieved by combined silencing of multiple genes, i.e., *asn1* (asparagine synthetase 1; for low asparagine formation), *ppo5* (polyphenol oxidase 5; reducing black spots), *R1* (starch-related R1 protein) and *PhL* (α-glucan phosphorylase) for reducing acrylamide by lowering sugar formation (Waltz 2015).

1.8.3 Genome Editing

Genome editing is the latest technology that has contributed significantly to plant improvement in recent years. This has been used to stably delete/mutate or replace a specific gene/genetic element from the genome and stands as a promising approach for generating novel crop varieties with controlled chances of genetic variations getting into the produced crops. The technology utilizes the targeted DNA double-strand breaks (DSBs) at a genomic locus with a consequent repair either by homologous or non-homologous DNA repair mechanisms, thereby inserting targeted sequence changes in plant genome (Wyman and

Kanaar 2006). The sequence-specific nuclease (SSN) recognizes and cleaves the target DNA sequences with high specificity. The main class of SSNs used for genetic modifications is named meganucleases, zinc-finger nucleases (ZFN) (Kim, Cha and Chandrasegaran 1996), transcription activator-like effector nucleases (TALENs) (Christian et al. 2010) and clustered regularly interspaced short palindromic repeats (CRISPRs)-Cas9. A comparative view of these different genome editing tools is summarized in Table 1.2.

1.8.3.1 Zinc-Finger Nucleases

These nucleases are zinc-finger proteins. Usually, transcription factors are fused to an endonuclease like Fok1 (Bibikova et al. 2003; Carroll 2011). These zinc-finger domains particularly identify DNA sequences that are trinucleotides. Likewise, a series of the zinc-finger domains recognizes longer DNA sequences that provide more sequence specificity. The endonuclease Fok1 endonuclease works as a dimer binding to the two opposite sides of DNA increasing the specificity of the ZFN. These ZFNs are precisely engineered to recognize closely related yet different DNA sequences on the target site limiting the off-target insertions.

1.8.3.2 Transcription Activator-Like Effector Nucleases

Similar to ZFN, TALENs are also fusion proteins comprising bacterial TALE proteins and Fok1 endonuclease (Bogdanove and Voytas 2011; Christian et al. 2010). These are quite similar to ZFN as these motifs are also linked to the Fok1 endonuclease requiring dimerization for successful cleavage of target DNA. A TALE motif identifies one nucleotide and an array of TALEs associates with a much longer nucleotide sequence (Table 1.2). The engineering of TALENs is much easier than ZFN as the activity of each TALEN is certainly limited to one nucleotide and does not affect the specificity of the adjoining TALE.

1.8.3.3 Clustered Regularly Interspaced Short Palindromic Repeats

This system naturally occurs as a bacterial immune system to combat foreign DNA that is engineered further so that it can be used for GE purposes. This consists of a nuclease named Cas9 and two RNAs, i.e. trans-activating crRNA (tracrRNA) and a single-guide RNA (sgRNA). This RNA pair recognizes the target sequence by simple Watson-Crick base pairing, followed by specific DNA motifs termed as the *protospacer adjacent motif* (PAM). Every Cas9 protein has a specific PAM sequence (Belhaj et al. 2015). For instance, for standard Cas9 the sequence is 5′-NGG-3′. The recognition is followed by Cas9-mediated DNA cleavage leading to a DSB in target DNA.

TABLE 1.2

A Comparative Analysis of the Different Gene Editing Tools Available for Crop Improvement

Feature	ZFNs	TALENs	CRISPR-Cas9
	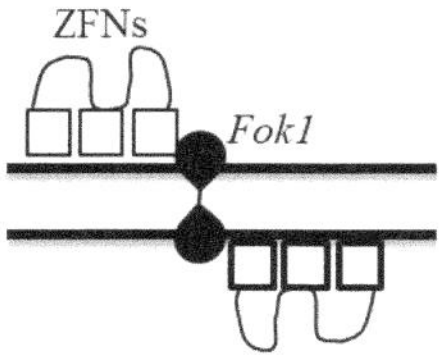	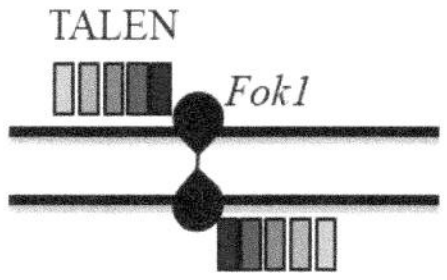	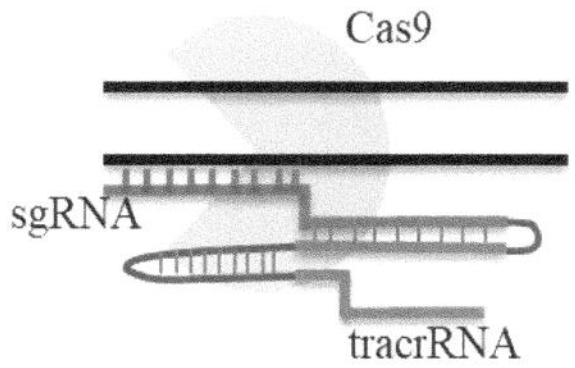
Size of target DNA	9–18 bp	30–40 bp	22 bp + protospacer adjacent motif (PAM) sequence
Mechanism of target DNA recognition	DNA–protein interaction	DNA–protein interaction	DNA–RNA interaction
Mechanism of DNA cleavage and repair	Fok1-induced double-strand break	Fok1-induced double-strand break	Cas9-induced single- or double-strand break
Design	Difficult and challenging.	Simple and easy	Simple and easy

TABLE 1.3
Plant Varieties Improved Through Gene-Editing Tools

S. No	Crop	Reference
Yield Performance		
1.	Rapeseed, Canola	Zheng et al. (2020); Yang et al. (2018)
2.	Rice	Zhou et al. (2019); Huang et al. (2018)
Disease Resistance		
3.	Tomato	Nekrasov et al. (2017)
4.	Wheat	Wang et al. (2014)
5.	Cocoa	Fister et al. (2018)
6.	Rice	Li et al. (2012); Oliva et al. (2019)
7.	Orange	Peng et al. (2017)
8.	Cucumber	Chandrasekaran et al. (2016)
9.	Cassava	Gomez et al. (2017)
Enhanced Food Quality		
10.	Gluten-free wheat	Sánchez-León et al. (2018)
11.	Apple	Dubois et al. (2015)
12.	Peanut	Dodo et al. (2008)
13.	Rice	Herman et al. (2003)
Enhanced Nutritional Profile		
14.	Lettuce	Zhang et al. (2018)
15.	Sorghum	Li et al. (2018)
16.	Soybean	Haun et al. (2014)

In this system recognition of the target DNA sequence is mediated by DNA-RNA interactions. The CRISPR-Cas9 system has various advantages over ZFN and TALEN such as it is easy to design and has the benefit of multiplexing (possibility to modify several DNA targets at the same time). The CRISPR system is the most used technique for genome editing and has revolutionized the gene-editing field; it is known as 'the biggest biotechnology discovery of the century' (Doudna and Charpentier 2014). All these gene editing tools can be used to improve different plant properties. Table 1.3 summarizes the plant properties that have been improved by applications of these tools.

1.9 Future Prospects and Conclusion

The increase in world population and ongoing climate change has initiated a need for developing crop varieties with high-yielding properties that are biotic and abiotic stress tolerant and are nutritionally rich. The introduction of GM technology has significantly increased the world's food productivity and has become the most adopted agricultural technology of the 20th century. This is supported by a study reporting a significant increase in the farmer's profit with a 22% increase in crop yield after applying transgenic technology (Klümper and Qaim 2014). Furthermore, it is also reported that the global area under transgenic crop production in 2018 was 191.7 million hectares with a 113-fold increase when compared with the production in 1996 of 1.7 million hectare (ISAAA 2018). The average area occupied by different transgenic crops worldwide in 2018 is given in Table 1.4. It clearly can be seen that the biggest area under GM production is occupied by crops with herbicide tolerance, i.e., 88.7 million hectares (47%), followed by crops with stacked traits and insect resistance with 41% and 14% area cover, respectively. According to the International Service for the Acquisition of Agri-biotech Applications (ISAAA), around 17 million farmers in 26 countries have positively adopted transgenic crops successfully, contributing to an increase in the global market with a value worth US$18.2 billion in 2018 (ISAAA 2018). It is evident from the reports that biotechnology has the potential to address the food scarcity problems worldwide. The employment of GM strategies in agriculture has significantly

TABLE 1.4
Percentage of Area Occupied by GM Crops Worldwide

GM Crop	Area Under Production (in Million Hectares)	Percentage of the Total Area Under Transgenic Crop Production
Soybean	95.9	50
Maize	58.9	31
Cotton	24.9	13
Canola	10.1	5.3
Other	1.9	<1

Source: https://www.isaaa.org.

minimized the crop losses due to different kinds of stresses and added to the value of the food crop via an increase in nutritional components like proteins, vitamins and so forth. These strategies also contributed to the longer shelf-life of perishable crops, thus reducing the post-harvest losses. Production of plant-based vaccines also contributed to the medicine industry. The global area producing GM crops is increasing every year, and it has not displayed any harmful effect on the environment with significantly increased consumption by humans. Furthermore, insect- and herbicide-resistant GM crops have significantly benefited farmers by minimizing the use of chemical insecticides and pesticides for crop management, increasing the overall crop yield. All these improved traits have proved the position of biotechnology in the agricultural sector. Besides this, improved species for the improved traits are also underway and are under extensive research in different parts of the world. It is pertinent to mention that the integration of traditional agricultural methods with modern biotechnology will be beneficial for achieving food security for generations to come. However, it is also necessary that the GM crops should undergo thorough safety assessment before being used for commercial cultivation and making it available to the common population for consumption. It is also pertinent to mention that improvement of agriculture through biotechnology has become an essential part of our lives, and this technology could be exploited for the future benefits of mankind.

REFERENCES

Aguirreburualde, María Sol Pérez, María Cristina Gómez, Agustín Ostachuk, Federico Wolman, Guillermo Albanesi, Andrea Pecora, Anselmo Odeon, Fernando Ardila, José M. Escribano, and María José Dus Santos. 2013. "Efficacy of a BVDV Subunit Vaccine Produced in Alfalfa Transgenic Plants." *Veterinary Immunology and Immunopathology* 151 (3–4): 315–324.

Arakawa, Takeshi, Daniel K.X. Chong, and William H.R. Langridge. 1998. "Efficacy of a Food Plant-Based Oral Cholera Toxin B Subunit Vaccine." *Nature Biotechnology* 16 (3): 292–297.

Bagla, Pallava. 2010. Hardy Cotton-Munching Pests Are Latest Blow to GM Crops. *Science*. 327 (5972): 1439.

Belhaj, Khaoula, Angela Chaparro-Garcia, Sophien Kamoun, Nicola J. Patron, and Vladimir Nekrasov. 2015. "Editing Plant Genomes with CRISPR/Cas9." *Current Opinion in Biotechnology* 32: 76–84.

Bhatnagar-Mathur, Pooja, V. Vadez, and Kiran K. Sharma. 2008. "Transgenic Approaches for Abiotic Stress Tolerance in Plants: Retrospect and Prospects." *Plant Cell Reports* 27 (3): 411–424.

Bibikova, Marina, Kelly Beumer, Jonathan K. Trautman, and Dana Carroll. 2003. "Enhancing Gene Targeting with Designed Zinc Finger Nucleases." *Science* 300 (5620): 764–764.

Bogdanove, Adam J., and Daniel F. Voytas. 2011. "TAL Effectors: Customizable Proteins for DNA Targeting." *Science* 333 (6051): 1843–1846.

Bravo, Javier A., Paul Forsythe, Marianne V. Chew, Emily Escaravage, Hélène M. Savignac, Timothy G. Dinan, John Bienenstock, and John F. Cryan. 2011. "Ingestion of Lactobacillus Strain Regulates Emotional Behavior and Central GABA Receptor Expression in a Mouse via the Vagus Nerve." *Proceedings of the National Academy of Sciences* 108 (38): 16050–16055.

Brennan, F. R., T. D. Jones, M. Longstaff, S. Chapman, T. Bellaby, H. Smith, F. Xu, W. D. O. Hamilton, and J.-I. Flock. 1999. "Immunogenicity of Peptides Derived from a Fibronectin-Binding Protein of S. Aureus Expressed on Two Different Plant Viruses." *Vaccine* 17 (15–16): 1846–1857.

Brooks, Matthew D., Jacopo Cirrone, Angelo V. Pasquino, Jose M. Alvarez, Joseph Swift, Shipra Mittal, Che-Lun Juang, Kranthi Varala, Rodrigo A. Gutiérrez, and Gabriel Krouk. 2019. "Network Walking Charts Transcriptional Dynamics of Nitrogen Signaling by Integrating Validated and Predicted Genome-Wide Interactions." *Nature Communications* 10 (1): 1–13.

Brower, Lincoln P., Orley R. Taylor, Ernest H. Williams, Daniel A. Slayback, Raul R. Zubieta, and M. Isabel Ramirez. 2012. "Decline of Monarch Butterflies Overwintering in Mexico: Is the Migratory Phenomenon at Risk?" *Insect Conservation and Diversity* 5 (2): 95–100.

Brown, D. C. W., and T. A. Thorpe. 1995. "Crop Improvement through Tissue Culture." *World Journal of Microbiology and Biotechnology* 11 (4): 409–415.

Bucchini, Luca and Lynn R. Goldman. 2002. "Starlink Corn: A Risk Analysis." *Environmental Health Perspectives* 110 (1): 5–13.

Bürling, Kathrin, Mauricio Hunsche, and Georg Noga. 2011. "Use of Blue-Green and Chlorophyll Fluorescence Measurements for Differentiation between Nitrogen Deficiency and Pathogen Infection in Winter Wheat." *Journal of Plant Physiology* 168 (14): 1641–1648. https://doi.org/10.1016/j.jplph.2011.03.016.

Carroll, Dana. 2011. "Genome Engineering with Zinc-Finger Nucleases." *Genetics* 188 (4): 773–782.

Chandrasekaran, Jeyabharathy, Marina Brumin, Dalia Wolf, Diana Leibman, Chen Klap, Mali Pearlsman, Amir Sherman, Tzahi Arazi, and Amit Gal-On. 2016. "Development of Broad Virus Resistance in Non-Transgenic Cucumber Using CRISPR/Cas9 Technology." *Molecular Plant Pathology* 17 (7): 1140–1153.

Chawla, Rekha, Roshani Shakya, and Caius M. Rommens. 2012. "Tuber-Specific Silencing of Asparagine Synthetase-1 Reduces the Acrylamide-Forming Potential of Potatoes Grown in the Field without Affecting Tuber Shape and Yield." *Plant Biotechnology Journal* 10 (8): 913–924.

Chebolu, Seethamahalakshmi and Henry Daniell. 2007. "Stable Expression of Gal/GalNAc Lectin of *Entamoeba histolytica* in Transgenic Chloroplasts and Immunogenicity in Mice towards Vaccine Development for Amoebiasis." *Plant Biotechnology Journal* 5 (2): 230–239. https://doi.org/10.1111/j.1467-7652.2006.00234.x.

Chikwamba, Rachel K., M. Paul Scott, Lorena B. Mejía, Hugh S. Mason, and Kan Wang. 2003. "Localization of a Bacterial Protein in Starch Granules of Transgenic Maize Kernels." *Proceedings of the National Academy of Sciences* 100 (19): 11127–11132. https://doi.org/10.1073/pnas.1836901100.

Chilton, Mary-Dell, Martin H. Drummond, Donald J. Merlo, Daniela Sciaky, Alice L. Montoya, Milton P. Gordon, and Eugene W. Nester. 1977. "Stable Incorporation of Plasmid DNA into Higher Plant Cells: The Molecular Basis of Crown Gall Tumorigenesis." *Cell* 11 (2): 263–271.

Christian, Michelle, Tomas Cermak, Erin L. Doyle, Clarice Schmidt, Feng Zhang, Aaron Hummel, Adam J. Bogdanove, and Daniel F. Voytas. 2010. "Targeting DNA Double-Strand Breaks with TAL Effector Nucleases." *Genetics* 186 (2): 757–761.

Christou, Paul. 1993. "Particle Gun Mediated Transformation." *Current Opinion in Biotechnology* 4 (2): 135–141.

Clarke, Tom. 2001. "Monarchs Safe from Bt." *Nature*, September, news010913-12. https://doi.org/10.1038/news010913-12.

Clemente, Marina, Roberto Curilovic, Alina Sassone, Alicia Zelada, Sergio O. Angel, and Alejandro N. Mentaberry. 2005. "Production of the Main Surface Antigen of Toxoplasma Gondii in Tobacco Leaves and Analysis of Its Antigenicity and Immunogenicity." *Molecular Biotechnology* 30 (1): 41–49.

Curtis, R. I., and C. A. Cardineau. 1990. "Oral Immunization by Transgenic Plants." World Patent Application, WO 90/02484." Patent Record Available from the World Intellectual Property Organization (WIPO), Geneva, Switzerland.

Daniell, Henry, Seung-Bum Lee, Tanvi Panchal, and Peter O. Wiebe. 2001. "Expression of the Native Cholera Toxin B Subunit Gene and Assembly as Functional Oligomers in Transgenic Tobacco Chloroplasts." *Journal of Molecular Biology* 311 (5): 1001–1009.

Das, Naren and Chirasree Roy Chaudhuri. 2015. "Reliability Study of Nanoporous Silicon Oxide Impedance Biosensor for Virus Detection: Influence of Surface Roughness." *IEEE Transactions on Device and Materials Reliability* 15 (3): 402–409.

Datta, Asis. 2012. "GM Crops: Dream to Bring Science to Society." *Agricultural Research* 1 (2): 95–99.

Davison, John. 2010. "GM Plants: Science, Politics and EC Regulations." *Plant Science* 178 (2): 94–98. https://doi.org/10.1016/j.plantsci.2009.12.005.

Dively, Galen P., Robyn Rose, Mark K. Sears, Richard L. Hellmich, Diane E. Stanley-Horn, Dennis D. Calvin, Joseph M. Russo, and Patricia L. Anderson. 2004. "Effects on Monarch Butterfly Larvae (*Lepidoptera*:

Danaidae) after Continuous Exposure to Cry1Ab-Expressing Corn during Anthesis." *Environmental Entomology* 33 (4): 1116–1125.

Dodo, Hortense W., Koffi N. Konan, Fur C. Chen, Marceline Egnin, and Olga M. Viquez. 2008. "Alleviating Peanut Allergy Using Genetic Engineering: The Silencing of the Immunodominant Allergen Ara h 2 Leads to Its Significant Reduction and a Decrease in Peanut Allergenicity." *Plant Biotechnology Journal* 6 (2): 135–145.

Dong, Jiang-Li, Ben-Guo Liang, Yong-Sheng Jin, Wan-Jun Zhang, and Tao Wang. 2005. "Oral Immunization with PBsVP6-Transgenic Alfalfa Protects Mice against Rotavirus Infection." *Virology* 339 (2): 153–163.

Doudna, Jennifer A. and Emmanuelle Charpentier. 2014. "The New Frontier of Genome Engineering with CRISPR-Cas9." *Science* 346 (6213).

Dubois, Anthony E. J., Giulia Pagliarani, Rixt M. Brouwer, Boudewijn J. Kollen, Lars O. Dragsted, Folmer Damsted Eriksen, Ole Callesen, Luud J. W. J Gilissen, Frans A. Krens, and Richard G. F. Visser. 2015. "First Successful Reduction of Clinical Allergenicity of Food by Genetic Modification: Mal d 1-Silenced Apples Cause Fewer Allergy Symptoms than the Wild-Type Cultivar." *Allergy* 70 (11): 1406–1412.

Ebinuma, H., K. Sugita, Endo Matsunaga, S. Endo, K. Yamada, and A. Komamine. 2001. "Systems for the Removal of a Selection Marker and Their Combination with a Positive Marker." *Plant Cell Reports* 20 (5): 383–392.

El-Shamy, M. M. 2011. "Management of Viral Disease in Banana Using Certified and Virus Tested Plant Material." *African Journal of Microbiology Research* 5 (32): 5923–5932.

Eun, A. J. and S. M. Wong. 2000. "Molecular Beacons: A New Approach to Plant Virus Detection." *Phytopathology* 90 (3): 269–275. https://doi.org/10.1094/PHYTO.2000.90.3.269.

Fang, Yi, Yogeswaran Umasankar, and Ramaraja P. Ramasamy. 2014. "Electrochemical Detection of P-Ethylguaiacol, a Fungi Infected Fruit Volatile Using Metal Oxide Nanoparticles." *The Analyst* 139 (15): 3804–3810. https://doi.org/10.1039/c4an00384e.

Fister, Andrew S., Lena Landherr, Siela N. Maximova, and Mark J. Guiltinan. 2018. "Transient Expression of CRISPR/Cas9 Machinery Targeting TcNPR3 Enhances Defense Response in *Theobroma Cacao*." *Frontiers in Plant Science* 9: 268.

Fraley, Robert T., Stephen G. Rogers, Robert B. Horsch, Patricia R. Sanders, Jeffery S. Flick, Steven P. Adams, Michael L. Bittner, Leslie A. Brand, Cynthia L. Fink, and Joyce S. Fry. 1983. "Expression of Bacterial Genes in Plant Cells." *Proceedings of the National Academy of Sciences* 80 (15): 4803–4807.

Franconi, Rosella, Paola Di Bonito, Francesco Dibello, Luisa Accardi, Antonio Muller, Alessia Cirilli, Paola Simeone, M. Gabriella Donà, Aldo Venuti, and Colomba Giorgi. 2002. "Plant-Derived Human Papillomavirus 16 E7 Oncoprotein Induces Immune Response and Specific Tumor Protection." *Cancer Research* 62 (13): 3654–3658.

Friso, Giulia, and Klaas J. van Wijk. 2015. "Posttranslational Protein Modifications in Plant Metabolism." *Plant Physiology* 169 (3): 1469–1487.

García-Gonzáles, Rolando, Karla Quiroz, Basilio Carrasco, and Peter Caligari. 2010. "Plant Tissue Culture: Current Status, Opportunities and Challenges." *International Journal of Agriculture and Natural Resources* 37 (3): 5–30.

Gassmann, Aaron J. 2012. "Field-Evolved Resistance to Bt Maize by Western Corn Rootworm: Predictions from the Laboratory and Effects in the Field." *Journal of Invertebrate Pathology* 110 (3): 287–293.

Ghosh, Sumit, Vijaykumar S. Meli, Anil Kumar, Archana Thakur, Niranjan Chakraborty, Subhra Chakraborty, and Asis Datta. 2011. "The N-Glycan Processing Enzymes α-Mannosidase and β-D-N-Acetylhexosaminidase Are Involved in Ripening-Associated Softening in the Non-Climacteric Fruits of Capsicum." *Journal of Experimental Botany* 62 (2): 571–582.

Gilleland Jr, Harry E., Linda B. Gilleland, John Staczek, Ronald N. Harty, Adolfo García-Sastre, Peter Palese, Frank R. Brennan, William DO Hamilton, Mohammed Bendahmane, and Roger N. Beachy. 2000. "Chimeric Animal and Plant Viruses Expressing Epitopes of Outer Membrane Protein F as a Combined Vaccine against *Pseudomonas aeruginosa* Lung Infection." *FEMS Immunology & Medical Microbiology* 27 (4): 291–297.

"Global Seeds Market Report 2018-2022 Featuring Bayer, DowDuPont, Groupe Limagrain, KWS, Land O' Lakes, Monsanto & Syngenta – ResearchAndMarkets.Com." In Buisness Wire. April 27, 2018. https://www.businesswire.com/news/home/20180427005463/en/Global-Seeds-Market-Report-2018-2022-Featuring-Bayer-DowDuPont-Groupe-Limagrain-KWS-Land-O-Lakes-Monsanto-Syngenta---ResearchAndMarkets.com.

Gómez, Evangelina, María Soledad Lucero, Silvina Chimeno Zoth, Juan Manuel Carballeda, María José Gravisaco, and Analía Berinstein. 2013. "Transient Expression of VP2 in *Nicotiana benthamiana* and Its Use as a Plant-Based Vaccine against Infectious Bursal Disease Virus." *Vaccine* 31 (23): 2623–2627. https://doi.org/10.1016/j.vaccine.2013.03.064.

Gomez, Michael A., Z. Daniel Lin, Theodore Moll, Collin Luebbert, Raj Deepika Chauhan, Anupama Vijayaraghavan, Renninger Kelley, Getu Beyene, Nigel J. Taylor, and J. Carrington. 2017. "Simultaneous CRISPR/Cas9-Mediated Editing of Cassava EIF4E Isoforms NCBP-1 and NCBP-2 Confers Elevated Resistance to Cassava Brown Streak Disease." *BioRxiv*, 209874.

Halpin, Claire. 2005. "Gene Stacking in Transgenic Plants–the Challenge for 21st Century Plant Biotechnology." *Plant Biotechnology Journal* 3 (2): 141–155.

Haq, Tariq A., Hugh S. Mason, John D. Clements, and Charles J. Arntzen. 1995. "Oral Immunization with a Recombinant Bacterial Antigen Produced in Transgenic Plants." *Science* 268 (5211): 714–716.

Haun, William, Andrew Coffman, Benjamin M. Clasen, Zachary L. Demorest, Anita Lowy, Erin Ray, Adam Retterath, Thomas Stoddard, Alexandre Juillerat, and Frederic Cedrone. 2014. "Improved Soybean Oil Quality by Targeted Mutagenesis of the Fatty Acid Desaturase 2 Gene Family." *Plant Biotechnology Journal* 12 (7): 934–940.

Haverkort, A. J., P. C. Struik, R. G. F. Visser, and E. J. P. R. Jacobsen. 2009. "Applied Biotechnology to Combat Late Blight in Potato Caused by *Phytophthora infestans*." *Potato Research* 52 (3): 249–264.

Heap, Ian. 2014. "Global Perspective of Herbicide-Resistant Weeds." *Pest Management Science* 70 (9): 1306–1315.

Heap, Ian and Stephen O. Duke. 2018. "Overview of Glyphosate-Resistant Weeds Worldwide." *Pest Management Science* 74 (5): 1040–1049.

Heritage, John. 2004. "The Fate of Transgenes in the Human Gut." *Nature Biotechnology* 22 (2): 170–172.

Herman, Eliot M., Ricki M. Helm, Rudolf Jung, and Anthony J. Kinney. 2003. "Genetic Modification Removes an Immunodominant Allergen from Soybean." *Plant Physiology* 132 (1): 36–43.

Herrera-Estrella, Luis, Marc De Block, E. H. J. P. Messens, J.-P. Hernalsteens, Marc Van Montagu, and Jeff Schell. 1983. "Chimeric Genes as Dominant Selectable Markers in Plant Cells." *The EMBO Journal* 2 (6): 987–995.

Hijri, Mohamed. 2009. "The Use of Fluorescent *In Situ* Hybridisation in Plant Fungal Identification and Genotyping." In *Plant Pathology: Techniques and Protocols*, edited by Robert Burns, 131–45. Methods in Molecular Biology. Totowa, NJ: Humana Press. https://doi.org/10.1007/978-1-59745-062-1_11.

Holme, Inger, Giuseppe Dionisio, Henrik Brinch-Pedersen, Éva Vincze, and Preben BachHolm. 2010. "Cisgenic Barley with Improved Phytase Activity." *In Vitro Cellular and Developmental Biology-Animal* 46 (Supplement 1): 187.

Huang, Hao, Elton Zeqiraj, Beihua Dong, Babal Kant Jha, Nicole M. Duffy, Stephen Orlicky, Neroshan Thevakumaran, et al. 2014. "Dimeric Structure of Pseudokinase RNase L Bound to 2-5A Reveals a Basis for Interferon-Induced Antiviral Activity." *Molecular Cell* 53 (2): 221–234. https://doi.org/10.1016/j.molcel.2013.12.025.

Huang, Liyu, Ru Zhang, Guangfu Huang, Yanxia Li, Getachew Melaku, Shilai Zhang, Haitao Chen, Yanjuan Zhao, Jing Zhang, and Yesheng Zhang. 2018. "Developing Superior Alleles of Yield Genes in Rice by Artificial Mutagenesis Using the CRISPR/Cas9 System." *The Crop Journal* 6 (5): 475–481.

Idowu, P. E., D. O. Ibitoye, and O. T. Ademoyegun. 2009. "Tissue Culture as a Plant Production Technique for Horticultural Crops." *African Journal of Biotechnology* 8 (16).

ISAAA (International Service for the Acquisition of Agri-biotech Applications). 2017. *Global Status of Commercialized Biotech/GM Crops in 2017: Biotech Crop Adoption Surges as Economic Benefits Accumulate in 22 Years. ISAAA Brief No. 53.* ISAAA: Ithaca, NY. https://www.isaaa.org/resources/publications/briefs/53/

ISAA (International Service for the Acquisition of Agri-biotech Applications database). 2019. *GM Approval Database retrieved on 17 Nov 2019.* https://www.isaaa.org/gmapprovaldatabase/

James, Clive. 1997. *Global Status of Transgenic Crops in 1997. ISAAA Brief 5-1977*, pp. 331. ISAAA: Ithaca, NY. https://www.isaaa.org/purchasepublications/itemdescription.asp?ItemType=BRIEF S&Control=IB005-1997

Ji, Hyeonso, Sung-Ryul Kim, Yul-Ho Kim, Jung-Pil Suh, Hyang-Mi Park, Nese Sreenivasulu, Gopal Misra, et al. 2016. "Map-Based Cloning and Characterization of the BPH18 Gene from Wild Rice Conferring Resistance to Brown Planthopper (BPH) Insect Pest." *Scientific Reports* 6 (September): 34376. https://doi.org/10.1038/srep34376.

Jiang, Xiao-Ling, Zhu-Mei He, Zhi-Qiang Peng, Yu Qi, Qing Chen, and Shou-Yi Yu. 2007. "Cholera Toxin B Protein in Transgenic Tomato Fruit Induces Systemic Immune Response in Mice." *Transgenic Research* 16 (2): 169–175.

Joelson, Thorleif, L. Akerblom, Per Oxelfelt, Bror Strandberg, Karin Tomenius, and T. Jack Morris. 1997. "Presentation of a Foreign Peptide on the Surface of Tomato Bushy Stunt Virus." *Journal of General Virology* 78 (6): 1213–1217.

Joensuu, J. J., Viola Niklander-Teeri, and J. E. Brandle. 2008. "Transgenic Plants for Animal Health: Plant-Made Vaccine Antigens for Animal Infectious Disease Control." *Phytochemistry Reviews* 7 (3): 553–577.

Johanson, A. and M. J. Jeger. 1993. "Use of PCR for Detection of *Mycosphaerella fijiensis* and *M. musicola*, the Causal Agents of Sigatoka Leaf Spots in Banana and Plantain." *Mycological Research* 97 (6): 670–674.

Jouanin, Aurélie, Theo Borm, Lesley A. Boyd, James Cockram, Fiona Leigh, Bruno A. C. M. Santos, Richard G. F. Visser, and Marinus J. M. Smulders. 2019. "Development of the GlutEnSeq Capture System for Sequencing Gluten Gene Families in Hexaploid Bread Wheat with Deletions or Mutations Induced by γ-Irradiation or CRISPR/Cas9." *Journal of Cereal Science* 88 (July): 157–166. https://doi.org/10.1016/j.jcs.2019.04.008.

Joung, Y. H., J. W. Youm, J. H. Jeon, B. C. Lee, C. J. Ryu, H. J. Hong, H. C. Kim, H. Joung, and H. S. Kim. 2004. "Expression of the Hepatitis B Surface S and PreS2 Antigens in Tubers of *Solanum tuberosum*." *Plant Cell Reports* 22 (12): 925–930. https://doi.org/10.1007/s00299-004-0775-1.

Jupe, Florian, Kamil Witek, Walter Verweij, Jadwiga Sliwka, Leighton Pritchard, Graham J. Etherington, Dan Maclean, et al. 2013. "Resistance Gene Enrichment Sequencing (RenSeq) Enables Reannotation of the NB-LRR Gene Family from Sequenced Plant Genomes and Rapid Mapping of Resistance Loci in Segregating Populations." *The Plant Journal: For Cell and Molecular Biology* 76 (3): 530–544. https://doi.org/10.1111/tpj.12307.

Kamthan, Ayushi, Mohan Kamthan, Mohammad Azam, Niranjan Chakraborty, Subhra Chakraborty, and Asis Datta. 2012. "Expression of a Fungal Sterol Desaturase Improves Tomato Drought Tolerance, Pathogen Resistance and Nutritional Quality." *Scientific Reports* 2: 951.

Kang, Tae-Jin, Nguyen-Hoang Loc, Mi-Ok Jang, Yong-Suk Jang, Young-Sook Kim, Jo-Eun Seo, and Moon-Sik Yang. 2003. "Expression of the B Subunit of *E. coli* Heat-Labile Enterotoxin in the Chloroplasts of Plants and Its Characterization." *Transgenic Research* 12 (6): 683–691. https://doi.org/10.1023/B:TRAG.0000005114.23991.bc.

Karasev, Alexander V., Scott Foulke, Candice Wellens, Amy Rich, Kyu J. Shon, Izabela Zwierzynski, David Hone, Hilary Koprowski, and Marvin Reitz. 2005. "Plant Based HIV-1 Vaccine Candidate: Tat Protein Produced in Spinach." *Vaccine* 23 (15): 1875–1880.

Keese, Paul. 2008. "Risks from GMOs Due to Horizontal Gene Transfer." *Environmental Biosafety Research* 7 (3): 123–149.

Kempf, Volkhard A. J., Karlheinz Trebesius, and Ingo B. Autenrieth. 2000. "Fluorescent *In Situ* Hybridization Allows Rapid Identification of Microorganisms in Blood Cultures." *Journal of Clinical Microbiology* 38 (2): 830–838. https://doi.org/10.1128/JCM.38.2.830-838.2000.

Key, Suzie, Julian K. C. Ma, and Pascal M. W. Drake. 2008. "Genetically Modified Plants and Human Health." *Journal of the Royal Society of Medicine* 101 (6): 290–298.

Kim, Yang-Gyun, Jooyeun Cha, and Srinivasan Chandrasegaran. 1996. "Hybrid Restriction Enzymes: Zinc Finger Fusions to Fok I Cleavage Domain." *Proceedings of the National Academy of Sciences* 93 (3): 1156–1160.

Klein, Theodore M., Elisabeth C. Harper, Zora Svab, John C. Sanford, Michael E. Fromm, and Pal Maliga. 1988. "Stable Genetic Transformation of Intact Nicotiana Cells by the Particle Bombardment Process." *Proceedings of the National Academy of Sciences* 85 (22): 8502–8505.

Kliot, Adi, Svetlana Kontsedalov, Galina Lebedev, Marina Brumin, Pakkianathan Britto Cathrin, Julio Massaharu Marubayashi, Marisa Skaljac, Eduard Belausov, Henryk Czosnek, and Murad Ghanim. 2014. "Fluorescence *In Situ* Hybridizations (FISH) for the Localization of Viruses and Endosymbiotic Bacteria in Plant and Insect Tissues." *Journal of Visualized Experiments: JoVE* 84 (February): e51030. https://doi.org/10.3791/51030.

Klöti, Andreas, Victor A. Iglesias, Joachim Wünn, Peter K. Burkhardt, Swapan K. Datta, and Ingo Potrykus. 1993. "Gene Transfer by Electroporation into Intact Scutellum Cells of Wheat Embryos." *Plant Cell Reports* 12 (12): 671–675.

Klümper, Wilhelm and Matin Qaim. 2014. "A Meta-Analysis of the Impacts of Genetically Modified Crops." *PLoS One* 9 (11): e111629.

Koenig, Renate. 1981. "Indirect ELISA Methods for the Broad Specificity Detection of Plant Viruses." *Journal of General Virology* 55 (1): 53–62.

Kolotilin, Igor, Angelo Kaldis, Bert Devriendt, Jussi Joensuu, Eric Cox, and Rima Menassa. 2012. "Production of a Subunit Vaccine Candidate against Porcine Post-Weaning Diarrhea in High-Biomass Transplastomic Tobacco." *PLoS One* 7 (8): e42405.

Komari, Toshihiko, Yukoh Hiei, Yasuhito Saito, Nobuhiko Murai, and Takashi Kumashiro. 1996. "Vectors Carrying Two Separate T-DNAs for Co-Transformation of Higher Plants Mediated by *Agrobacterium tumefaciens* and Segregation of Transformants Free from Selection Markers." *The Plant Journal* 10 (1): 165–174.

Kuckenberg, Jan, Iryna Tartachnyk, and Georg Noga. 2009. "Temporal and Spatial Changes of Chlorophyll Fluorescence as a Basis for Early and Precise Detection of Leaf Rust and Powdery Mildew Infections in Wheat Leaves." *Precision Agriculture* 10 (1): 34–44. https://doi.org/10.1007/s11119-008-9082-0.

Kumar, G. B. Sunil, T. R. Ganapathi, L. Srinivas, C. J. Revathi, and V. A. Bapat. 2006. "Expression of Hepatitis B Surface Antigen in Potato Hairy Roots." *Plant Science* 170 (5): 918–925.

Kumar, Krishan, Chetana Aggarwal, Ishwar Singh, and Pranjal Yadava. 2018. "Microbial Genes in Crop Improvement." In *Crop Improvement Through Microbial Biotechnology*, 39–56. Amsterdam: Elsevier.

Kumar, Krishan, Geetika Gambhir, Abhishek Dass, Amit Kumar Tripathi, Alla Singh, Abhishek Kumar Jha, Pranjal Yadava, Mukesh Choudhary, and Sujay Rakshit. 2020. "Genetically Modified Crops: Current Status and Future Prospects." *Planta* 251: 1–27.

Lamphear, Barry J., Joseph M. Jilka, Lyle Kesl, Mark Welter, John A. Howard, and Stephen J. Streatfield. 2004. "A Corn-Based Delivery System for Animal Vaccines: An Oral Transmissible Gastroenteritis Virus Vaccine Boosts Lactogenic Immunity in Swine." *Vaccine* 22 (19): 2420–2424.

Lee, Tak, Sunmo Yang, Eiru Kim, Younhee Ko, Sohyun Hwang, Junha Shin, Jung Eun Shim, Hongseok Shim, Hyojin Kim, and Chanyoung Kim. 2015. "AraNet v2: An Improved Database of Co-Functional Gene Networks for the Study of *Arabidopsis thaliana* and 27 Other Nonmodel Plant Species." *Nucleic Acids Research* 43 (D1): D996–D1002.

Li, Aixia, Shangang Jia, Abou Yobi, Zhengxiang Ge, Shirley J. Sato, Chi Zhang, Ruthie Angelovici, Thomas E. Clemente, and David R. Holding. 2018. "Editing of an Alpha-Kafirin Gene Family Increases, Digestibility and Protein Quality in Sorghum." *Plant Physiology* 177 (4): 1425–1438.

Li, Tian, Jing Kuan Sun, Zhao Hua Lu, and Qing Liu. 2011. "Transformation of HBsAg (Hepatitis B Surface Antigen) Gene into Tomato Mediated by *Agrobacterium tumefaciens*." *Czech Journal of Genetics and Plant Breeding* 47 (2): 69–77.

Li, Ting, Bo Liu, Martin H. Spalding, Donald P. Weeks, and Bing Yang. 2012. "High-Efficiency TALEN-Based Gene Editing Produces Disease-Resistant Rice." *Nature Biotechnology* 30 (5): 390.

Lövei, Gabor L., Thomas Bøhn, and Angelika Hilbeck. 2010. *Biodiversity, Ecosystem Services, and Genetically Modified Organisms*. Malaysia: Third World Network.

MacGregor, James T., John A. Heddle, Mark Hite, Barry H. Margolin, Clacs Ramel, Michael F. Salamone, Raymond R. Tice, and Dieter Wild. 1987. "Guidelines for the Conduct of Micronucleus Assays in Mammalian Bone Marrow Erythrocytes." *Mutation Research/Genetic Toxicology* 189 (2): 103–112.

Mahlein, Anne-Katrin, Erich-Christian Oerke, Ulrike Steiner, and Heinz-Wilhelm Dehne. 2012. "Recent Advances in Sensing Plant Diseases for Precision Crop Protection." *European Journal of Plant Pathology* 133 (1): 197–209. https://doi.org/10.1007/s10658-011-9878-z.

Mason, Hugh S. and Charles J. Arntzen. 1995. "Transgenic Plants as Vaccine Production Systems." *Trends in Biotechnology* 13 (9): 388–392.

Mason, Hugh S., D. M. Lam, and Charles J. Arntzen. 1992. "Expression of Hepatitis B Surface Antigen in Transgenic Plants." *Proceedings of the National Academy of Sciences* 89 (24): 11745–11749.

McCabe, Dennis E., William F. Swain, Brian J. Martinell, and Paul Christou. 1988. "Stable Transformation of Soybean (*Glycine max*) by Particle Acceleration." *Bio/Technology* 6 (8): 923–926.

McWhite, Claire D., Ophelia Papoulas, Kevin Drew, Rachael M. Cox, Viviana June, Oliver Xiaoou Dong, Taejoon Kwon, Cuihong Wan, Mari L. Salmi, and Stanley J. Roux. 2020. "A Pan-Plant Protein Complex Map Reveals Deep Conservation and Novel Assemblies." *Cell* 181 (2): 460–474.e14.

Meli, Vijaykumar S., Sumit Ghosh, T. N. Prabha, Niranjan Chakraborty, Subhra Chakraborty, and Asis Datta. 2010. "Enhancement of Fruit Shelf Life by Suppressing N-Glycan Processing Enzymes." *Proceedings of the National Academy of Sciences* 107 (6): 2413–2418.

Mertens, Martha. 2008. "Assessment of Environmental Impacts of Genetically Modified Plants." *Implementation of the Biosafety Protocol Development of Assessment Bases*, FKZ 201 67 430/07, 234. Federal Agency for Nature Conservation, Germany.

Miller, Jamie K. and Kent J. Bradford. 2010. "The Regulatory Bottleneck for Biotech Specialty Crops." *Nature Biotechnology* 28 (10): 1012–1014. https://doi.org/10.1038/nbt1010-1012.

Moravec, František and Jean-Lou Justine. 2007. "A New Species of Ascarophis (*Nematoda, Cystidicolidae*) from the Stomach of the Marine Scorpaeniform Fish *Hoplichthys citrinus* from a Seamount off the Chesterfield Islands, New Caledonia." *Acta Parasitologica* 52 (3): 238–246.

Moricca, S., A. Ragazzi, T. Kasuga, and K. R. Mitchelson. 1998. "Detection of *Fusarium oxysporum* f. *Sp. vasinfectum* in Cotton Tissue by Polymerase Chain Reaction." *Plant Pathology* 47 (4): 486–494.

Moshou, D., C. Bravo, R. Oberti, J. West, L. Bodria, A. McCartney, and H. Ramon. 2005. "Plant Disease Detection Based on Data Fusion of Hyper-Spectral and Multi-Spectral Fluorescence Imaging Using Kohonen Maps." *Real-Time Imaging, Spectral Imaging II* 11 (2): 75–83. https://doi.org/10.1016/j.rti.2005.03.003.

Nekrasov, Vladimir, Congmao Wang, Joe Win, Christa Lanz, Detlef Weigel, and Sophien Kamoun. 2017. "Rapid Generation of a Transgene-Free Powdery Mildew Resistant Tomato by Genome Deletion." *Scientific Reports* 7 (1): 1–6.

Nochi, Tomonori, Hidenori Takagi, Yoshikazu Yuki, Lijun Yang, Takehiro Masumura, Mio Mejima, Ushio Nakanishi, Akiko Matsumura, Akihiro Uozumi, and Takachika Hiroi. 2007. "Rice-Based Mucosal Vaccine as a Global Strategy for Cold-Chain-and Needle-Free Vaccination." *Proceedings of the National Academy of Sciences* 104 (26): 10986–10991.

Oerke, E.-C., U. Steiner, H.-W. Dehne, and M. Lindenthal. 2006. "Thermal Imaging of Cucumber Leaves Affected by Downy Mildew and Environmental Conditions." *Journal of Experimental Botany* 57 (9): 2121–2132. https://doi.org/10.1093/jxb/erj170.

Oliva, Ricardo, Chonghui Ji, Genelou Atienza-Grande, José C. Huguet-Tapia, Alvaro Perez-Quintero, Ting Li, Joon-Seob Eom, Chenhao Li, Hanna Nguyen, and Bo Liu. 2019. "Broad-Spectrum Resistance to Bacterial Blight in Rice Using Genome Editing." *Nature Biotechnology* 37 (11): 1344–1350.

Ow, David W. 2005. "2004 SIVB Congress Symposium Proceeding: Transgene Management via Multiple Site-Specific Recombination Systems." *In Vitro Cellular & Developmental Biology-Plant* 41 (3): 213–219.

Ow, David W. 2007. "GM Maize from Site-Specific Recombination Technology, What Next?" *Current Opinion in Biotechnology* 18 (2): 115–120.

Peng, Aihong, Shanchun Chen, Tiangang Lei, Lanzhen Xu, Yongrui He, Liu Wu, Lixiao Yao, and Xiuping Zou. 2017. "Engineering Canker-Resistant Plants through CRISPR/Cas9-Targeted Editing of the Susceptibility Gene Cs LOB 1 Promoter in Citrus." *Plant Biotechnology Journal* 15 (12): 1509–1519.

Perdikaris, Antonios, Nikon Vassilakos, Iakovos Yiakoumettis, Oxana Kektsidou, and Spiridon Kintzios. 2011. "Development of a Portable, High Throughput Biosensor System for Rapid Plant Virus Detection." *Journal of Virological Methods* 177 (1): 94–99.

Potrykus, Ingo. 2007. "Gene Transfer Methods for Plants and Cell Cultures." In *Ciba Foundation Symposium 154 - Bioactive Compounds from Plants*, 198–212. John Wiley & Sons, Ltd. https://doi.org/10.1002/9780470514009.ch14.

Richter, Liz J., Yasmin Thanavala, Charles J. Arntzen, and Hugh S. Mason. 2000. "Production of Hepatitis B Surface Antigen in Transgenic Plants for Oral Immunization." *Nature Biotechnology* 18 (11): 1167–1171. https://doi.org/10.1038/81153.

Rigano, M. Manuela, M. Lucrecia Alvarez, Julia Pinkhasov, Y. Jin, Francesco Sala, C. J. Arntzen, and Amanda Maree Walmsley. 2004. "Production of a Fusion Protein Consisting of the Enterotoxigenic *Escherichia coli* Heat-Labile Toxin B Subunit and a Tuberculosis Antigen in *Arabidopsis thaliana*." *Plant Cell Reports* 22 (7): 502–508.

Rommens, Caius M., Michel A. Haring, Kathy Swords, Howard V. Davies, and William R. Belknap. 2007. "The Intragenic Approach as a New Extension to Traditional Plant Breeding." *Trends in Plant Science* 12 (9): 397–403.

Sailaja, M., M. Tarakeswari, and M. Sujatha. 2008. "Stable Genetic Transformation of Castor (*Ricinus communis* L.) via Particle Gun-Mediated Gene Transfer Using Embryo Axes from Mature Seeds." *Plant Cell Reports* 27 (9): 1509.

Sánchez-León, Susana, Javier Gil-Humanes, Carmen V. Ozuna, María J. Giménez, Carolina Sousa, Daniel F. Voytas, and Francisco Barro. 2018. "Low-Gluten, Nontransgenic Wheat Engineered with CRISPR/Cas9." *Plant Biotechnology Journal* 16 (4): 902–910.

Saunders, James A. and Benjamin F. Matthews. 1997. "Plant transformation by gene transfer into pollen." World Patent Application, US5629183, issued May 13, 1997.

Schaad, Norman W. and Reid D. Frederick. 2002. "Real-Time PCR and Its Application for Rapid Plant Disease Diagnostics." *Canadian Journal of Plant Pathology* 24 (3): 250–258. https://doi.org/10.1080/07060660209507006.

Schouten, Henk J., Frans A. Krens, and Evert Jacobsen. 2006. "Cisgenic Plants Are Similar to Traditionally Bred Plants: International Regulations for Genetically Modified Organisms Should Be Altered to Exempt Cisgenesis." *EMBO Reports* 7 (8): 750–753.

Sears, Mark K., Richard L. Hellmich, Diane E. Stanley-Horn, Karen S. Oberhauser, John M. Pleasants, Heather R. Mattila, Blair D. Siegfried, and Galen P. Dively. 2001. "Impact of Bt Corn Pollen on Monarch Butterfly Populations: A Risk Assessment." *Proceedings of the National Academy of Sciences* 98 (21): 11937–11942.

Shoji, Yoko, Hong Bi, Konstantin Musiychuk, Amy Rhee, April Horsey, Gourgopal Roy, Brian Green, Moneim Shamloul, Christine E. Farrance, and Barbara Taggart. 2009. "Plant-Derived Hemagglutinin Protects Ferrets against Challenge Infection with the A/Indonesia/05/05 Strain of Avian Influenza." *Vaccine* 27 (7): 1087–1092.

Song, Wen-Yuan, Guo-Liang Wang, Li-Li Chen, Han-Suk Kim, Li-Ya Pi, Tom Holsten, J. Gardner, et al. 1995. "A Receptor Kinase-Like Protein Encoded by the Rice Disease Resistance Gene, Xa21." *Science* 270 (5243): 1804–1806. https://doi.org/10.1126/science.270.5243.1804.

Staczek, John, Mohammed Bendahmane, Linda B. Gilleland, Roger N. Beachy, and Harry E. Gilleland Jr. 2000. "Immunization with a Chimeric Tobacco Mosaic Virus Containing an Epitope of Outer Membrane Protein F of *Pseudomonas aeruginosa* Provides Protection against Challenge with *P. aeruginosa*." *Vaccine* 18 (21): 2266–2274.

Steuernagel, Burkhard, Sambasivam K. Periyannan, Inmaculada Hernández-Pinzón, Kamil Witek, Matthew N. Rouse, Guotai Yu, Asyraf Hatta, et al. 2016. "Rapid Cloning of Disease-Resistance Genes in Plants Using Mutagenesis and Sequence Capture." *Nature Biotechnology* 34 (6): 652–655. https://doi.org/10.1038/nbt.3543.

Stewart, James McD. 1981. "*In Vitro* Fertilization and Embryo Rescue." *Environmental and Experimental Botany* 21 (3–4): 301–315.

Thorpe, Trevor A. 2007. "History of Plant Tissue Culture." *Molecular Biotechnology* 37 (2): 169–180.

Toriyama, Kinya, Youichi Arimoto, Hirofumi Uchimiya, and Kokichi Hinata. 1988. "Transgenic Rice Plants after Direct Gene Transfer into Protoplasts." *Bio/Technology* 6 (9): 1072–1074.

Tregoning, John S., Peter Nixon, Hiroshi Kuroda, Zora Svab, Simon Clare, Frances Bowe, Neil Fairweather, Jimmy Ytterberg, Klaas J. van Wijk, and Gordon Dougan. 2003. "Expression of Tetanus Toxin Fragment C in Tobacco Chloroplasts." *Nucleic Acids Research* 31 (4): 1174–1179.

Trout, C. L., J. B. Ristaino, M. Madritch, and T. Wangsomboondee. 1997. "Rapid Detection of *Phytophthora infestans* in Late Blight-Infected Potato and Tomato Using PCR." *Plant Disease* 81 (9): 1042–1048.

Tsatsakis, Aristidis M., Muhammad Amjad Nawaz, Demetrios Kouretas, Georgios Balias, Kai Savolainen, Victor A. Tutelyan, Kirill S. Golokhvast, Jeong Dong Lee, Seung Hwan Yang, and Gyuhwa Chung. 2017. "Environmental Impacts of Genetically Modified Plants: A Review." *Environmental Research* 156: 818–833.

Turpen, Thomas H., Stephen J. Rein, Yupin Charoenvit, Stephen L. Hoffman, Victoria Fallarme, and Laurence K. Grill. 1995. "Malaria Epitopes Expressed on the Surface of Recombinant Tobacco Mosaic Virus." *Bio/Technology* 13 (1): 53–57.

Tuteja, Narendra and Sarvajeet S. Gill. 2013. *Climate Change and Plant Abiotic Stress Tolerance*. Weinheim: John Wiley & Sons.

Van den Berg, Johnnie, and P. Campagne. 2015. "Resistance of Busseola Fusca to Cry1Ab Bt Maize Plants in South Africa and Challenges to Insect Resistance Management in Africa." *Bt Resistance: Characterization and Strategies for GM Crops Producing Bacillus Thuringiensis Toxins*, January, 36–48.

Van Rensburg, J. B. J. 2007. "First Report of Field Resistance by the Stem Borer, *Busseola fusca* (Fuller) to Bt-Transgenic Maize." *South African Journal of Plant and Soil* 24 (3): 147–151.

Vanblaere, Thalia, Iris Szankowski, Jan Schaart, Henk Schouten, Henryk Flachowsky, Giovanni AL Broggini, and Cesare Gessler. 2011. "The Development of a Cisgenic Apple Plant." *Journal of Biotechnology* 154 (4): 304–311.

Vos, Clazien J. de and Manon Swanenburg. 2018. "Health Effects of Feeding Genetically Modified (GM) Crops to Livestock Animals: A Review." *Food and Chemical Toxicology* 117: 3–12.

Waltz, Emily. 2015. *USDA Approves Next-Generation GM Potato.* Nature Publishing Group.

Wang, Yanpeng, Xi Cheng, Qiwei Shan, Yi Zhang, Jinxing Liu, Caixia Gao, and Jin-Long Qiu. 2014. "Simultaneous Editing of Three Homoeoalleles in Hexaploid Bread Wheat Confers Heritable Resistance to Powdery Mildew." *Nature Biotechnology* 32 (9): 947–951.

Wang, Yuanyuan, Hanqing Deng, Xiaobo Zhang, Hailin Xiao, Yunbo Jiang, Yunfeng Song, Liurong Fang, Shaobo Xiao, Yonglian Zhen, and Huanchun Chen. 2009. "Generation and Immunogenicity of Japanese Encephalitis Virus Envelope Protein Expressed in Transgenic Rice." *Biochemical and Biophysical Research Communications* 380 (2): 292–297.

West, Jonathan S., Cedric Bravo, Roberto Oberti, Dimitri Lemaire, Dimitrios Moshou, and H. Alastair McCartney. 2003. "The Potential of Optical Canopy Measurement for Targeted Control of Field Crop Diseases." *Annual Review of Phytopathology* 41 (1): 593–614. https://doi.org/10.1146/annurev.phyto.41.121702.103726.

Wyman, Claire and Roland Kanaar. 2006. "DNA Double-Strand Break Repair: All's Well That Ends Well." *Annual Review of Genetics*40: 363–383.

Yang, Chung-Da, Jia-Teh Liao, Chen-Yen Lai, Ming-Hwa Jong, Chi-Ming Liang, Yeou-Liang Lin, Na-Sheng Lin, Yau-Heiu Hsu, and Shu-Mei Liang. 2007. "Induction of Protective Immunity in Swine by Recombinant Bamboo Mosaic Virus Expressing Foot-and-Mouth Disease Virus Epitopes." *BMC Biotechnology* 7 (1): 62.

Yang, Yang, Kaiyu Zhu, Huailin Li, Shaoqing Han, Qingwei Meng, Shahid Ullah Khan, Chuchuan Fan, Kabin Xie, and Yongming Zhou. 2018. "Precise Editing of CLAVATA Genes in *Brassica napus* L. Regulates Multilocular Silique Development." *Plant Biotechnology Journal* 16 (7): 1322–1335.

Yusibov, V., D. C. Hooper, S. V. Spitsin, N. Fleysh, R. B. Kean, T. Mikheeva, D. Deka, A. Karasev, S. Cox, and J. Randall. 2002. "Expression in Plants and Immunogenicity of Plant Virus-Based Experimental Rabies Vaccine." *Vaccine* 20 (25–26): 3155–3164.

Yusibov, Vidadi, Anna Modelska, Klaudia Steplewski, Michail Agadjanyan, David Weiner, D. Craig Hooper, and Hilary Koprowski. 1997. "Antigens Produced in Plants by Infection with Chimeric Plant Viruses Immunize against Rabies Virus and HIV-1." *Proceedings of the National Academy of Sciences* 94 (11): 5784–5788.

Zhang, G. Gary, Lauren Rodrigues, Benjamin Rovinski, and K. Andrew White. 2002. "Production of HIV-1 P24 Protein in Transgenic Tobacco Plants." *Molecular Biotechnology* 20 (2): 131–136.

Zhang, Huawei, Xiaomin Si, Xiang Ji, Rong Fan, Jinxing Liu, Kunling Chen, Daowen Wang, and Caixia Gao. 2018. "Genome Editing of Upstream Open Reading Frames Enables Translational Control in Plants." *Nature Biotechnology* 36 (9): 894–898.

Zheng, Ming, Liang Zhang, Min Tang, Jinglin Liu, Hongfang Liu, Hongli Yang, Shihang Fan, William Terzaghi, Hanzhong Wang, and Wei Hua. 2020. "Knockout of Two Bna MAX 1 Homologs by CRISPR/Cas9-Targeted Mutagenesis Improves Plant Architecture and Increases Yield in Rapeseed (*Brassica napus* L.)." *Plant Biotechnology Journal* 18 (3): 644–654.

Zhou, Haisheng, Meihong Zhang, Xue Tian, Hongxia Shao, Kun Qian, Jianqiang Ye, and Aijian Qin. 2017. "Identification of a Novel Recombinant Virulent Avian Infectious Bronchitis Virus." *Veterinary Microbiology* 199: 120–127.

Zhou, Jianping, Xuhui Xin, Yao He, Hongqiao Chen, Qian Li, Xu Tang, Zhaohui Zhong, Kejun Deng, Xuelian Zheng, and Sayed Abdul Akher. 2019. "Multiplex QTL Editing of Grain-Related Genes Improves Yield in Elite Rice Varieties." *Plant Cell Reports* 38 (4): 475–485.

Zhu, Guangtao, Junbo Gou, Harry Klee, and Sanwen Huang. 2019. "Next-Gen Approaches to Flavor-Related Metabolism." *Annual Review of Plant Biology* 70: 187–212.

2

Advances in Genomics and Proteomics in Agriculture

Manas Mathur and Rakesh Kumar Prajapat
Suresh Gyan Vihar University

Tarun Kumar Upadhyay
Parul Institute of Applied Sciences, Parul University

Dalpat Lal
Jagannath University

Nisha Khatik
Maharshi Dayanand Saraswati University

Deepak Sharma
JECRC University

CONTENTS

2.1 Introduction

Today the world population is increasing and climate change is the major challenge of food security. The challenges include a growing global population, climate change and environmental pressure for accelerating breeding work for high production. However, the impediments are extensive; there is a massive agreement that swelling of the potentiality and productivity of recent agricultural production is important. Researchers are searching for convenient and modern techniques for crop production. A new era of omics techniques in molecular biology which include genomics, transcriptomics, proteomics, metabolomics and phenomics have a wide role in accelerating the biotechnology work in the future. The word genomics was invented by the geneticist Thomas Roderick in 1986 during the mapping of the human genome. It deals with an innovative scientific chastisement of sequencing, analysing and mapping genomes or it is the experimental approach that requires molecular characterisation and cloning of complete genomes to comprehend the morphology, application and synthesis of genes. After sequencing of the complete genome to genotyping for genome-wide annotation research to genomic assumption, innovations in technology and importance have resulted in revolutions in crop improvement (Liu and Zhang 2019). To date, more than 50 crops have been sequenced (Table 2.1).

Most of the plants selected to be sequenced are based on investigations of the scientific world; prototypical beings or economically viable, small genomes; diploid and homozygous lines; access to molecular mapping; expressed sequence tags (ESTs)/transcriptome and different genomic implements. Most of the plant genome (73%) publications have been on crop species such as *Arabidopsis thaliana*, *A. lyrata*,

TABLE 2.1
List of Some Sequenced Common Plant Genomes

Serial No.	Common Name	Scientific Name	Year	Genome Size	Gene (#)
1	Tulsi	*Ocimum tenuiflorum*	2015	612 Mb	
2	Pearl millet	*Pennisetum glaucum*	2017	179 Gb	38,579
3	Jojoba	*Simmondsia chinensis*	2020	887 Mb	23,490
4	*Chlorella*	*Chlorella sorokiniana*	2018	61.4 Mb	16,697
5	*Dunaliella*	*Dunaliella salina*	2017	343.7 Mb	
6	Wheel Tree	*Trochodendron aralioides*	2019	1.614 Gb	35,328
7	Pigweed	*Amaranthus hypochondriacus*	2016	403.9 Mb	23,847
8	Silver birch	*Betula pendula*	2017	435 Mbp	28,399
9	Hyacinth bean	*Lablab purpureus*	2018	397 Mbp	20,946
10	Chinese chestnut)	*Castanea mollissima*	2019	785.53 Mb	36,479
11	Loquat	*Eriobotrya japonica*	2020	760.1 Mb	45,743
12	Yellowhorn	*Xanthoceras sorbifolium*	2019	504.2 Mb	24,672
13	Agarwood	*Aquilaria sinensis*	2020	726.5 Mb	29,203
14	Dodder	*Cuscuta australis*	2018	265 Mbp	19,671

Brachypodium distachyon, *Physcomitrella patens* (moss) and *Selaginella moellendorffii* (spike moss). Almost 95% genomes sequenced to date are angiosperms, among them 36 belong to dicots and 16 to monocots, whereas spruce which is an exceptional gymnosperm, 1 bryophyte (moss) and 1 lycophyta (clubmoss) have been studied in detail after genome sequencing (Michael and Jackson 2017). Genomics also has a role in increasing the genomic resources, utility, diversity, yield and other agronomic traits. The application of genomics with next-generation sequencing (NGS) can change the molecular plant breeding and will accordingly allow it to be achieved with precision (Lal 2020).

2.2 The "Omics" Techniques

Generally, *omics* means an area of research or large-scale studies, such as comprehensive biological data sets in biology concluding in *-omics*, viz., genomics, proteomics or metabolomics. The associated suffix -ome is denoted to refer to the substantial research of such areas, like genome, proteome or metabolome, respectively. Plant molecular biology targets research of the metabolic pathway of cellular processes, their genetic switch and associations with ecological variations. This broad-scale and thorough research needs large-scale trials involving complete genetic, morphological and functional mechanisms. The achievement of molecular breeding relies on the different tools which are applied in the potent manipulation of genetic changes. Whole "omics", arrays and high-throughput inventions have been conceived to conduct optimum broad-spectrum genetic analyses and breeding trials. These innovations have been unified into many innovative genetic and breeding procedures (Deshmukh et al. 2014).

2.3 Branches of Genomics

Genomics is defined as a field in a range of genetics that deals with the organisation, sequencing and investigation of an organism's genome. Genomics is often divided into two main branches, structural genomics and functional genomics, but other branches are also studied under genomics.

2.3.1 Structural Genomics

Structural genomics is a branch of science which deals with the organisation and sequence of the DNA in the whole genome. It represents initial steps of genome analysis including (1) the construction of a high-resolution genetic map which is based on recombination frequency (cM), physical maps of an organism where genes are placed in relation to distance measure in base pairs; (2) sequencing of the genome and (3) determining the complete set of proteins. There are several tools of structural genomic studies like DNA-based markers and genome sequencing techniques. A molecular marker is a sequence of DNA that is willingly sensed and whose legacy can effortlessly be scrutinised. They are phenotypically unbiased and factually uncountable, have permitted perusing of the whole genome and convey landmarks in bulk compactness on each chromosome in several plant species. The molecular process of most DNA markers and the genetic diversity can be engendered by restriction site or mutation at a polymerase chain reaction (PCR) priming site, insertion, deletion or by modifying the number of replicate units between two restriction or PCR priming sites and nucleotide mutation leading to a single nucleotide polymorphism (SNP).

2.3.1.1 Some Properties of Molecular Markers

Properties of molecular markers include (1) vibrant different allele structures (so that different alleles can be recognised easily), (2) codominance (so that heterozygotes can be differenced from homozygotes), (3) free from environment factors, (4) easy diagnosis (so that the complete pathway can be automated), (5) high copy number (so that the outcomes can be amassed and shared between laboratories), (6) stable inheritance/Mendelian segregation, (7) economical marker development and genotyping, (8) large-scale genetic polymorphism, (9) uniform dispersal on the whole genome and (10) neutral selection (without pleiotropic effect) (Lal 2015).

Several types of DNA markers depend upon how the polymorphism is revealed: hybridisation-based polymorphisms [random fragment length polymorphism (RFLP)] and PCR-based polymorphisms [random amplified polymorphic DNA (RAPD), amplified fragment length polymorphism (AFLP), simple sequence repeats (SSR), single nucleotide polymorphism (SNP), etc.] and sequenced based markers [cleaved amplified polymorphic sequences (CAPS)], sequence characterised amplified regions (SCARs), sequence target sites (STS) and so forth.

2.3.1.2 Sequencing of the Genome

The process of genome sequencing is highly sophisticated and technical. The genome size is variable (depends on species) and obtained in different sizes of pieces. These pieces have to be then assembled into a sequence for a genome. The pieces for the sequence are generated by breaking the genomic DNA into random sites and these are further cloned into a suitable vector. Cloning of the fragments is essential to produce a large number of copies of each fragment required for sequencing. Further, these fragments are sequenced by using the sequencing method.

2.3.1.2.1 Method of Sequencing

2.3.1.2.1.1 Maxam and Gilbert Method The Maxam and Gilbert Method is the first DNA sequencing technique where the DNA fragment to be sequenced is end-labelled by the addition of ^{32}p-dATP either at the 5′ end or at the 3′ end of its two strands. It is also called the chemical degradation method (at sites G, C, G+C/C+T) of single end-labelled DNA strands.

2.3.1.2.1.2 Sanger Sequencing Sanger sequencing is the first cohort sequencing method based on the chain termination/dideoxy method/enzymatic procedure of sequencing. Generally, read length of this method is 500–1000 bp; however, the acceptable read length is 200 bp, but it has some disadvantages like amplifying the DNA fragments before sequencing and electrophoresis for separating the DNA ladders that are generated. This sequencing technique was used in the Human Genome Project in 2003.

Application of Sanger sequencing

- Bacterial identification – 16s RNA sequence
- Microsatellite detection
- Clone confirmation
- Bacterial purity testing
- Genetically modified organism (GMO) testing/screening for transgenic plant using PCR
- SNP identification

2.3.1.2.1.3 Shotgun Sequencing Whole-genome shotgun sequencing is the most commonly used technique for sequencing of prokaryotic genomes. Genomes which recur less can be simply sequenced using this protocol. First of all, high-quality DNA is isolated from the target organisms and then DNA is cleaved into random various tiny segments of 2–10 kb. These segments are sequenced using the Sanger sequencing method to attain reads. Several reads for the target DNA are attained by conducting many repeated cycles of this disintegration and sequencing. Computer programs then use the overlying ends of various reads to accumulate them into an unceasing sequence. A large number of shotgun clones are sequenced by using automated DNA sequencing machines. Sequenced data are assembled using computational methods (Kaur et al. 2015).

2.3.1.2.1.4 Clone-by-Clone Sequencing Clone-by-clone sequencing is a hierarchical sequencing method used for the sequencing of large genomes. Using this method, there is an urge to design genetic maps of an individual. So, accessibility of genomic assets in terms of DNA markers and the bacterial artificial chromosome (BAC)/yeast artificial chromosome (YAC) library are very crucial to designing maps of plants. Once a genetic linkage map is available, it is used for the construction of physical maps.

TABLE 2.2
Next-Generation Sequencing Platforms

Platforms	PCR type	Mode Working	Read Length (bp)
Second Generation of Sequencing			
454/ Roche	Emulsion	Pyrosequencing	250–400
Illumina/Solexa	Bridge PCR	Reversible terminator	85–100
SOLiD	Emulsion	Sequencing by ligation	35
Third Generation of Sequencing			
Heliscope	Single molecule	Reversible terminator	35
SMRT (Single molecule real time)	Single molecule	Real time sequencing by synthesis	>1000
Ion torrent	Emulsion PCR	Detection of H^+	25–50

Abbreviation: PCR, polymerase chain reaction.

2.3.1.2.1.5 Next-Generation Sequencing NGS, also called high-throughput sequencing, refers to different current sequencing procedures that empower researchers to sequence DNA and RNA at a broad level, along with corresponding sequencing, and permit millions of DNA fragments to be sequenced instantaneously more rapidly and economically than Sanger sequencing. These protocols of DNA sequencing allow the precise sequencing of nucleotides within a DNA molecule. So, NGS has transfigured the research of molecular biology and genomics. Different novel procedures for DNA sequencing were innovated in the mid to late 1990s and were executed in profitable DNA sequences in past years (Liu et al. 2020) (Table 2.2).

Application of NGS

- Whole-genome/transcriptome sequencing.
- Chip and methylation sequencing.
- Small RNA/miRNA sequencing.
- Mitochondria/chloroplast DNA sequencing.
- Microarray-based NGS sequencing validation.
- The human genome is being re-sequenced to screen genetic causes that result in disorders.
- Mutation detection.
- Digital gene expression.

Genome Organisation

Genomic organisation refers to the linear arrangements of nucleotides and their structure of chromosomes in the nucleus. Eukaryotes possess linear chromosomes, huge genomes with telomeres and centromeres and small gene flow with introns and large monotonous sequences, whereas prokaryotes have tiny genomes, single and circular chromosomes (few linear) without centromeres or telomeres, large gene bulk without introns and scanty monotonous sequences. The genome extent means a single set of chromosomes (haploid) or C-value and is calculated by re-association kinetics. However, various cells inside a single organism can possess different ploidy. Followed by denaturation, the velocity of re-association relies on the size of the genome. The more complex the genome, the more repetitive DNA sequences and more duration to re-anneal is needed for more C-value. $Cot^{1/2}$ is the output of the DNA quantity and duration needed to carry forward halfway to re-association. It is unswervingly connected to the quantity of DNA in the genome. Genome size is generally associated with plant development and ecology, and enormously large genomes may be inadequate both ecologically and evolutionarily. The assorted physiological and cellular outcomes of complex genomes may be a key attribute of the collection of the major machinery that subsidises genome size like gene duplication and transposable elements (Meyerson et al. 2020).

2.3.2 Functional Genomics

Functional genomics is defined as a branch of genomics that focusses on gene transcription, translation and protein functions and interfaces with the help of genome information and high-throughput tools. It aims to understand genome functions at various progressive phases and under the ecological environment. It is featured by high-throughput or large-scale investigated mutual protocols with computational (bioinformatics) and statistical evaluation of the outcomes. The novel evidence proved by the entire omics field will lead the researchers from plant sciences to *in silico* recreations of plant growth, development and retort to environmental variation. To assess a gene function by using *in silico* prediction of gene function (annotation), expression analysis and forward and reverse genetics studies are being used. Two types of studies are used: up-regulation (activation tagging-promoter control/enhancer control, overexpression of gene, multiple copy gene insertion or gain in function mutagenesis) and down-regulation [insertion mutagenesis (T-DNA tagging transpose on tagging), PTGS (post-transcriptional gene silencing), VIGS (virus induced gene silencing) and chemical mutagenesis; Dwivedi and Rautela (2018).

2.3.3 Comparative Genomics

Comparative genomics compares the structure and functions of genomes of different organisms. The information gained in one organism can have applications in other even distantly related organisms. It allows us to achieve a greater understanding of evolution with information about what is common and what is unique between different species at the genome level. It helps to identify both coding and non-coding genes and regulatory elements. Comparative genomics is an extensive tool in screening variations in the structure of the genome caused by segmental duplications, rearrangements and polyploidy. This tool has been recommended to interpret homologous genes and consequently to recognise conserved *cis*-regulatory motifs. Serval studies of comparative genomics have shown that, in related species, there is synteny (conserved gene location within large blocks of the genome in different species). Furthermore, comparative genomics is also used to comprehend the evolution of novel traits (Susičet al. 2020).

2.3.4 Evolutionary Genomics

The genomic sequence of organisms from different species has been recommended to research how the genome varies during evolution. Partial or complete genomes modify structure during evolution through deletions and duplications. Once genome sequence has blown out through the community, then modification in genome size occurs at an instant rate. The genome sequence of a recent species can be utilised to conjecture when duplications and deletions have fallen out in the past and to assume that the evolutionary phases are allied with variations in a genome site (Van Tassel et al. 2020).

2.3.5 Epigenomics

Epigenomics is a branch of molecular biology which deals with a whole set of epigenetic variations on the genetic constituent of a cell, called an epigenome. Epigenetic variations are reversible on a cell's DNA or histones that disturb the expression of genes without fluctuating the DNA sequence. Two of the most featured epigenetic alterations are histone modification and DNA methylation. Epigenetic variations play a crucial role in gene expression and regulation and are engaged in different cellular progressions like development or differentiation and tumorigenesis (Kapazoglou et al. 2018).

2.3.6 Metagenomics

Metagenomics is defined as a branch of biology which deals with studies related to metagenomes, genetic material recuperated unswervingly from environmental tissues. It is also recommended as environmental genomics, eco-genomics or community genomics. However, conventional microbiology and microbial genome sequencing depends upon sophisticated clonal cultures, initial environmental gene sequencing and cloned particular genes (16S rRNA gene) to yield a shape of diversity in a naturally occurring species (Kumar et al. 2020).

2.4 Applications of Genomics

- Gene identification and cloning.
- Gene prediction/discovery.
- Genome sequence gives the structure/architecture of the chromosome, genetic mapping and location of the genes.
- Genome sequencing identified the mutated region in sequence.
- Genome manipulation in genetic features like crop yield, disease resistance, growth abilities, nutritive qualities and drought tolerance.
- Quantitative trait locus (QTL) analysis and marker-assisted selection.
- Comparative genomics.
- Gene banks and chromosome stocks.
- Understanding expression profiles, responses and interactions.
- Oral plant vaccines against hepatitis B where transgenes create surface antigens that stimulate immunity.
- Compare the genomic sequence of species and understand how the genome has been remodelled in the course of evolution.
- Genomics also engages the research of intragenomic methodologies such as heterosis, epistasis and pleiotropy along with the interfaces between loci and alleles within the genome.

Genomics is the study of the genome and it has contributed to enhancement in agriculture by using structural, functional and comparative information. Molecular markers and sequencing technologies provided insight at the structural level. Further development of molecular markers has enhanced due to advances in sequencing and genotyping technologies. This also leads to crop genotyping on a large scale, which can further be used to develop high-density genetic as well as physical maps. Functional genomics is used to know the function of the gene. Furthermore, it is used to understand the genome of most crops and we can ensure a better future with the capability of more perfect and precise genetic modification for improving yield and survival in stress conditions. The combination of genomics with other omics technology can be used in SMART (Simple Modular Architecture Research Tool) breeding or breeding by design.

2.5 Proteomics

Proteomics is the hefty gage of research of proteins, predominantly their task and assembly. It is an exceptional way to carry out research and variations in metabolism in retort to various unfavourable environments. The term "proteomics" came into existence in 1995, recommended as the extensive feature of the whole protein accompaniment of a cell, tissue, or organism.

2.6 Categories of Proteomics

2.6.1 Expression Proteins

Expression proteomics is applied in the field of qualitative and quantifiable display of whole proteins under two environments. Naturally, expression proteomics research is spoked to the research of the expression protein decorations in anomalous cells. Two-dimensional (2D) gel electrophoresis along mass spectrometry (MS) analysis has been recommended to visualise the protein expressional behaviours, and whether or not it is expressed in the cancerous cell, when associated it can be recognised and featured as protein accomplishments, multi-protein complexes and signalling routes (Kwok et al. 2020). Identification of these proteins provides cherished knowledge about the molecular biology of the development of oncogenes in cells and the particular state of disease application in diagnostic markers or therapeutic agents.

2.6.2 Structural Proteomics

Structural proteomics assists in comprehending three-dimensional (3D) morphology and the physical intricacies of functional proteins. This assumption of a protein when its amino acid sequence is resolute unswervingly is done by sequencing or through a process known as homology modelling. This technique provides elaborated evidence about the morphology and role of protein complexes in a particular organelle of the cell. It is conceivable to know all the proteins existing in composite structures like membranes, ribosomes and cell organelles and to analyse all the protein complexities that can be among these proteins and protein complexes. Various techniques like nuclear magnetic resonance (NMR) spectroscopy and X-ray crystallography were particularly applied for structure fortitude (Beale 2020).

2.6.3 Functional Proteomics

This technique elucidates the empathetic role of protein along with evasive molecular contrivances within the cell that relies on the recognition of the interrelating protein complexes. The complexity of a mysterious protein with its complexes fitting a particular protein engaged in a specific manner will be strappingly redolent of its biological function. Therefore, elaborated explanation of the various intracellular signalling mechanism might be a vital tool to understand the description of protein-protein interactions *in vivo* (Liu et al. 2018; Popp and Maquat 2018; Heusel et al. 2019; Schaffer et al. 2019; Bludau and Aebersold 2020; Salas et al. 2020).

2.6.4 Techniques Involved in Proteomics

In this area of biological sciences, both logical and bioinformatics tools have been recommended to recognise protein morphology and its role. These tools include 2D gel electrophoresis and matrix-assisted laser desorption/ionisation time-of-flight MS (MALDI-TOF-MS) as recommended in recent scenarios along with other software.

2.6.4.1 2DGel Electrophoresis

With 2D gel electrophoresis, protein tissues are determined centred on the charge, termed as isoelectric focussing (IEF), and followed by molecular weight. The output is the shadow in thousands of minute spots, each denoting a protein. A prominent 2D gel can determine 1000–2000 protein spots, which can be visualised after staining, like dots in the gel. This method is particularly recommended to compare two of the same samples to determine particular protein variation. Further, the protein solution is prepared at a concentration and then applied for IEF. Some specific factors to keep in mind are the relative existence, solubility and native charge of key proteins. Proteins are separated based on isoelectric point (pI). An optimum immobilized pH gradient (IPG) strip of a particular length is selected and the sample is loaded as a pH gradient, and there is loading of differently expressed proteins in accordance to size by sodium dodecyl sulphate-polyacrylamide gel electrophoresis (SDS-PAGE). Further using SDS-PAGE, proteins are separated based on size after which proteins of optimum size are selected. Proteins are screened using fluorescent protein tags. After that, digital images are captured by 2D configurations using optimum imaging machinery and software. After that, the expression ways are studied using 2D software. Isolated protein of prime importance are cleaved from the gel and digested.

2.6.4.2 *MALDI-TOF*

MALDI-TOF, an innovative method that is a soft ionisation technology applied in spectrometry, is confined to investigate biomolecules like protein, peptides and DNA. These biologically active molecules along with artificial polymers have minute instability and are thermally wobbly, which is a drawback for the application of MS as a means of identification of a specific protein of interest. These glitches have been sorted by innovations from MALDI-TOF-MS, which is generally applied to the molecular weight examination of bioactive compounds by vaporisation and ionisation. It does not inactivate the compound and it remains in its native form thereafter; a laser beam has been recommended to ionise the sample (Hou et al. 2019).

The protein sample is then carried for purification by high-performance liquid chromatography (HPLC) or SDS-PAGE by spawning peptide fates, which are the key bases for fingerprints of protein as a device to identify the transparency of a recognised protein in a known sample. MS provides an idea about a peptide fate when proteins are cleaved with trypsin, which is a proteolytic enzyme. Based on these peptide maps, a sequence database can be designed to search a perfect similarity from the prevailing records.

2.7 Advanced Methods in Proteomics

2.7.1 Isotope-Coded Affinity Tags (ICAT)

The isotope-coded affinity tag (ICAT) is free from gel and is applied to determine measurable proteomics that depends on synthetic labelling reagents. These probes possess three general constituents, i.e., an isotopically coded linker, defined amino acid side chain and a docket for the empathy separated of tagged proteins/peptides. For quantitative research of two proteomes, each sample is tagged with isotopic light and the other sample with bulky report. Both samples were shared with isotope-coded tagging mixtures. These peptides are studied by LC-MS. Deuterium tags are used generally. The process is particularly applied for the virtual measurement of proteins existing in two or more biological samples. Observable ICATs can be applied in this method. A visible tag that tolerates the electrophoresis hierarchy of tagged peptides can be conveniently studied (Bąchor et al. 2019).

2.7.2 Isobaric Tags for Relative and Absolute Quantification (iTRAQ)

Isobaric tags for relative and absolute quantification (iTRAQ) is also a non–gel-based technique recommended to enumerate proteins. It is applied in proteomics to research measurable variations in the proteome. The reagents 4-plex and 8-plex can be applied to tag all peptides from various samples based on the covalent bonding of the N-terminus and side-chain amines of peptides from protein cleaved with tags of different weight. Further samples can be run on MS/MS. Various types of software are recommended for interpretation of the MS/MS spectra, i.e., j-Tracker and j-TraqX 20 (Morales et al. 2017).

2.7.3 Absolute Quantification (AQUA)

The absolute quantification (AQUA) technique is recommended for studies of the unqualified measurement of proteins and their variable forms. Covalent variations are done to design *in vitro* proteins. Such variations are chemically similar to logically existing posttranslational mechanisms. These types of peptides are applied to enumerate the posttranslational altered proteins after complete digestion with the assistance of a tandem MS (Schnatbaum et al. 2020).

2.7.4 ESI-Q-IT-MS

Electrospray ionisation-quadrupole-ion trap-mass spectroscopy (ESI-Q-IT-MS) has excellent applications in proteomics. In ESI, ionisation proteins are ionised in solution and possess different charges. The benefit of implementing this technique for the investigation for studies on the mass of protein is that because of the large charge state of proteins their m/z dimensions are stereotypically less than 2000 and the TOF detector has excellent mass precision in this examination area. The outcomes are more precise mass determinations for proteins in ESI-Q-TOF (Bian et al. 2020).

2.7.5 SELDI (Surface-Enhanced Laser Desorption/Ionisation) TOF-MS

The surface-enhanced laser desorption/ionisation (SELDI) TOF-MS technique is applied for the determination of protein concoctions; it is an ionisation assay applied in MS. SELDI is particularly applied with TOF mass spectrometers and is applied to screen proteins in clinical trials along with a comparison of protein amount with and without a disorder that can be recommended for biomarker research (Hill et al. 2020).

2.8 Applications of Proteomics

The initial protocol is to examine the sequence of the peptides after fragmentation spectra data from MS has been received. Two various approaches can be used including *de novo* peptide sequencing and probing against the fragmentation spectra databases (Chen et al. 2020). In the latter techniques, a target database is reputable from the *in silico* cleaved pattern of all expressed or predicted protein sequences. Then a peptide spectrum match (PSM) score is determined for each fragmentation spectra and all experimental fragmentation spectra evidence from the target database. The peptide that has the maximum PSM score can be kept on priority for the query peptide. It is continually a tedious assignment to select optimum probing algorithms that give high eminence peptide spectrum matching consequences from databases. Conventional protein database search engines, like SEQUEST (Timp and Timp 2020), execute a recording purpose based on standardised cross-correlation of the mass-to-charge ratio assumed from the identified sequence of amino acid in databases and the fragment ions recognised from the tandem mass spectrum. MASCOT (Timp and Timp 2020) is an additional famous software innovated thereafter, which assimilates frame capacities with protein sequence data, peptide molecular weights from protein absorption and tandem MS data to load a probability-based score for protein recognition.

2.8.1 Gene Ontology

Gene ontology (GO) is the most broadly applied technology in upgrading analysis. It is defined as a set of predefined clusters to which various genes are allocated based on their practical features, thus assisting to decrease the severance in expressions. The GO terms are positionally bunched with three chief classes of terms: "molecular function", "the biological process" and "cellular component". Each term has a different recogniser and is associated with each other. The AmiGO database (Munoz-Torres and Carbon 2017) delivers GO term footnotes for different species, but not all proteins have a complete and exact annotation. Enlightening GO terms from proteins with the same database can be applied to those proteins which have incomplete interpretations. GO term assumption algorithms such as ProLoc-GO (Lande et al. 2020), PFP (Wang et al. 2020) and IGNA (Piovesanet al. 2015) are designed to combat this tedious task. The most generalised statistical trials recommended in GO enrichment are Fisher's exact test and the hypergeometric test. Statistically noteworthy GO terms are those that seem to recur more often in the effort protein slope than would be anticipated conditionally, and they may have denoted amazing biological mechanisms for additional research. Therefore, since GO terms frequently characterise ORF products, rather than established protocols, scientists should cautiously test the GO terms in the improved outcomes to guarantee that they have prominent associations between the proteoforms and related genes. Similar to GO terms, former acquaintances with steadied mechanism networks and disorders can also be used to achieve an augmentation study (Welzenbach et al. 2016).

Biological routes indicate biological schedules and chemical responses among molecules within a cell that revealed a convinced biological mechanism. Databases such as PANTHER and Reactome (Aslam et al. 2017) curate the interface maps for various mechanisms along with a set of enhancement tools to achieve enhancement unswervingly on the database webpage. For example, Pathview (Luo and

Brouwer 2013) is an R package for the KEGG pathway (Kanehisa et al. 2017) constructed data amalgamation and picturing. Protein set enrichment analysis (PSEA) is an alternative famous enrichment tactic and is obtained from gene set enrichment analysis (GSEA). In PSEA, the augmentation output is concluded from a weighted running sum statistic and proteins without adequate tremendous variations may damagingly affect the upgrading score. While many types of GSEA software can also execute PSEA, tools innovated particularly for protein quantification data, like PSEA-Quant (Lavallée-Adam et al. 2014), may deliver a more comfortable and justified protocol for proteomics.

2.8.2 Oncology

Oncology is the science of the study of tumour cells. Behind tumour metastasis, the major encounter with drugs is to pronounce molecular and cellular phenomena. Proteomics is used to predict the protein expressions connected to the dangerous effect of this tumour, which assists us in appreciating the process of metastasis, enabling the innovation of approaches for the beneficial involvement and clinical administration of cancer for innovation of novel drugs which can combat its effects. Proteomics is an organised branch of science, which focusses on identifying the protein expressions, the role of cancerous cells and broad application in biomarker inventions (Macklin et al. 2020).

2.8.3 Applications in Biomedical Science

The area of research of the association between microbial pathogens and their hosts is defined as "interactomics". It is an amazing field in proteomics. Interactomics assists in understanding the basics of the contagion's source and its consequence on organs. The key target of these investigations is to avoid or prevent disease at the initial stage. Advanced diagnostic concerns are associated with developing contaminations, elevation in finicky bacteria, and a group of patient-tailored phenotypes (Katsani and Sakellari 2019).

2.8.4 Applications in Agriculture

The bids of plant proteomics are immobile at a promising level. Plant proteomics is also used to determine plant-insect associations that assist in recognising key genes engaged in the defensive stimuli of plants to herbivores. The population growth and the effect of global atmosphere variations strike harsh restrictions on the stability of agricultural crop yields (Chin and Tan 2018).

2.8.5 Food Microbiology

The application of proteomics in food technology is mainly for identification and optimisation of resources, course innovation and revealing batch-to-batch differences and the eminence switch of the final product. Further study is devoted to the features of food safety, particularly concerning biological and microbial security and the application of genetically modified crops (Wu et al. 2017).

2.9 Conclusion and Future Perspectives

Thus, based on the above key findings and applications of proteomics, this chapter has been designed to focus on the applications of proteomics in different aspects of science and delivers a way to apply the expressed protein resource in a more controlled manner. Coalescing the proteome data with other omics processes is vital to implement as a novel prototype in bioinformatics. In a well-defined manner, judgments between proteomics and transcriptomics can be innovated since the key denoting of genes and proteins can be linearly arranged between the two omics spaces. Proteins/genes with conflicting tendencies may denote the participation of substantial transcriptional and posttranslational control process. Measurable proteomics using *in vivo* SILAC (stable isotope labelling by/with amino acids in cell culture) mouse technology has been effectively used in investigating the posttranscriptional phenomena in several aspects.

The contrast design of protein signalling grids can also be advantageous from other omics data. The simulated effect of virtual interference of protein-activity by enriched regulon analysis (VIPER) algorithm can complete computational studies of protein studies based on the association between transcription factors and their probable goals recognised from transcriptome data. Based on the data available, various reforms of omics data are frequently opposed to each other, the MS-based proteomics will have gained a great deal of attention when analysed from a multi-omics perspective. As the MS-based proteomics process endures to generate, the connected bioinformatics tools need to be modernised consequently. To date, most of the scrutiny protocols described in this chapter have been cited with an expedient and user-friendly line.

REFERENCES

Aslam, B., Basit, M., Nisar, M.A., Khurshid, M., and Rasool, M.H. 2017. Proteomics: technologies and their applications. *Journal of Chromatographic Science* 55(2): 182–196.

Bąchor, R., Waliczek, M., Stefanowicz, P., and Szewczuk, Z. 2019. Trends in the design of new isobaric labelling reagents for quantitative proteomics. *Molecules* 24(4): 701.

Beale, J.H. 2020. Macromolecular X-ray crystallography: soon to be a road less travelled? *Acta Crystallographica Section D: Structural Biology* 76(5): 400–405.

Bian, Y., Zheng, R., Bayer, F. P., et al. 2020. Robust, reproducible and quantitative analysis of thousands of proteomes by micro-flow LC–MS/MS. *Nature Communications* 11(1): 1–12.

Bludau, I., and Aebersold, R. 2020. Proteomic and interactomic insights into the molecular basis of cell functional diversity. *Nature Reviews Molecular Cell Biology* 21(6):1–14.

Chen, C., Hou, J., Tanner, J. J., and Cheng, J. 2020. Bioinformatics methods for mass spectrometry-based proteomics data analysis. *International Journal of Molecular Sciences* 21(8): 2873.

Chin, C.F., and Tan, H.S. 2018. The use of proteomic tools to address challenges faced in clonal propagation of tropical crops through somatic embryogenesis. *Proteomes* 6(2): 21.

Deshmukh, R., Sonah, H., Patil, G., Chen, W., Prince, S., Mutava, R., and Nguyen, H. T. 2014. Integrating omic approaches for abiotic stress tolerance in soybean. *Frontiers in Plant Science* 3(5): 244.

Dwivedi, M., and Rautela, A. 2018. Application of functional genomics in agriculture. *International Journal of Chemical Studies* 6(1): 462–465.

Heusel, M., Bludau, I., Rosenberger, G., et al. 2019. Complex centric proteome profiling by SEC SWATH MS. *Molecular Systems Biology* 15(1): e8438.

Hill, V., Kuhnert, P., Erb, M., and Machado, R. A. 2020. Identification of *Photorhabdus* symbionts by MALDI-TOF mass spectrometry. *bioRxiv*. https://doi.org/10.1101/2020.01.10.901900

Hou, T.Y., Chiang-Ni, C., and Teng, S.H. 2019. Current status of MALDI-TOF mass spectrometry in clinical microbiology. *Journal of Food and Drug Analysis* 27(2): 404–14.

Kanehisa, M., Furumichi, M., Tanabe, M., Sato, Y., and Morishima, K. 2017. KEGG: new perspectives on genomes, pathways, diseases and drugs. *Nucleic Acids Research* 45(1): 353–361.

Kapazoglou, A., Ganopoulos, I., Tani, E., and Tsaftaris, A.2018. Epigenetics, epigenomics and crop improvement. In *Advances in Botanical Research* 86,287–324. Academic Press, London.

Katsani, K. R., and Sakellari, D. 2019. Saliva proteomics updates in biomedicine. *Journal of Biological Research-Thessaloniki* 26(1): 17.

Kaur, R., Kumar, K., and Sharma, N. 2015. Genomics in agriculture. *CAB Reviews* 10(046): 1–25.

Kumar, A., Ravindran, M., Sarsaiya, B., Chen, S., Wainaina, H., Singh, S., and Zhang, Z. 2020. Metagenomics for taxonomy profiling: tools and approaches. *Bioengineered* 11(1): 356–374.

Kwok, C.S.N., Lai, K.K.Y., Lam, S.W., Chan, K.K., Xu, S.J.L., and Lee, F.W.F. 2020. Production of high-quality two-dimensional gel electrophoresis profile for marine medaka samples by using trizol-based protein extraction approaches. *Proteome Science* 18: 1–13.

Lal, D. 2015. An overview on molecular marker for crop improvement. *Agrobios Newsletter* 13(9): 110–11.

Lal, D. 2020. Trends in genomics: branches and application. *Biotica Research Today* 2(5): 153–55.

Lande, N.V., Barua, P., Gayen, D., Kumar, S., Chakraborty, S., and Chakraborty, N. 2020. Proteomic dissection of the chloroplast: Moving beyond photosynthesis. *Journal of Proteomics* 212: 103542.

Lavallée-Adam, M., Rauniyar, N., McClatchy, D. B., and Yates III, J. R.2014. PSEA-Quant: a protein set enrichment analysis on label-free and label-based protein quantification data. *Journal of Proteome Research* 13(12): 5496–5509.

Liu, D. D., and Zhang, L. 2019. Trends in the characteristics of human functional genomic data on the gene expression omnibus, 2001–2017. *Laboratory Investigation* 99(1): 118–127.

Liu, F., Zhan, W., Xie, Q., Chen, H., Lou, B., and Xu, W. 2020. A first genetic linage map construction and QTL mapping for growth traits in Larimichthys polyactis. *Scientific Reports* 10(1): 1–9.

Liu, X., Salokas, K., Tamene, F., et al.2018. An AP-MS-and BioID-compatible MAC-tag enables comprehensive mapping of protein interactions and subcellular localizations. *Nature Communications* 9(1): 1–16.

Luo, W., and Brouwer, C. 2013. Pathview: an R/Bioconductor package for pathway-based data integration and visualization. *Bioinformatics* 29(14): 1830–1831.

Macklin, A., Khan, S., and Kislinger, T. 2020. Recent advances in mass spectrometry based clinical proteomics: applications to cancer research. *Clinical Proteomics* 17: 1–25.

Meyerson, L. A., Pyšek, P., Lučanová, M., Wigginton, S., Tran, C. T.,and Cronin, J. T. 2020. Plant genome size influences stress tolerance of invasive and native plants via plasticity. *Ecosphere* 11(5): e03145.

Michael, T. P., and Jackson, S. 2017. The first 50 plant genomes. *The Plant Genome* 6(2).

Morales, A. G., Lachén-Montes, M., Ibáñez-Vea, M., Santamaría, E., and Fernández-Irigoyen, J. 2017. Application of isobaric tags for relative and absolute quantitation (iTRAQ) to monitor olfactory proteomes during Alzheimer's disease progression. In *Current Proteomic Approaches Applied to Brain Function*, 29–42. Humana Press, New York.

Munoz-Torres, M., and Carbon, S. 2017. Get GO! retrieving GO data using AmiGO, QuickGO, API, files, and tools. In *The Gene Ontology Handbook*, 149–160. Humana Press, New York.

Piovesan, D., Giollo, M., Leonardi, E., Ferrari, C., and Tosatto, S. C. 2015. INGA: protein function prediction combining interaction networks, domain assignments and sequence similarity. *Nucleic Acids Research* 43(W1): W134–140.

Popp, M. W., and Maquat, L. E. 2018. Nonsense-mediated mRNA decay and cancer. *Current Opinion in Genetics & Development* 48: 44–50.

Salas, D., Stacey, R.G., Akinlaja, M., and Foster, L.J. 2020. Next-generation interactomics: considerations for the use of co-elution to measure protein interaction networks. *Molecular and Cellular Proteomics* 19(1): 1–10.

Schaffer, L. V., Millikin, R. J., and Miller, R. M., et al. 2019. Identification and quantification of proteoforms by mass spectrometry. *Proteomics* 19(10): 1800361.

Schnatbaum, K., Solis-Mezarino, V., Pokrovsky, D., et al. 2020. Front cover: new approaches for absolute quantification of stable-isotope-labelled peptide standards for targeted proteomics based on a UV active tag. *Proteomics* 20(10): 2070081.

Susič, N., Janežič, S., Rupnik, M., and Stare, B.G. 2020. Whole genome sequencing and comparative genomics of two nematicidal *Bacillus* strains reveals a wide range of possible virulence factors. *G3: Genes, Genomes, Genetics* 10(3): 881–890.

Timp, W., and Timp, G. 2020. Beyond mass spectrometry, the next step in proteomics. *Science Advances* 6(2): eaax8978.

Van Tassel, D. L., Tesdell, O., Schlautman, B., Rubin, M. J., DeHaan, L. R., and Crews, T. E., et al. 2020. New food crop domestication in the age of gene editing: genetic, agronomic and cultural change remain co-evolutionarily entangled. *Frontiers in Plant Science* 11.

Wang, L., Law, J., Murali, T.M.N., and Pandey, G. 2020. Data integration through heterogeneous ensembles for protein function prediction. *bioRxiv*. https://doi.org/10.1101/2020.05.29.123497

Welzenbach, J., Neuhoff, C., and Heidt, H., et al. 2016. Integrative analysis of metabolomic, proteomic and genomic data to reveal functional pathways and candidate genes for drip loss in pigs. *International Journal of Molecular Sciences* 17(9): 1426.

Wu, F., Zhong, F., and He, F. 2016. Microbial proteomics: approaches, advances, and applications. *Journal of Bioinformatics, Proteomics and Imaging Analysis* 2(1): 85–91.

3

Integrated Metabolome and Transcriptome Analysis: A New Platform/Technology for Functional Biology and Natural Products Research

Rohit Jain and Swati Gupta
Manipal University Jaipur

Sumita Kachhwaha
University of Rajasthan

S.L. Kothari
Amity University Rajasthan

CONTENTS

3.1 Introduction

Plants produce a diverse array of metabolites not only important for plant growth, and defence mechanisms, but they also account for the various applications as food, cosmetics, medicine, industrial raw materials and biofuels (Hong, Yang and Shi 2016; Hirai et al. 2004). Metabolites directly involved in plant growth and development are classified as primary metabolites, whereas those involved in plant-environment interaction, and defence against various biotic and abiotic stresses and do not have any known direct influence on the plant developmental process are classified as secondary metabolites or natural products (Djande et al. 2020; Agarrwal and Nair 2020).

The metabolic composition of plants is very much dependent on the evolutionary process and environmental adaptation at both the inter- and intra-species levels (Oksman-Caldentey and Saito 2005; Delfin,

Watanabe and Tohge 2019). Genetic diversity and the inherent property of certain metabolites to switch between different chemical structures and functions also contribute to the metabolite diversity (Delfin et al. 2019). Plant metabolic diversity is largely contributed by the secondary metabolism that enables the formation of complex mixtures of unique secondary metabolites (Delfin et al. 2019; Hong et al. 2016). These secondary metabolites possess different biological activities such as pharmaceuticals, dyes, food additives, coagulants and so forth (Delfin et al. 2019). In fact, the nutritional and edible value of the plants is also governed by their metabolite composition (Christ et al. 2018), showing the importance of natural products for the survival of living beings.

The collective metabolite pool of plants is more abundant, diverse and complex compared with other organisms, which can be attributed to the structural and physicochemical differences due to various environmental and intrinsic factors (Saito and Matsuda 2010). Secondary metabolites are low molecular weight compounds specifically produced at different events of the plant lifecycle, usually in response to external stresses (Yang et al. 2014). Plants produce approximately $>2 \times 10^5$ secondary metabolites, however, only half of them have been identified and characterised for various applications (Yang et al. 2014). On the contrary, the non-toxic and biodegradable nature of these metabolites led to an exponential increase in the global demand for multiple uses, thereby increasing the burden over natural flora to meet the industrial demands of raw materials for various products, including pharmaceutics, flavours, insecticides and aromas. Applications of plants as "chemical factories" for such bioactive metabolites are likely to pose a great danger to the existence of the natural flora (Oksman-Caldentey and Saito 2005). Moreover, the variations in metabolite composition due to environmental factors and production of comparatively lesser quantities after post-harvest processing are some of the major bottlenecks that withhold use of plant-derived products. Therefore, detailed studies related to structural and functional characterisation of natural products along with the development protocols for enhanced production of these metabolites independent of the external factors is the need of the hour.

Over the last few decades, biotechnological research and development in the omics-technology have opened new avenues for a better understanding of the metabolic processes involved in the biosynthesis of secondary metabolites, which could further be used for metabolic pathway engineering followed by targeted and enhanced biosynthesis of natural products with desired functional characteristics. Omics-technology including genomics, proteomics and transcriptomics has emerged as a powerful tool for deciphering secondary metabolism in recent years (Figure 3.1). Although secondary metabolism in plants is a complex process involving biological interactions between multiple metabolic pathways at gene, protein and metabolite levels, understanding of gene regulation, expression and interaction are still far-fetched with the use of a single omics-tool (Yang et al. 2014; Maroli et al. 2018). Genomics aims to study the genome functions, while transcriptomics and proteomics perform gene expression and protein (quantitative and qualitative) analysis, respectively. Being the end products of gene expression and protein activity, metabolites have been lately regarded as an additional component of the "Central Dogma" (Srinivasan and Kannan 2019). Analysis of metabolite mixtures (metabolomics) alone provides a precise picture of possible metabolisms involved, whilst simultaneous metabolome, gene expression and proteome analysis have proven to be a fruitful approach in the characterisation of genes involved in secondary metabolism (Maroli et al. 2018; Turumtay et al. 2016). Therefore, integration of multiple-omics tools provides the missing links between genotypic and phenotypic changes, which are vital for the elucidation of gene-to-gene and gene-to-trait networks (Figure 3.1).

Biochemical-omics techniques have been widely used to create a snapshot of different metabolic processes and to understand cross-linking among different metabolisms. Comprised of metabolomics, transcriptomics and proteomics, biochemical-omics bridges the gap between regulatory genes and related genetic markers by identification of genes and metabolic pathways having direct association(s) with the final traits (Teh et al. 2017). This approach has widely been adopted to gain deeper insights into the complex metabolic framework and associated phenotypic traits and to characterise the functionally active/bioactive products of different cellular processes. The knowledge generated through such studies has been widely utilised for the development of novel plant breeding and selection strategies aimed at crop improvement (Teh et al. 2017; Srinivasan and Kannan 2019; Maroli et al. 2018) (Figure 3.1).

Therefore, this chapter summarises various methods adopted for metabolome and transcriptome analysis along with their applications in an integrated-omics approach.

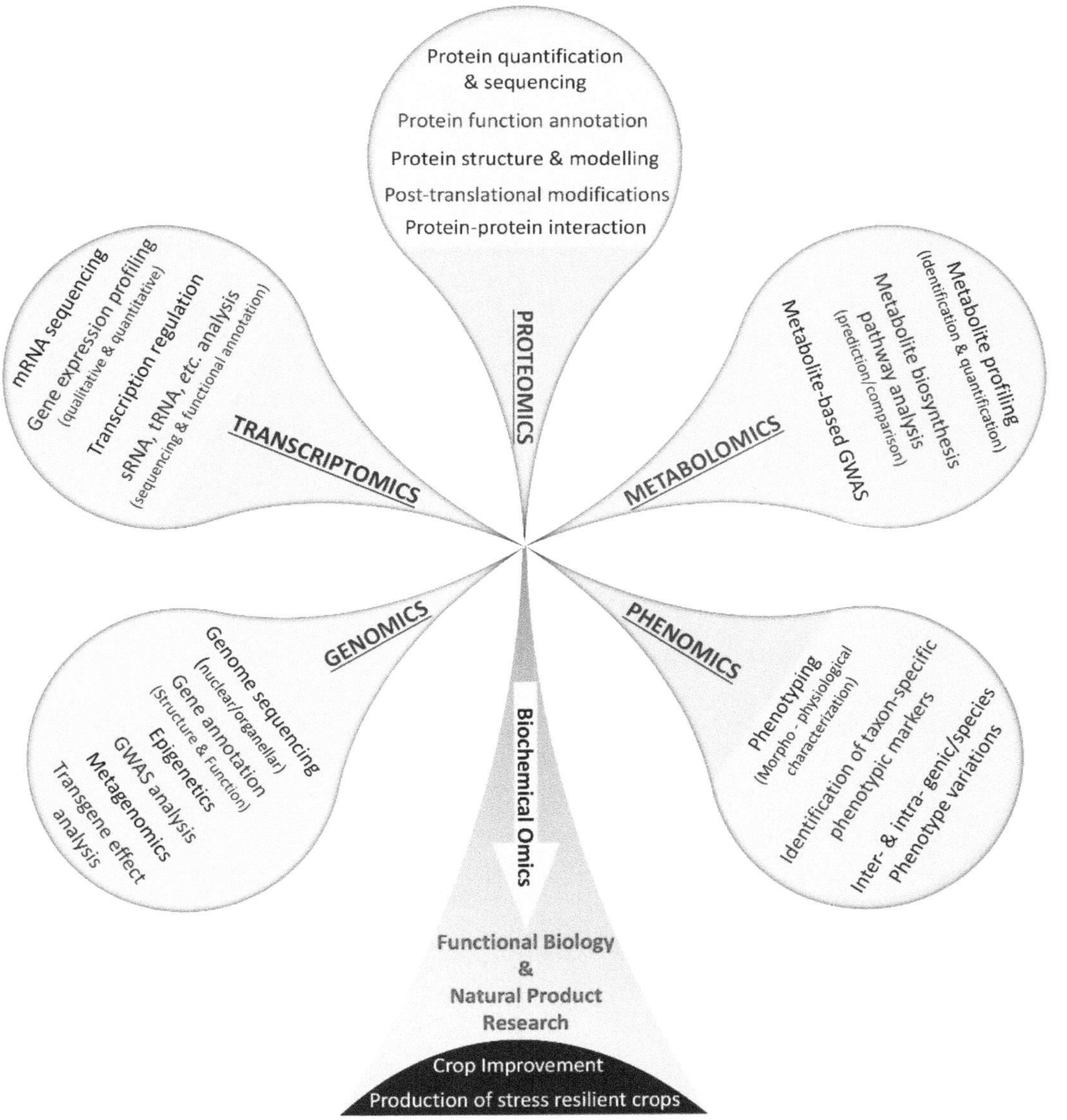

FIGURE 3.1 Role of different omics-platforms in functional biology and natural product research. GWAS, genome-wide association study; mRNA, messenger RNA; sRNA, small RNA; tRNA, transfer RNA.

3.2 Metabolomics: Methods and Applications

The process of detection, quantification and structural and functional characterisation of known and novel metabolites produced by plants followed by identification of their biosynthesis pathway is termed as "metabolomics" (Turumtay et al. 2016). Precisely, metabolomics provides an overview of the various metabolic events that mediate different plant physiological and developmental functions, i.e., it provides the information about why, when and how the plant metabolic machinery works at both cellular and organismal levels. Considering the inherent chemical complexity, diversity and dynamic nature of plant metabolomes, a holistic approach to provide a comprehensive overview of the metabolome is required

(Tugizimana et al. 2018). Based on the objectives of the study, the following strategies are currently being adopted for metabolite analysis:

- ***Targeted analysis:*** Detection and quantitative estimation of known metabolites using authentic chemical standards (Agarrwal and Nair 2020; Turumtay et al. 2016). This approach is restricted to a narrow range of metabolites, as preliminary knowledge of the metabolites to be identified is required. Since detection is based on a well-defined hypothesis and the preliminary knowledge of these metabolites fails to offer a comprehensive picture of the metabolic pathways, it is not considered very effective in deciphering the functions of unknown metabolites. Targeted analysis is usually employed for the detection of common metabolites, e.g., phytohormones, sugars, amino acids, alkaloids, flavonoids and terpenes.
- ***Metabolic profiling:*** Identification and quantification of a large group of predefined targets of specific metabolic pathways and structurally related natural products. It is a semi-targeted analysis, in which the metabolites are quantitatively and tentatively identified (Djande et al. 2020).
- ***Metabolomic fingerprinting:*** Rapid, high-throughput screening and identification of metabolites present in a sample (specific plant parts or extracts). It is a non-targeted approach that provides comprehensive information about the structural and chemical characteristics along with the relative concentrations of different metabolites identified (Djande et al. 2020). This strategy is often used to identify novel metabolites biosynthesised during different biological processes.

3.2.1 Overview of the Metabolomics Workflow

Practically, metabolome analysis is a complex process that includes a diverse array of data acquisition and mining tools. The workflow of metabolomics includes sample preparation, data acquisition using either single or a combination of analytical platforms and data mining using various statistical and bioinformatics tools (Figure 3.2). The selection of an appropriate strategy for metabolite extraction and analytical technique(s) for analysis are critical steps as their combined sensitivity and selectivity govern the outcomes of a metabolomics experiment (Djande et al. 2020). An overview of all the steps involved in a metabolomics study has been provided in subsequent sections.

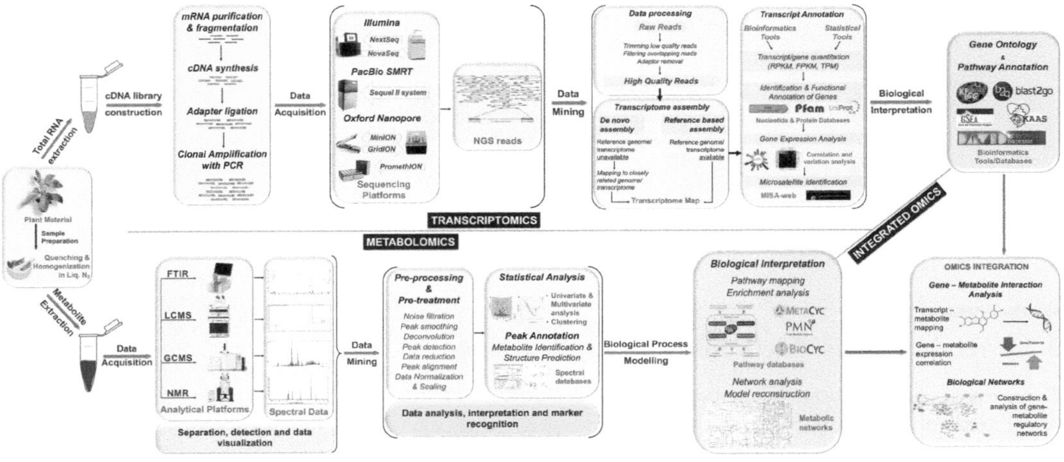

FIGURE 3.2 Overview of the steps involved in transcriptomics, metabolomics and integrated omics analysis. cDNA, complementary DNA; FT-IR, Fourier transform infrared spectroscopy; GC-MS, gas chromatography-mass spectrometry; LC-MS, liquid chromatography mass spectrometry; NGS, next generation sequencing; NMR, nuclear magnetic resonance spectroscopy.

3.2.1.1 Sample Preparation and Metabolite Extraction

The method of sample preparation should be easy, quick, reproducible and non-selective such that it could extract maximum metabolites from the plant tissue(s). Harvesting, quenching and extraction are the basic steps of sample preparation; however, the efficacy of each step is determined by the plant material studied, solvent system(s) used and metabolites of interest (Razzaq et al. 2019; Djande et al. 2020). The selection of appropriate plant material depends upon its developmental stage, physiological functions and metabolite composition (Pérez-Alonso et al. 2018). Followed by the selection, utmost caution is required for harvesting and quenching as metabolites are sensitive to external stress. However, to avoid post-harvest changes in its metabolome, plant material is generally preserved in liquid nitrogen immediately after harvesting (Razzaq et al. 2019).

Various extraction methods have been developed for the maximum recovery of metabolites from different plants. For extraction, the type of solvent is very important as it should not alter the chemical or structural integrity of the metabolites. Also, dissolution and solubility of metabolites in different solvents is taken into account while optimising the solvent system (Razzaq et al. 2019). A wide range of polar (water, methanol, ethanol, acetone) and non-polar (n-hexane, chloroform, petroleum ether) systems are used either as single-phase or multi-phase systems for extraction of different metabolites (Pérez-Alonso et al. 2018). Extraction of metabolites in these solvents is done using (1) solid-phase microextraction (SPME), (2) temperature- or pressure-assisted liquid extraction and (3) microwave-assisted selection (Djande et al. 2020). Supercritical fluid extraction (SPFE), swiss rolling technique, ultrasound-assisted extraction and enzyme-assisted extraction are other recent methods used (Djande et al. 2020).

3.2.1.2 Analytical Platforms for Data Acquisition

With the advancement in analytical tools, multiple analytical platforms have been developed for metabolomics study including chromatography and spectrometry techniques such as Fourier transform infrared spectroscopy (FT-IR), gas chromatography-mass spectrometry (GC-MS), liquid chromatography (LC)-MS and nuclear magnetic resonance (NMR) spectroscopy (Kumar et al. 2017). The selection of appropriate techniques predominantly relies on the aim of the study and solvent system used for extraction, along with other factors such as sensitivity, reproducibility and versatility of the analytical technique(s) (Piasecka, Kachlicki and Stobiecki 2019). The biochemical complexity of the metabolites results in differences in their chemical properties, solubility and sensitivity to different analytical techniques; therefore, a single technique is not able to analyse the entire metabolome (Kumar et al. 2017). Details of various techniques currently in use for different types of metabolome studies are summarised in Table 3.1.

3.2.1.3 Data Mining

Metabolome analyses generate a deluge of complex data sets that require highly advanced and powerful tools for data processing and interpretation. The use of statistical and high-throughput computational tools facilitated rapid and precise identification (Razzaq et al. 2019). Interpretation of the metabolomics data is a multilevel process involving pre-processing and pre-treatment followed by compound annotation and identification (Djande et al. 2020). The raw data obtained from the metabolomic analysis are processed to remove redundant data sets, eliminating the risk of false identification. Both pre-processing (noise filtration, peak detection and peak alignment) and pre-treatment (data normalisation, scaling, batch effect correction and integrity validation) methods are critical to generating high-quality information (Djande et al. 2020; Razzaq et al. 2019). Various statistical and bioinformatics tools used for the analysis of the processed metabolomics data have been summarised in Table 3.2.

3.2.2 Recent Applications and Advances

Metabolomics has emerged as a novel approach to understanding the underlying mechanisms of different physiological and developmental processes in plants. Over the past few years, extensive research on the

TABLE 3.1
Commonly Used Analytical Platforms in Metabolomics

Analytical Technique	Advantages	Disadvantages	References
FT-IR	✓ Inexpensive ✓ Functional groups identification and structure prediction	× Not suitable for liquid samples × Suitable for pure samples than the mixtures × Low chemical specificity	Pérez-Alonso et al. (2018); Turumtay et al. (2016); Srinivasan and Kannan (2019); Razzaq et al. (2019)
LC-MS	✓ High sensitivity and resolution ✓ Suitable for complex mixtures ✓ Minimal sample preparation required ✓ Enables quantitative profiling	× Retention time is dependent on various factors × No structural information × Highly purified samples are required	Pérez-Alonso et al. (2018); Turumtay et al. (2016); Srinivasan and Kannan (2019); Razzaq et al. (2019)
GC-MS	✓ High resolution ✓ Consistent and reproducible retention time(s) ✓ Cost-effective ✓ Large databases available for comparison ✓ Fast and accurate compound identification	× Suitable only for volatile and low molecular weight compounds × Analytes should be heat stable × Complicated derivatisation process × Artefacts due to derivatisation	Pérez-Alonso et al. (2018); Turumtay et al. (2016); Srinivasan and Kannan (2019); Razzaq et al. (2019)
NMR	✓ High sensitivity and selectivity ✓ Reproducible spectrum ✓ Non-destructive in nature ✓ Suitable for quantitative profiling ✓ Facilitates structure determination	× Cost intensive × Complicated spectral data × Tedious process of spectral analysis × Less sensitive and requires highly purified compounds	Pérez-Alonso et al. (2018); Turumtay et al. (2016); Srinivasan and Kannan (2019); Pontes et al. (2017); Razzaq et al. (2019)

metabolome of different medicinally and other agro-economically important plants has been carried out to identify the accumulation of bioactive compounds and their regulatory mechanism(s) under different stresses. The knowledge has widely been used for numerous applications such as crop improvement, chemotaxonomy (i.e., species and cultivar identification based on biochemical markers), development of natural product(s) formulations and so forth (Pontes et al. 2017; Foito and Stewart 2018; Parida, Panda and Rangani 2018).

Comprehensive and non-targeted metabolomics studies have been carried out for identification of metabolites involved in different biological processes such as senescence (Watanabe et al. 2018), shoot growth and development (Arkorful et al. 2020) and cyanogenesis (Diaz-Vivancos et al. 2017) using high-performance liquid chromatography (HPLC), GC/time-of-flight (TOF)-MS, LC/electrospray ionisation (ESI)-MS, and ultra-high-performance liquid chromatography (UHPLC)-MS, respectively. Comparative metabolite profiling and fingerprinting using UHPLC-HRMS, H^1-NMR, GC-MS has also been employed as a powerful tool for chemotaxonomic distinction of plants belonging to different habitats, species and genera (Vanderplanck and Glauser 2018; Grauso et al. 2019; Skubel et al. 2020; Chen et al. 2017b). Similar metabolomics platforms have also been used to decipher the mechanism of metabolic changes associated with plant-microbe interactions (Song et al. 2017). Studies pertaining to metabolomics-assisted identification of natural products having anti-viral (Prinsloo and Vervoort 2018), anti-diabetic (Hasanpour, Iranshahy and Iranshahi 2020) and anti-tuberculosis (Tuyiringire et al. 2018) activities have recently been reported. Moreover, applications of metabolomics in the early detection of plant diseases such as late blight in *Solanum lycopersicum* have also been established (Garcia et al. 2018). The effects of various environmental factors on plant metabolome have also been used to assess their ecotoxicological effect on plant physiological and developmental processes (Matich et al. 2019; Sampaio et al. 2016). Further, a relatively new approach of metabolomic analysis using radiolabelled isotopes for identification of whole or missing links in metabolic pathways has also been reported in recent years (Freund and Hegeman 2017; Nakabayashi and Saito 2020).

TABLE 3.2

Commonly Used Statistical and Bioinformatics Tools for Metabolomics

Tool	Weblink
Statistical Analysis	
MetaboAnalyst	www.metaboanalyst.ca/
MetAlign	www.metalign.nl
Babelomics 5	http://www.babelomics.org/
COVAIN	http://www.univie.ac.at/mosys/software.html
GenePattern	http://software.broadinstitute.org/cancer/software/genepattern/
Data Annotation and Processing	
MetaboSearch	http://omics.georgetown.edu/metabosearch.html
MeltDB 2.0	https://meltdb.cebitec.uni-bielefeld.de
MetiTree	http://www.metitree.nl/
MetaCrop 2.0	http://metacrop.ipk-gatersleben.de
MetAssign	http://mzmatch.sourceforge.net/
MET-COFEA	http://bioinfo.noble.org/manuscript-support/met-cofea/
iMet-Q	http://ms.iis.sinica.edu.tw/comics/Software_iMet-Q.html
XCMS	https://xcmsonline.scripps.edu
MZedDB	http://maltese.dbs.aber.ac.uk:8888/hrmet/index.html
MaxQuant	https://www.maxquant.org/
Maven	https://maven.apache.org/
MZmine2	http://mzmine.github.io/
MS-DIAL	http://prime.psc.riken.jp/Metabolomics_Software/MS-DIAL/
CAMERA	https://bioconductor.org/packages/release/bioc/html/CAMERA.html
ADAP	http://www.du-lab.org/software.html
Metabolite Annotation and Data Analysis	
METLIN	https://metlin.scripps.edu/
MetFrag	https://ipb-halle.github.io/MetFrag/
MetaGeneAlyse	http://metagenealyse.mpimp-golm.mpg.de/
MassBank	http://www.massbank.jp/
MarVis	http://marvis.gobics.de/
CFM-ID	http://cfmid.wishartlab.com
Pathway Analysis	
KEGG	http://www.genome.jp/kegg/
Mummichog	http://mummichog.org
MSEA	http://www.metaboanalyst.ca/
MetaboAnalyst (MetPA)	http://metpa.metabolomics.ca
MetExplore	http://metexplore.toulouse.inra.fr
AraCyc	https://www.plantcyc.org/typeofpublication/aracyc
MetaCyc	http://www.metacyc.org
PMN/PlantCyc	http://www.plantcyc.org
SMPDB	http://www.smpdb.ca
WikiPathways	http://wikipathways.org

3.3 Transcriptomics: Methods and Applications

Transcriptome refers to the entire pool of transcripts present in an organism or single cell under a given condition or at a certain developmental stage (Han et al. 2016). Being an indispensable connecting link between genomics and proteomics, transcriptomics plays a key role in unravelling the connection between different genotypic and phenotypic traits, thereby making it an excellent tool for functional characterisation of genes and identification of the molecular basis of biochemicals processes (Han et al. 2016; Abbai et al. 2017). Transcriptomics is a versatile approach for identification of functional genes as it (1) archives all types of RNA transcripts present in the target organism or cell, (2) deals with the structural and other related characteristics of genes, (3) measures the gene expression level and (4) quantifies impacts of environmental factors (Han et al. 2016). Since the advent of the post-genomic era, a quantum leap in developing research platforms for comprehensive and high-throughput analysis of transcripts and genomes has been witnessed (Jain et al. 2019). With the evolution of sequencing methods such as next-generation sequencing (NGS) over the past few decades, generation and analysis of multiple data sets of gene expression profiles has now become an inevitable step in functional biology research (Jain et al. 2019; Abbai et al. 2017). The use of NGS platform(s) for transcriptomics has facilitated high-throughput gene expression analysis in an unprecedentedly rapid, cost-effective, tractable and reproducible manner (Han et al. 2016). The methods for interpretation of transcript information into corresponding functions are broadly divided into two categories: (1) hybridisation-based and (2) sequencing-based platforms, the details of which are discussed in subsequent sections. However, since the inception of sequencing technologies, use of hybridisation methods has significantly diminished in the past few years.

3.3.1 Hybridisation-Based Platforms

The expression levels of each transcript are determined based on the intensities of signals generated through DNA/RNA hybridisation and the pre-selected library of probes (Jain et al. 2019). The probe library consists of short nucleotide sequences selected based on genes to be identified (target genes) in the sample sequence. DNA microarray is one of the most common hybridisation methods, in which thousands of labelled complementary DNA (cDNA) target sequences are simultaneously allowed to hybridise with a series of immobilised cDNA probes, specific to the gene(s) of interest (Abbai et al. 2017; Jain et al. 2019). The fluorescent signals emitted as a result of probe-target hybridisation are then converted into digital outputs for qualitative and quantitative gene expression analysis. Fluorescent signals are used for labelling the target cDNA; however, radioisotopes and chemiluminescent molecules can also be used for labelling of the same.

Suppression subtractive hybridisation (SSH) is another method, in which the cDNA target sequences (tester) are allowed to hybridise with the cDNA sequences of the genes of interest (driver) (Abbai et al. 2017). The signal processing and data analysis are performed similarly to that of the DNA microarray technique. This method has commonly been used to study the differences in gene expression profiles under different biotic and abiotic stresses, in which cDNA from the treated group is allowed to hybridise with that of the control group, such that the genes uniquely synthesised in response to a specific set of conditions given to the treated group remain unhybridised (Abbai et al. 2017).

3.3.2 Sequencing-Based Platforms

Advances in biotechnological and computational tools have led to the emergence of a diverse range of sequencing methods and platforms, in a very short span of time. Expressed sequence tags (ESTs) are typically about 500- to 800-bp short sequences from the cDNA library, which represent genes expressed in a cell under specific conditions (Abbai et al. 2017; Jain et al. 2019). Transcriptome profiles are generated through sequencing of ESTs followed by their cataloguing into transcript clusters, which are then mapped on target cDNA sequences and used for gene discovery, gene annotation, proteome analysis and so forth (Abbai et al. 2017; Jain et al. 2019). Serial analysis of gene expression (SAGE) is a deep sequencing approach in which multiple short (10–14 bp) cDNA reads unique to each gene are used to identify the transcripts (if present) in the target sequence (Abbai et al. 2017; Jain et al. 2019). This strategy also enables a quantitative comparison of transcriptomes under different stress conditions and predicts genes directly involved in different metabolic processes.

RNA-Seq is the most recent and extensively used strategy for transcriptome analysis. The cDNA library constructed using RNAs isolated from different tissues or plants under different treatments are sequenced and assembled either *de novo* or using a reference genome or transcriptome. The assembled transcripts are then mapped against various gene and protein databases to identify the key compounds involved in metabolism (Abbai et al. 2017). Transcriptome sequencing and analysis is a multi-step process that involves the preparation of plant material, RNA extraction, synthesis of cDNA libraries, deep sequencing, assembly of fragments generated and extraction of relevant information from the assembled sequences using computational tools and databases (Han et al. 2016) (Figure 3.2). Currently, BGISEQ-500 and Illumina HiSeq 4000 are the two most efficient and advanced transcriptome sequencing platforms that have been used in several studies (Zhu et al. 2018a). Tzfadia et al. (2018) reported a novel TranSeq 3′ end sequencing method for high-throughput transcriptomics and gene space refinement in plant genomes.

3.3.3 Recent Applications and Advances

With the emergence of the NGS technique, transcriptome analysis has become an important step in crop improvement and breeding programs. Transcriptomics has facilitated the identification of regulatory genes for various physiological processes, which in turn has opened avenues for engineering specific metabolic pathways. Many studies have been conducted to identify candidate genes responsible for stress tolerance among plants including identification of regulatory genes and elucidation of underlying defence mechanism(s) of *Arabidopsis* against *Pseudomonas aeruginosa* (Iizasa et al. 2017), *Triticum aestivum* against *Puccinia striiformis* (Hao et al. 2016), *Camellia sinensis* against *es obliqua* (Wang et al. 2016) and *Brassica napus* against *Leptosphaeria maculans* (Becker et al. 2017). Transcriptomics has also been employed to get insights into various plant developmental and physiological processes such as identification of regulatory networks involved in sex determination in *Jatropha curcas* (Chen et al. 2017a), pre-harvest and post-harvest decay mechanism in *Fragaria* (Landi et al. 2017), zig-zag stem formation in *Camellia sinensis* (Cao et al. 2020) and nitrogen fixation in *Medicago truncatula* (Liese, Schulze and Cabeza 2017). Comparative transcriptomics has recently been used to study the differential gene expression patterns for varietal discrimination such as wild and domesticated plant species (Huang et al. 2018). Transcriptomics-assisted determination of evolutionary relationships among different plants belonging to similar or different taxa has also been performed (Wu et al. 2017). Microarray-assisted transcriptome-wide analysis has been reported to provide information about genes expressed differentially in *Solanum lycopersicum* in response to some external stimuli (Ertani, Schiavon and Nardi 2017), whilst single-cell transcriptome profiling of different plant cell types has also been performed to understand differential or cell-specific regulatory mechanisms of various metabolisms.

3.4 Integrated Omics for Functional Biology Research

The primary goal of functional biology research is to gain in-depth understanding of the operational and regulatory framework of different biological processes by decoding the information stored in genes, identifying different key factors and establishing gene to phenotype relation (Rai, Yamazaki and Saito 2019). With the huge leap of advancements in omics-technology, researchers are able to gain deeper and better insights on the molecular, biochemical and physiological aspects of various biological processes, which could further be used to develop more efficient crop improvement strategies (Rai et al. 2019). To accomplish this, comprehensive knowledge of regulatory determinants of metabolic processes leading to desired traits is the first prerequisite. The regulatory framework of any biological process is governed by complex interactions between genes, transcripts, proteins and metabolites (Kumar et al. 2017).

Use of multi-omics as a tool to obtain a complete snapshot of the complexly intertwined metabolisms leading to various physiological, phenotypic and developmental changes has emerged as a promising approach in functional biology research (Pinu et al. 2019; Rai et al. 2019). The pool of data generated through multi-omics provides a more holistic understanding of the physiological and developmental processes in plants, which can further be used to enhance the expression of genes and production of metabolites of interest (Rai et al. 2019). Being the ultimate downstream products of the complex interactions between genes, transcripts and proteins, metabolites have become an indispensable part of multi-omics studies, and thus metabolomics serves as a common platform in all the multi-omics studies, enabling identification of genetic determinants of targeted metabolites (Kumar et al. 2017). Further, integration of metabolomics data with transcriptome can collectively aid in the identification of genes directly affecting levels of specific metabolite(s) and determination of the relation between metabolism and gene functions (Agarrwal and Nair 2020).

Due to their inherent large complex data sets and data differences, the integration of multiple-omics platforms requires high-end statistical and computational tools (Pinu et al. 2019). Thus, acquisition, processing and correlation analysis of multi-omics data are critical, which determine the significance and quality of the information retrieved from such data-rich pools (Pinu et al. 2019; Jamil et al. 2020). Moreover, due to enormously complex metabolic diversity, poorly annotated genomes of non-model species and various plant symbiotic associations, integration of data obtained from multi-omics approaches has comparatively been more challenging in plants than other organisms (Jamil et al. 2020). Over the years, the applicability of numerous bioinformatics tools and platforms for multi-omics integration has been reviewed in several studies (Pinu et al. 2019). Jamil et al. (2020), recently proposed a three-level approach for multi-omics data integration as discussed below (Figure 3.2):

1. ***Level 1 (element-based):*** Unbiased approach in which correlation, clustering and multivariate analyses of two or more omics data sets is performed to identify the nature (positive or negative) of the relation between them.
2. ***Level 2 (pathway-based):*** Knowledge-based approach in which pathway mapping against existing pathways available in databases and co-expression analysis to assess the strength of relationships among expressed molecules is performed.
3. ***Level 3 (mathematical-based):*** Differential and genome-scale analyses are performed to develop functional mathematical models and well-defined differential equation for prediction and measurement of different responses at the gene, transcript, protein or metabolome level.

Consequently, integration of metabolome data, with that generated through other omics-platforms has not only provided avenues for accelerated research in identifying novel bioactive compounds but has also generated scope for crop improvement through the development of omics-assisted breeding (Kumar et al. 2017; Christ et al. 2018; Alseekh et al. 2018; Abdelrahman et al. 2018) and production of genetically engineered crops by metabolic pathways engineering (Gaston et al. 2020; Zhu et al. 2018b; Ichihashi et al. 2018).

3.4.1 Advances in Integrated Metabolomics and Transcriptomics Research

NGS and other technological advances have revolutionised the functional biology and natural products research in plants. A sudden hike in the studies aimed at understanding a variety of biological responses of several models as well as non-model plants has been performed in the past few decades. Susceptibility of plants to various biotic and abiotic factors greatly influences their metabolic processes (particularly secondary metabolism) such that it may either promote/trigger or suppress biosynthesis of natural products, which in turn affects their quality and yield. Therefore, to develop efficient and reproducible breeding, thorough knowledge of various processes involved in response to different stresses that lead to the production of desired compounds is an essential requirement.

The majority of the commercially and medicinally important chemicals are produced during secondary metabolism, therefore extensive studies have been performed to elucidate biosynthesis pathway(s) of flavonoid (Gutierrez et al. 2017; Li et al. 2018; Chen et al. 2020), volatile organic compounds (Dhandapani et al. 2017), rotenoids (Gu et al. 2018), saponins (Zhang et al. 2020), phenylpropanoids (Meng et al. 2019), phenolics (Han et al. 2020) and pigment (anthocyanins)-producing compounds (Rothenberg et al. 2019; Dong et al. 2019). Integrated metabolome and transcriptome studies have also been performed to identify regulatory networks of lipid synthesis (Tan et al. 2019) and carbon and nitrogen metabolisms (Xin et al. 2019; Tahir et al. 2020). Multi-omics have also been utilised to identify the molecular and biochemical factors involved in various plants' developmental and physiological processes such as nitrogen fixation (Abdelrahman et al. 2018), tiller production (Wang et al. 2019), seed germination (Qu et al. 2019), photosynthesis (Lallemand et al. 2019) and aroma production (Xu et al. 2018).

Comparative omics-analysis has been used as an effective approach to differentiate among plant varieties and to identify the factors responsible for differential metabolite synthesis in the same parts of different plants (Guo et al. 2020; Tsaballa et al. 2020; Li et al. 2019). Multi-omics-guided crop improvement through metabolic modelling (Lakshmanan et al. 2016) and engineering (Gaston et al. 2020; Zhu et al. 2018b; Ichihashi et al. 2018) has also been reported.

3.5 Conclusion and Future Prospects

The past decade has witnessed a paradigm shift from conventional breeding to modern omics-assisted strategies for crop improvement. Although genetic marker–assisted or metabolite marker–assisted breeding has now become a normal practice, these methods have several limitations, the most important of which is the missing link between genotype and phenotype, which often leads to unwanted genetic or metabolic repercussions, thereby declining the overall potency of the breeding and selection programs. Metabolome provides a complete overview of the biochemical composition of the plants, while transcriptome provides information about various regulatory and non-regulatory factors involved in metabolite biosynthesis. Hence, it would be appropriate to say that metabolomics and transcriptomics complement each other in such a way that their co-analysis can provide a complete overview of the functional characteristics of plant metabolisms. Apart from being able to provide an absolute picture of the target metabolic process(s), integration of metabolomics with transcriptomics has also made the discovery of novel genes and products, identification of gene function, detection of novel pathways, regulation and reconstruction of metabolic networks possible, which has generated avenues for the development of rapid, cost-effective and efficient methods for crop improvement. Therefore, with technological advances in genetic engineering and functional biology working hand in hand, the production of improved crop varieties can be attained without causing any damage to their natural gene pool, which indeed could be the first step of mankind towards sustainable agriculture and development.

REFERENCES

Abbai, R., S. Su bramaniyam, R. Mathiyalagan, and D. Yang. 2017. Functional genomic approaches in plant research. In *Plant Bioinformatics*, ed., 215–239. Springer.

Abdelrahman, M., M. El-Sayed, A. Hashem, E. Abd_Allah, A. Alqarawi, D. Burritt, and L. Tran. 2018. Metabolomics and transcriptomics in legumes under phosphate deficiency in relation to nitrogen fixation by root nodules. *Frontiers in Plant Science* 9: 922.

Agarrwal, R., and S. Nair. 2020. Metabolomics-assisted crop improvement. In *Advancement in Crop Improvement Techniques*, ed., 263–274. Elsevier.

Alseekh, S., L. Bermudez, L. De Haro, A. Fernie, and F. Carrari. 2018. Crop metabolomics: from diagnostics to assisted breeding. *Metabolomics* 14: 148.

Arkorful, E., Y. Yu, C. Chen, L. Lu, S. Hu, H. Yu, Q. Ma, K. Thangaraj, R. Periakaruppan, and A. Jeyaraj. 2020. Untargeted metabolomic analysis using UPLC-MS/MS identifies metabolites involved in shoot growth and development in pruned tea plants (*Camellia sinensis* (L.) O. Kuntz). *Scientia Horticulturae* 264: 159–164.

Becker, M., X. Zhang, P. Walker, J. Wan, J. Millar, D. Khan, M. Granger, J. Cavers, A. Chan, and D. Fernando. 2017. Transcriptome analysis of the *Brassica napus–Leptosphaeria maculans* pathosystem identifies receptor, signaling and structural genes underlying plant resistance. *The Plant Journal* 90: 573–586.

Cao, H., F. Wang, H. Lin, Y. Ye, Y. Zheng, J. Li, Z. Hao, N. Ye, and C. Yue. 2020. Transcriptome and metabolite analyses provide insights into zigzag-shaped stem formation in tea plants (*Camellia sinensis*). *BMC Plant Biology* 20: 1–14.

Chen, M., B. Pan, Q. Fu, Y. Tao, J. Martínez-Herrera, L. Niu, J. Ni, Y. Dong, M. Zhao, and Z. Xu. 2017a. Comparative transcriptome analysis between gynoecious and monoecious plants identifies regulatory networks controlling sex determination in *Jatropha curcas*. *Frontiers in Plant Science* 7: 1953.

Chen, Q., X. Lu, X. Guo, Q. Guo, and D. Li. 2017b. Metabolomics characterization of two *Apocynaceae* plants, *Catharanthus roseus* and *Vinca minor*, using GC-MS and LC-MS methods in combination. *Molecules* 22: 997.

Chen, Y., W. Pan, S. Jin, and S. Lin. 2020. Combined metabolomic and transcriptomic analysis reveals key candidate genes involved in the regulation of flavonoid accumulation in *Anoectochilus roxburghii. Process Biochemistry* 91: 339–351.

Christ, B., T. Pluskal, S. Aubry, and J. Weng. 2018. Contribution of untargeted metabolomics for future assessment of biotech crops. *Trends in Plant Science* 23: 1047–1056.

Delfin, J., M. Watanabe, and T. Tohge. 2019. Understanding the function and regulation of plant secondary metabolism through metabolomics approaches. *Theoretical and Experimental Plant Physiology* 31: 127–138.

Dhandapani, S., J. Jin, V. Sridhar, R. Sarojam, N. Chua, and I. Jang. 2017. Integrated metabolome and transcriptome analysis of *Magnolia champac*a identifies biosynthetic pathways for floral volatile organic compounds. *BMC Genomics* 18: 463.

Diaz-Vivancos, P., A. Bernal-Vicente, D. Cantabella, C. Petri, and J. Hernández. 2017. Metabolomics and biochemical approaches link salicylic acid biosynthesis to cyanogenesis in peach plants. *Plant and Cell Physiology* 58: 2057–2066.

Djande, C., C. Pretorius, F. Tugizimana, L. Piater, and I. Dubery. 2020. Metabolomics: A tool for cultivar phenotyping and investigation of grain crops. *Agronomy* 10: 831.

Dong, T., R. Han, J. Yu, M. Zhu, Y. Zhang, Y. Gong, and Z. Li. 2019. Anthocyanins accumulation and molecular analysis of correlated genes by metabolome and transcriptome in green and purple asparaguses (*Asparagus officinalis*, L.). *Food Chemistry* 271: 18–28.

Ertani, A., M. Schiavon, and S. Nardi. 2017. Transcriptome-wide identification of differentially expressed genes in *Solanum lycopersicon* L. in response to an *Alfalfa*-protein hydrolysate using microarrays. *Frontiers in Plant Science* 8: 1159.

Foito, A., and D. Stewart. 2018. Metabolomics: a high-throughput screen for biochemical and bioactivity diversity in plants and crops. *Current Pharmaceutical Design* 24: 2043–2054.

Freund, D., and A. Hegeman. 2017. Recent advances in stable isotope-enabled mass spectrometry-based plant metabolomics. *Current Opinion in Biotechnology* 43: 41–48.

Garcia, P., F. dos Santos, S. Zanotta, M. Eberlin, and C. Carazzone. 2018. Metabolomics of *Solanum lycopersicum* infected with *Phytophthora infestans* leads to early detection of late blight in asymptomatic plants. *Molecules* 23: 3330.

Gaston, A., S. Osorio, B. Denoyes, and C. Rothan. 2020. Applying the *Solanaceae* strategies to strawberry crop improvement. *Trends in Plant Science* 25: 130–140.

Grauso, L., M. Zotti, W. Sun, B. de Falco, V. Lanzotti, and G. Bonanomi. 2019. Spectroscopic and multivariate data-based method to assess the metabolomic fingerprint of Mediterranean plants. *Phytochemical Analysis* 30: 572–581.

Gu, L., Z. Zhang, H. Quan, M. Li, F. Zhao, Y. Xu, J. Liu, M. Sai, W. Zheng, and X. Lan. 2018. Integrated analysis of transcriptomic and metabolomic data reveals critical metabolic pathways involved in rotenoid biosynthesis in the medicinal plant *Mirabilis himalaica. Molecular Genetics and Genomics* 293: 635–647.

Guo, J., Y. Wu, G. Wang, T. Wang, and F. Cao. 2020. Integrated analysis of the transcriptome and metabolome in young and mature leaves of *Ginkgo biloba* L. *Industrial Crops and Products* 143: 111906.

Gutierrez, E., A. García-Villaraco, J. Lucas, A. Gradillas, F. Gutierrez-Mañero, and B. Ramos-Solano. 2017. Transcriptomics, targeted metabolomics and gene expression of blackberry leaves and fruits indicate flavonoid metabolic flux from leaf to red fruit. *Frontiers in Plant Science* 8: 472.

Han, R., A. Rai, M. Nakamura, H. Suzuki, H. Takahashi, M. Yamazaki, and K. Saito. 2016. *De novo* deep transcriptome analysis of medicinal plants for gene discovery in biosynthesis of plant natural products. In *Methods in Enzymology*, ed., 19–45. Elsevier.

Han, Z., M. Ahsan, M. Adil, X. Chen, M. Nazir, I. Shamsi, F. Zeng, and G. Zhang. 2020. Identification of the gene network modules highly associated with the synthesis of phenolics compounds in barley by transcriptome and metabolome analysis. *Food Chemistry* 126862.

Hao, Y., T. Wang, K. Wang, X. Wang, Y. Fu, L. Huang, and Z. Kang. 2016. Transcriptome analysis provides insights into the mechanisms underlying wheat plant resistance to stripe rust at the adult plant stage. *PLoS One* 11: e0150717.

Hasanpour, M., M. Iranshahy, and M. Iranshahi. 2020. The application of metabolomics in investigating antidiabetic activity of medicinal plants. *Biomedicine & Pharmacotherapy* 128: 110263.

Hirai, M., M. Yano, D. Goodenowe, S. Kanaya, T. Kimura, M. Awazuhara, M. Arita, T. Fujiwara, and K. Saito. 2004. Integration of transcriptomics and metabolomics for understanding of global responses to nutritional stresses in *Arabidopsis thaliana*. *Proceedings of the National Academy of Sciences* 101: 10205–10210.

Hong, J., L. Yang, D. Zhang, and J. Shi. 2016. Plant metabolomics: an indispensable system biology tool for plant science. *International Journal of Molecular Sciences* 17: 767.

Huang, X., B. Wang, J. Xi, Y. Zhang, C. He, J. Zheng, J. Gao, H. Chen, S. Zhang, and W. Wu. 2018. Transcriptome comparison reveals distinct selection patterns in domesticated and wild Agave species, the important CAM plants. *International Journal of Genomics* 2018: 5716518.

Ichihashi, Y., M. Kusano, M. Kobayashi, K. Suetsugu, S. Yoshida, T. Wakatake, K. Kumaishi, A. Shibata, K. Saito, and K. Shirasu. 2018. Transcriptomic and metabolomic reprogramming from roots to haustoria in the parasitic plant *Thesium chinense*. *Plant and Cell Physiology* 59: 729–738.

Iizasa, S., E. Iizasa, K. Watanabe, and Y. Nagano. 2017. Transcriptome analysis reveals key roles of AtLBR-2 in LPS-induced defense responses in plants. *BMC Genomics* 18: 995.

Jain, D., N. Ashraf, J. Khurana, and M. Kameshwari. 2019. The 'Omics' approach for crop improvement against drought stress. In *Genetic Enhancement of Crops for Tolerance to Abiotic Stress: Mechanisms and Approaches*, ed., 183–204. Springer.

Jamil, I., J. Remali, K. Azizan, N. Muhammad, M. Arita, H. Goh, and W. Aizat. 2020. Systematic multi-omics integration (MOI) approach in plant systems biology. *Frontiers in Plant Science* 11: 944.

Kumar, R., A. Bohra, A. Pandey, M. Pandey, and A. Kumar. 2017. Metabolomics for plant improvement: status and prospects. *Frontiers in Plant Science* 8: 1302.

Lakshmanan, M., C. Cheung, B. Mohanty, and D. Lee. 2016. Modeling rice metabolism: from elucidating environmental effects on cellular phenotype to guiding crop improvement. *Frontiers in Plant Science* 7: 1795.

Lallemand, F., M. Martin-Magniette, F. Gilard, B. Gakière, A. Launay-Avon, É. Delannoy, and M. Selosse. 2019. *In situ* transcriptomic and metabolomic study of the loss of photosynthesis in the leaves of mixotrophic plants exploiting fungi. *The Plant Journal* 98: 826–841.

Landi, L., R. M. De Miccolis Angelini, S. Pollastro, E. Feliziani, F. Faretra, and G. Romanazzi. 2017. Global transcriptome analysis and identification of differentially expressed genes in strawberry after preharvest application of benzothiadiazole and chitosan. *Frontiers in Plant Science* 8: 235.

Li, Y., J. Fang, X. Qi, M. Lin, Y. Zhong, L. Sun, and W. Cui. 2018. Combined analysis of the fruit metabolome and transcriptome reveals candidate genes involved in flavonoid biosynthesis in *Actinidia arguta*. *International Journal of Molecular Sciences* 19: 1471.

Li, Y., W. Wang, Y. Feng, M. Tu, P. Wittich, N. Bate, and J. Messing. 2019. Transcriptome and metabolome reveal distinct carbon allocation patterns during internode sugar accumulation in different sorghum genotypes. *Plant Biotechnology Journal* 17: 472–487.

Liese, R., J. Schulze, and R. Cabeza. 2017. Nitrate application or P deficiency induce a decline in *Medicago truncatula* N_2-fixation by similar changes in the nodule transcriptome. *Scientific Reports* 7: 46264.

Maroli, A., T. Gaines, M. Foley, S. Duke, M. Doğramacı, J. Anderson, D. Horvath, W. Chao, and N. Tharayil. 2018. Omics in weed science: A perspective from genomics, transcriptomics, and metabolomics approaches. *Weed Science* 66: 681–695.

Matich, E., N. Soria, D. Aga, and G. Atilla-Gokcumen. 2019. Applications of metabolomics in assessing ecological effects of emerging contaminants and pollutants on plants. *Journal of Hazardous Materials* 373: 527–535.

Meng, J., B. Wang, G. He, Y. Wang, X. Tang, S. Wang, Y. Ma, C. Fu, G. Chai, and G. Zhou. 2019. Metabolomics integrated with transcriptomics reveals redirection of the phenylpropanoids metabolic flux in *Ginkgo biloba*. *Journal of Agricultural and Food Chemistry* 67: 3284–3291.

Nakabayashi, R., and K. Saito. 2020. Higher dimensional metabolomics using stable isotope labeling for identifying the missing specialized metabolism in plants. *Current Opinion in Plant Biology* 55: 84–92.

Oksman-Caldentey, K., and K. Saito. 2005a. Integrating genomics and metabolomics for engineering plant metabolic pathways. *Current Opinion in Biotechnology* 16: 174–179.

Parida, A., A. Panda, and J. Rangani. 2018. Metabolomics-guided elucidation of abiotic stress tolerance mechanisms in plants. In *Plant Metabolites and Regulation Under Environmental Stress*, ed., 89–131. Elsevier.

Pérez-Alonso, M., V. Carrasco-Loba, and S. Pollmann. 2018. Advances in plant metabolomics. *Annual Plant Reviews Online* 1: 557–588.

Piasecka, A., P. Kachlicki, and M. Stobiecki. 2019. Analytical methods for detection of plant metabolomes changes in response to biotic and abiotic stresses. *International Journal of Molecular Sciences* 20: 379.

Pinu, F., D. Beale, A. Paten, K. Kouremenos, S. Swarup, H. Schirra, and D. Wishart. 2019. Systems biology and multi-omics integration: Viewpoints from the metabolomics research community. *Metabolites* 9: 76.

Pontes, J., A. Brasil, G. Cruz, R. de Souza, and L. Tasic. 2017. NMR-based metabolomics strategies: plants, animals and humans. *Analytical Methods* 9: 1078–1096.

Prinsloo, G., and J. Vervoort. 2018. Identifying anti-HSV compounds from unrelated plants using NMR and LC–MS metabolomic analysis. *Metabolomics* 14: 134.

Qu, C., Z. Zuo, L. Cao, J. Huang, X. Sun, P. Zhang, C. Yang, L. Li, Z. Xu, and G. Liu. 2019. Comprehensive dissection of transcript and metabolite shifts during seed germination and post-germination stages in poplar. *BMC Plant Biology* 19: 279.

Rai, A., M. Yamazaki, and K. Saito. 2019. A new era in plant functional genomics. *Current Opinion in Systems Biology* 15: 58–67.

Razzaq, A., B. Sadia, A. Raza, M. Hameed, and F. Saleem. 2019. Metabolomics: A way forward for crop improvement. *Metabolites* 9: 303.

Rothenberg, D., H. Yang, M. Chen, W. Zhang, and L. Zhang. 2019. Metabolome and transcriptome sequencing analysis reveals anthocyanin metabolism in pink flowers of anthocyanin-rich tea (*Camellia sinensis*). *Molecules* 24: 1064.

Saito, K., and F. Matsuda. 2010. Metabolomics for functional genomics, systems biology, and biotechnology. *Annual Review of Plant Biology* 61: 463–489.

Sampaio, B., R. Edrada-Ebel, and F. Da Costa. 2016. Effect of the environment on the secondary metabolic profile of *Tithonia diversifolia*: a model for environmental metabolomics of plants. *Scientific Reports* 6: 29265.

Skubel, S., X. Su, A. Poulev, L. Foxcroft, V. Dushenkov, and I. Raskin. 2020. Metabolomic differences between invasive alien plants from native and invaded habitats. *Scientific Reports* 10: 1–9.

Song, T., H. Xu, N. Sun, L. Jiang, P. Tian, Y. Yong, W. Yang, H. Cai, and G. Cui. 2017. Metabolomic analysis of alfalfa (*Medicago sativa* L.) root-symbiotic rhizobia responses under alkali stress. *Frontiers in Plant Science* 8: 1208.

Srinivasan, T., and R. Kannan. 2019. Single-cell-type metabolomics for crop improvement. In *Single-Cell Omics*, ed., 315–339. Elsevier.

Tahir, A., J. Kang, F. Choulet, C. Ravel, I. Romeuf, F. Rasouli, A. Nosheen, and G. Branlard. 2020. Deciphering carbohydrate metabolism during wheat grain development via integrated transcriptome and proteome dynamics. *Molecular Biology Reports* 47: 5439–5449.

Tan, H., J. Zhang, X. Qi, X. Shi, J. Zhou, X. Wang, and X. Xiang. 2019. Correlation analysis of the transcriptome and metabolome reveals the regulatory network for lipid synthesis in developing *Brassica napus* embryos. *Plant Molecular Biology* 99: 31–44.

Teh, H., B. Neoh, N. Ithnin, L. Daim, T. Ooi, and D. Appleton. 2017. Omics and strategic yield improvement in oil crops. *Journal of the American Oil Chemists' Society* 94: 1225–1244.

Tsaballa, A., E. Sarrou, A. Xanthopoulou, E. Tsaliki, C. Kissoudis, E. Karagiannis, M. Michailidis, S. Martens, E. Sperdouli, and Z. Hilioti. 2020. Comprehensive approaches reveal key transcripts and metabolites highlighting metabolic diversity among three oriental tobacco varieties. *Industrial Crops and Products* 143: 111933.

Tugizimana, F., M. Mhlongo, L. Piater, and I. Dubery. 2018. Metabolomics in plant priming research: the way forward. *International Journal of Molecular Sciences* 19: 1759.

Turumtay, H., C. Sandallı, and E. Turumtay. 2016. Plant metabolomics and strategies. In *Plant Omics: Trends and Applications*, ed., 399–406. Springer.

Tuyiringire, N., D. Tusubira, J. Munyampundu, C. Tolo, C. Muvunyi, and P. Ogwang. 2018. Application of metabolomics to drug discovery and understanding the mechanisms of action of medicinal plants with anti-tuberculosis activity. *Clinical and Translational Medicine* 7: 1–12.

Tzfadia, O., S. Bocobza, J. Defoort, E. Almekias-Siegl, S. Panda, M. Levy, V. Storme, S. Rombauts, D. Jaitin, and H. Keren-Shaul. 2018. The 'TranSeq'3′-end sequencing method for high-throughput transcriptomics and gene space refinement in plant genomes. *The Plant Journal* 96: 223–232.

Vanderplanck, M., and G. Glauser. 2018. Integration of non-targeted metabolomics and automated determination of elemental compositions for comprehensive alkaloid profiling in plants. *Phytochemistry* 154: 1–9.

Wang, Y., L. Tang, Y. Hou, P. Wang, H. Yang, and C. Wei. 2016. Differential transcriptome analysis of leaves of tea plant (*Camellia sinensis*) provides comprehensive insights into the defense responses to *Ectropis oblique* attack using RNA-Seq. *Functional & Integrative Genomics* 16: 383–398.

Wang, Z., H. Shi, S. Yu, W. Zhou, J. Li, S. Liu, M. Deng, J. Ma, Y. Wei, and Y. Zheng. 2019. Comprehensive transcriptomics, proteomics, and metabolomics analyses of the mechanisms regulating tiller production in low-tillering wheat. *Theoretical and Applied Genetics* 132: 2181–2193.

Watanabe, M., T. Tohge, S. Balazadeh, A. Erban, P. Giavalisco, J. Kopka, B. Mueller-Roeber, A. Fernie, and R. Hoefgen. 2018. Comprehensive metabolomics studies of plant developmental senescence. In *Plant Senescence*, ed., 339–358. Springer.

Wu, F., C. Tang, Y. Guo, Z. Bian, J. Fu, G. Lu, J. Qi, Y. Pang, and Y. Yang. 2017. Transcriptome analysis explores genes related to shikonin biosynthesis in *Lithospermeae* plants and provides insights into Boraginales' evolutionary history. *Scientific Reports* 7: 1–11.

Xin, W., L. Zhang, W. Zhang, J. Gao, J. Yi, X. Zhen, Z. Li, Y. Zhao, C. Peng, and C. Zhao. 2019. An integrated analysis of the rice transcriptome and metabolome reveals differential regulation of carbon and nitrogen metabolism in response to nitrogen availability. *International Journal of Molecular Sciences* 20: 2349.

Xu, Q., Y. He, X. Yan, S. Zhao, J. Zhu, and C. Wei. 2018. Unraveling a crosstalk regulatory network of temporal aroma accumulation in tea plant (*Camellia sinensis*) leaves by integration of metabolomics and transcriptomics. *Environmental and Experimental Botany* 149: 81–94.

Yang, D., X. Du, Z. Yang, Z. Liang, Z. Guo, and Y. Liu. 2014. Transcriptomics, proteomics, and metabolomics to reveal mechanisms underlying plant secondary metabolism. *Engineering in Life Sciences* 14: 456–466.

Zhang, J., Z. Cun, H. Wu, and J. Chen. 2020. Integrated analysis on biochemical profiling and transcriptome revealed nitrogen-driven difference in accumulation of saponins in a medicinal plant *Panax notoginseng*. *Plant Physiology and Biochemistry* 154: 564–580.

Zhu, F., M. Chen, N. Ye, W. Qiao, B. Gao, W. Law, Y. Tian, D. Zhang, D. Zhang, and T. Liu. 2018a. Comparative performance of the BGISEQ-500 and Illumina HiSeq4000 sequencing platforms for transcriptome analysis in plants. *Plant Methods* 14: 69.

Zhu, G., S. Wang, Z. Huang, S. Zhang, Q. Liao, C. Zhang, T. Lin, M. Qin, M. Peng, and C. Yang. 2018b. Rewiring of the fruit metabolome in tomato breeding. *Cell* 172: 249–261.e12.

4

Recent Advances in Protein Bioinformatics

Mahak Tufchi
GCW Parade Jammu

Rashmi
Govind Ballabh Pant University of Agriculture and Technology

Naveen Kumar
ICAR-Indian Institute of Maize Research

CONTENTS

4.1 Introduction

Life sustenance on earth is possible due to the biological processes that occur at the molecular level. The study of molecular biology unveils the trapped information in the nucleotide sequence that constitutes the genetic material. This flow of information encoded in the messenger RNA in the form of a triplet codon ultimately leads to the synthesis of a polypeptide chain or protein at the expressional level. The discovery of DNA structure and the incipience of various sequencing techniques resulted in the sequencing of the whole genome of various organisms leading to the explosion of biological data. This created the need for several biological data banks which can store this huge information of nucleotides, protein or even motif sequences and a new advanced branch of biology came into existence called Bioinformatics.

Bioinformatics branch deals with the application of high-order computational and analytical tools to capture and analyze biological information. Applying this to the branch of protein study or proteomics offers the management, data elaboration and integration of new software packages and algorithms. This chapter thus deals with the current scenario of protein bioinformatics in terms of recent databases. A database is the collection of sequence and structure information that is featured, annotated and retrievable. The data can be searchable using a search engine, updated periodically and even cross-referenced. The phenomenon of sequencing of the insulin protein led to the creation of the first database by Margaret Dayhoff et al. in 1965. The most important aspect of all databases is the retrieval system which is available for obtaining the requisite information as and when required by the researcher. Commonly used retrieval systems are SRS (Sequence Retrieval System) and Entrez. The SRS is the most common type of retrieval system in which the data information and its acquisition is made accessible to the searcher. Entrez, on the other hand, is a World Wide Web based data retrieval tool developed by the

NCBI (National Center for Biotechnology Information), simplest to use and does not even require any specific browser requirements as it directly accesses through PubMed.

4.2 Protein Sequence Databases

The protein sequence database was initiated in the year 1986 and was called Swiss-Prot. The primary databases include the data directly submitted by the researcher with relatively little validation. Swiss-Prot, PIR (Protein Information Resource), UniProt and TrEMBL (Translation of the European Molecular Biology Laboratory) are primary protein sequence databases, while PDB (Protein Data Bank) and NDB (Nucleic Acid Database) are the primary structure databases. The secondary databases are manually curated or automatically generated through analysis of entries from the primary databases. They contain information of conserved sequences, signature sequences and active site residues of the protein families. Swiss-Prot is the common example of a curated sequence database. The third kind of database is the Composite Databases, where the data from multiple primary resources is gathered together, making querying and searching more efficient. BioSilico, NRDB (Non-redundant Database) and OWL (Web Ontology Language) are examples of composite databases. Different types of protein sequence databases with their functional significance are represented in a tabulated form in Table 4.1. A combination of three protein sequence databases (Swiss-Prot, TrEMBL and PIR-International Protein Sequence Database)

TABLE 4.1
An Update on Protein Sequence Databases

Protein Sequence Databases	Links	Description
Swiss-Prot	https://www.uniprot.org/statistics/Swiss-Prot	It provides a high level of annotation
PIR	https://proteininformationresource.org	Resource for protein sequences and functional information
TrEMBL	https://www.uniprot.org/statistics/TrEMBL	It is a computer-annotated protein sequence database
GenPept	https://in.mathworks.com/help/bioinfo/ref/getgenpept.html	It consists of translation of all CDS (coding sequence) features with a translation qualifier
RefSeq	http://www.ncbi.nlm.nih.gov/RefSeq	It is collection of a non-redundant set of nucleotide and protein sequences with bibliographic annotation
UniProt	https://www.uniprot.org	Universal protein resources
UniProt Knowledgebase (UniProtKB)	https://www.uniprot.org/help/uniprotkb	It is the central hub for a collection of functional information on proteins
UniProt Archive UniParc)	https://www.uniprot.org/help/uniparc	It is a non-redundant database that contains protein sequences
UniProt Reference Clusters (UniRef)	https://www.uniprot.org/uniref/	It contains all UniprotKnowldege bases record plus selected UniPrac records
UniProt Proteomes	uniport.org/help/proteomes	Linked to one or more proteomes
PDBe	http://www.ebi.ac.uk/pdbe	Available PDB data to identify important structural features
PDB	https://www.rcsb.org	Protein data bank for 3D structure of proteins
InterPro	http://www.ebi.ac.uk/interpro/	Protein sequence analysis and classification
Pfam	http://pfam.xfam.org	Family and domain databases
PIRSF	http://pir.georgetown.edu/pirsf	Family classification system for protein functional and evolutionary analysis
PROSITE	https://prosite.expasy.org	Consists of entries describing protein families, domains and functional sites as well as amino acid patterns and profiles
PRIDE	http://www.ebi.ac.uk/pride	Data repository of mass spectrometry-based proteomics data

yielded a single resource database as "UniProt". It comprises two sections known as UniProtKB/Swiss-Prot (reviewed annotated entries) and UniProtKB/TrEMBL (unreviewed annotated entries). UniProt provides many datasets including Literature Citations, Taxonomy, Keywords, Subcellular locations, etc., besides sequence search, alignment and retrieval tools like BLAST (Basic Local Alignment Search Tool), Align and ID (Identifier) mapping (Pundir et al. 2017). A compliment of UniProtKB is neXtProt (https://www.nextprot.org), a human protein knowledge base generating data on proteomics (85%) and genetic variations in humans. The neXtProt is a continuously evolving database that provides comprehensive and organised up-to-date information to the researchers involved in human protein studies (Gaudet et al. 2017). A major advancement in human transmembrane proteins (TMPs) is the development of MutHTP (http://www.iitm.ac.in/bioinfo/MutHTP/), the database which contains information on 183,395 disease-associated and 17,827 neutral mutations. It provides the mutation site location around the TMP topology and notifies the structural environment. This integrated database will help researchers to design and develop personalised treatment strategies for human diseases due to mutations (Kulandaisamy et al. 2018). The recent development in the field of protein-degrading enzymes is the MEROPS database publicly available at http://www.ebi.ac.uk/merops/, an integrated source of all sets of information about peptidases. This database provides an in-depth look into the evolution of peptidase families concerning origin and complexity in organisms as well as the recent structures for proteasome complexes in archaea. Recently, peptidase homologs have been identified and compared to other pro and eukaryote proteomes (Rawlings et al. 2018).

4.2.1 Family and Domain Databases

Pfam is a database of protein families that utilises seed alignment to build a profile based on the hidden Markov model (HMM) using the software available at http://hmmer.org/. It is intended to include as much of the protein sequences with functional information as possible with the utilisation of the fewest number of HMMs. New additions include domain boundary clarifications and the formation of Pfam clans, along with functional annotation of a total of 17,929 numbers of families. Scaling up is being controlled in *pfamseq* as it is an offshoot only from the UniProtKB sequences and each entry is linked to an ORCID identifier (El-Gebali et al. 2019). Domain databases like ProDom available at http://prodom.prabi.fr/prodom/current/html/home.php comprise a comprehensive set of protein domain families automatically generated from the UniProt Knowledge Database (Mishra et al. 2019) while CATH (Class Architecture Topology and Homologous Superfamily) is a type of database used for hierarchical classification of protein domains and identifies structural similarities or homology relationships based on similarity and dissimilarity among different domains. Divergence in protein domains from the same family leads to the formation of groups under as SuperFamilies. Sillitoe et al. (2015) established that a particular protein structure search can be feasible through the use of an algorithm called CATHEDRAL specifically dedicated to domain recognition. SCOP (The Structural Classification of Proteins), another domain database, creates curated manual domain data along with a hierarchal pattern of the three-dimensional (3D) structure of proteins to determine the evolutionary links between proteins. SCOPe stands for structural classification of proteins extended, a database that was manually curated to retain unambiguous protein structures in 2014. There is also another system for classification dedicated to the study of evolution heterogeneity known as SCOP2 available at http://scop.berkeley.edu/help/ver=2.05#scopchanges (Khandelwal et al. 2016). The TOPDOM database available at http://topdom.enzim.hu extends the incorporation of the collection of domains and motifs from globular proteins besides transmembrane by utilising the CCTOP algorithm which determines the type and topology of TMPs (Varga et al. 2016). Uniclust server at https://uniclust.mmseqs.com comprises a triplet of clustered protein sequence databases and multiple sequence alignments as a stock for sequence search, analysis and function prediction. The Uniclust database annotates 17% more Pfam domains than UniProt. Besides it, the associated MMseqs2 software has fast and sensitive protein sequence searching and clustering, making Uniclust database a unique and useful database (Mirdita et al. 2017).

Similarly, based on homologies (both distant and close) and sequence similarities the protein domains in ECOD database are classified and the family groups are formed. Comparison to Pfam and CDD reveals that 27% and 16% of ECOD families are new and 35% and 48% of these families had modified boundaries (Liao et al. 2018). Certain family databases like PDZscape (http://www.actrec.gov.in:8080/pdzscape) comprise more than 58,648 PDZ-containing protein families having significance in signal transduction cellular pathways through localisation and membrane receptor clustering. The comprehensive compilation includes mutations and diseases associated with PDZ-containing proteins with their known and putative binding partners. It also features tools like PDZ-BLAST and identifiers for PDZ-binding proteins (Doshi et al. 2018).

4.3 Protein Structure Databases

Proteins when expressed undergo posttranslational modification to resume their functional 3D structure. Proteomics is the branch of omics that involves sequencing of the protein, 3D structure determination and configuring its functional annotations.

The data from X-ray crystallography, nuclear magnetic resonance (NMR) spectroscopy, 3D electron microscopy (3DEM), including cryo-electron diffraction and tomography, serial femtosecond crystallography, and X-ray free electron laser studies, have been used as an input for running various software to elucidate the structure prediction of unknown proteins either by homology modelling, in which an already deciphered structure of a related protein is used as a template to model the unknown structure, or template-free modelling, which accesses novel protein folds through conformational sampling and physics-based energy changes (Kuhlman and Bradley 2019). Utilising the available crystal structure, details from various databases have led to the structure prediction of almost two-thirds of the protein families known to date. PDB is the first protein structure database that was established (https://www.rcsb.org) and is integrated with more than 40 external data resources providing rich structural views. Recent advancement in it is the PDBFlex database (http://pdbflex.org) which provides information on the flexibility of protein structures by analyzing their structural differences and clustering them according to their similarities. The main feature tools include two-dimensional (2D) scaling for root-mean-square deviation (RMSD) calculations, creation of distance maps among coordinates, graphical representation through structural disorder in secondary structure and 3D views using JSMol visualisation software. A multiple-sequence alignment of the target protein and related sequences is the key step in the prediction of structure in template-free modelling. The various predicted features include a secondary structure with torsion angles and inter-residue distances across the polypeptide chain that guide for building 3D models, their refinement and final predictions (Hrabe et al. 2016). Similarly, BOCTOPUS2 webserver available at (https://github.com/NBISweden/predictprotein-webserver-boctopus2), a devoted method for topology prediction and identification of residue orientations in β-domain TMPs, utilises the dyad–repeat pattern. BOCTOPUS2 successfully has predicted the topology in 69% of the 42 proteins used in the study, an improvement of more than 10% over the best earlier method used (Hayat et al. 2016). SPIDER2 a webserver package available at http://sparks-lab.org on the other end utilises deep neural networks to predict the secondary structure of proteins with 82% accuracy. With this, it predicted the actual solvent accessible surface area correlation as 0.76 along with main-chain torsional angles (Yang et al. 2017). Indicating a major contribution in the large-scale annotation of proteomes with authenticated intrinsic disorder information, MobiDB-lite available at http://protein.bio.unipd.it/mobidblite/ predicted few false positives based on consensus prediction using eight different predictors. MobiDB-lite is indeed useful and thus integrated into the MobiDB, DisProt and InterPro databases (Necci et al. 2017). High-speed homology-driven function annotation of proteins (HFSP) is a computational method that uses the MMseqs2 algorithm along with similarity in alignment length and sequence identity as criteria to predict functional annotations of unknown proteins (Mahlich et al. 2018). Recently MUFOLD-SS available at http://dslsrv8.cs.missouri.edu/~cf797/MUFoldSS/download.html outperformed all other deep neural networks significantly by exploiting PSI-BLAST and HHBlits profiles for accurate secondary structure prediction of proteins along with the physical/chemical properties of amino acids (Fang et al. 2018). SPOT-1D prediction of protein secondary structure tool available at https://sparks-lab.org/server/spot-1d/ works by exploiting the convolutional neural networks and contact maps for determining the backbone angles, solvent accessibility and contact numbers. The method is expected to be useful for advancing protein secondary structure and even prediction of function. Prediction of helix secondary structure from protein sequences in polyproline II utilises bidirectional recurrent neural networks laid on dihedral angle filtering (Hanson et al. 2019). PiPred available at https://toolkit.tuebingen.mpg.de/#/tools/quick2d can predict π-helices with individual amino acid precision (48%) with sensitivity (46%). The functional aspect in the near future will be modelling the 3D structure of transmembrane domains, suggesting the protein-ligand interactions specifically ligand-binding pockets (Ludwiczak et al. 2019). A PDB database needs enrichment as only 2% of TMP information is available, although 25% of the human proteome is predicted to be composed of them. TMCrys, available at https://github.com/brgenzim/tmcrys, is a crystallisation prediction method that improves target selection of TMPs. Besides it, the major problem encountered while predicting the secondary structures is the intrinsic disorder which remains unrecognised in X-ray crystallography (Varga and Tusnady 2018). The data from the DisProt database available at http://www.disprot.org/assessment/ established the long intrinsic protein disorders through benchmarking in new protein sequences. However, a large fraction remains undetected and the ranking of methods remains different (Necci et al. 2018). Validation of the predicted intrinsic disorders has emphasised the need for quality checks through scores. The probability of correct prediction at a residue level in protein structure is accounted for by QA (Quality Assessment) score and has been well practiced in other branches of bioinformatics. The QUARTER (QUality Assessment for pRotein in Trinsic disordEr pRedictions) method available at http://biomine.cs.vcu.edu/servers/QUARTER/ accommodates a diverse set of ten disorder predictors and contributes to the overall predictive performance of this tool. Recently the QA score-based predictions were utilised to interpret the intrinsic disorders in the proteome of humans (Hu et al. 2019). PPIIPRED software available at http://bioware.ucd.ie/PPIIPRED site favors even the amino acids whose side chains extend from the backbone. More specifically in protein sequences, π-helices prediction is suggested by a deep learning neural network-based method which makes it a standalone tool (O'Brien et al. 2020).

Other than the neural network–based approach, the use of fragment libraries has recently been used to predict the protein structure and has substantially proved to improve the quality of the predicted structures with few drawbacks. The selection of the fragment size length from the different fragment sizes for accurate prediction of the native structure is done through the process of model building. It was indicated that fragments that shorten create the more accurate native structures than the long fragments. Models obtained from the libraries which used the sequence similarity criterion for structure prediction were better than others. Certain critical guidelines are essential to be followed in protein structure prediction using the fragment libraries (Trevizani et al. 2017).

There is no single tool that predicts secondary structures consistently and thus a combined use of the above discussed prediction tools can enhance the efficacy of *in silico* protein structure prediction. Need for a new database that will fill the knowledge gap and predict the protein functions precisely resulted in COMBREX-DB (http://combrex.bu.edu), an online microbial repository of information related to protein structure prediction and their phenotypes based on experimental data ensuing accelerated rate of gene function validation. The database includes ~3.3 million knowns and predicted proteins of bacterial and archaeal genomes to help in the function prediction of unknown proteins within families (Chang et al. 2016). Similarly, e23D is a database where modeled protein structure from A-to-I RNA editing sites from human, mouse and fly are visualised. The 3D structures are generally derived from PDB or theoretical models from ModBase (Solomon et al. 2016). Secondary structure predictions in the virus world have been a dramatic issue that needed apprehension as it unveils benign from pathogenic variants along with molecular consequence. Recently, the JNetpred algorithm was used to predict the structure of p7 protein of the hepatitis C virus genotype 4 having one α-helix and three β-sheets along with two coiled coils. However, to determine the exact location of a genomic coordinate is still a matter of concern (Mohamed et al. 2017). VarMap, a web-based tool, is a master blend that annotates as well as validates the protein 3D structures through the utilisation of UniProt data. It can even map the chromosome coordinates and is better than the other tools established recently which further reflect the advances made in the field of protein bioinformatics (Stephenson et al. 2019).

4.4 Protein-Protein Interaction Databases

Enhanced computational power with tremendous data on protein sequence and structure has ignited the development of numerous novice approaches for structure predictions, designing protein folds and protein-protein interfaces. Proteins with therapeutic and signaling properties have been designed along with fluorescent proteins with the novel or enhanced utility using these approaches (Li-Ping et al. 2018). The protein-protein interactions (PPIs) are vital for understanding the mechanism of action that governs various biological processes including the prediction of protein function. PDB has only 50% of structures in the form of complexes and thus limits its use for PPI predictions paving the way towards *in silico* approaches (Northey et al. 2018). The recent 2019 update on one of the pioneer PPI databases BioGRID (Biological General Repository for Interaction Datasets) is the annotation of genome-wide CRISPR/Cas9-based observations for human drug and bioactive compounds interaction that determine relationships between gene and phenotype. Apart from being an open-access database available at https://thebiogrid.org, it is meant for storage of genetic, protein and chemical interactions from almost all experimental organisms including humans, it also records for more than 700,000 posttranslational modification sites (Oughtred et al. 2019). Second, a structure-based PPI prediction software MEGADOCK-Web is freely available for use at http://www.bi.cs.titech.ac.jp/megadock-web/ and enables archival of a comprehensive set of all possible predicted PPIs through docking calculations with biochemical pathways. MEGADOCK-Web records more than 28,331,628 predicted PPIs with all possible amalgamations of proteins along with annotated structures through PDB and other 3D molecular viewers (Hayashi et al. 2018). Similarly, another structure-based predictor IntPred available at site https://bio.tools/intpred works on the machine learning principle and has shown superiority in its performance when a criterion was specificity, over many other predictors (Northey et al. 2018). A sequence-based deep learning approach is another way to enhance the high-quality data for PPI prediction. DPPI is freely available software at site https://github.com/hashemifar/DPPI/ and it can predict homodimeric interactions and cytokine-receptor binding affinities more accurately than any other sequence-aided approach (Hashemifar et al. 2018). PCLPred, an amalgamation method involving a position-specific scoring matrix (PSSM) of protein with the robust relevance vector machine (RVM), resulted in the development of a new sequence-based approach to predict PPIs in the *Saccharomyces cerevisiae* dataset and yielded results with nearly 95% precision and sensitivity. It emerged as one of the predominant proteomics tools with high-performance potency (Li-Ping et al. 2018). Some of the well-established facts about these protein structures are their utility in drug discovery and functionality determination in proteins through the inception of ligand binding site detection. Considering the two-prime criterion of stability and speed, many webservers and template-based tools have failed to upstand the expectations of the researchers, except recently P2Rank, a stand-alone tool based on machine learning and ability to predict ligand of local chemical neighborhoods. P2Rank is an independent tool outperforming several other existing tools by capturing the likely binding sites consisting of residues from multiple chains. P2Rank is a fast, accurate and stable predictor of ligand binding sites in proteins and in the near future can be used for novel allosteric site prediction (Krivák and Hoksza 2018). BIPSPI webserver dedicated to partner-specific PPI prediction is freely available at http://bipspi.cnb.csic.es. The uniqueness of this server is its approach to consider a pair of interacting proteins rather than a single protein for PPI predictions. An interesting feature of this is that both sequence and structural features are considered while compiling the PPI prediction data from the interacting protein partners (Sanchez-Garcia et al. 2019).

The STRING resource available at https://string-db.org/ is a comprehensive compendium of global networks of proteins through direct and indirect means. It brings the easy retrieval of information from a set of different sources available along with their computational predictions. The latest version of STRING (11.0) features a genome dataset as input with easy visualisation of the interaction network (interactome) and thus performs gene-set enrichments. STRING offers high-throughput text mining as an additional classification system, besides GO and KEGG, thereby featuring the enrichment process. This protein-protein association network has an overall enhanced coverage, and predictive capacity of associations even on large datasets (Szklarczyk et al. 2019). Short linear motifs have been recently used for virus-host PPI detections. A computational approach for predicting motif-mediated PPIs between humans and the human immunodeficiency virus 1 (HIV-1) using protein sequences as input has been devised. The futuristic prospect of this software so developed will be the detection of mechanisms of virus attacks. Further, the efficacy of a particular PPI determined through different approaches discussed above needs to be evaluated on the accuracy of prediction (Becerra et al. 2017). PRODIGY (PROtein binDIng enerGY prediction) is a webserver that aids in predicting the binding affinity of protein-protein complexes. It exploits the tertiary structural information of the predicted model and is freely available at https://bianca.science.uu.nl/prodigy/lig site. The server is fast, performing the prediction in a few seconds for the largest complexes and can contribute to the scale-up of new predictive approaches (Xue et al. 2016). A slight change in the structure due to a sudden mutation adversely affects the accuracy of predicted PPI and thus the impact of mutations on kinetics, thermodynamics and binding energy of protein complexes needs to be addressable. SKEMPI 2.0 available at https://life.bsc.es/pid/skempi2/ is a database dedicated to manually curated data pertaining to the changes in binding free energy, entropy and enthalpy upon mutation and its influence in detecting binding (Jankauskaite et al. 2019). APID is a comprehensive repository of curated 'protein interactomes' accessible at http://apid.dep.usal.es. It includes 500 experimentally detected PPIs of more than 1100 organisms from nearly 30 species. The five primary databases of molecular interactions including BioGRID, along with the structural data from PDB, act as a feedstock for APID search. It provides information of PPIs detected through valid experimental interaction methods and text mining, pointing to the interactions as 'binary' or not and leading to the creation of a new collection of 'binary interactomes' for multiple species with allocated proportions (Alonso-López et al. 2019).

4.4.1 PPIs and Model Organisms

Developments in the field of protein bioinformatics have led to the establishment of new databases specified to PPIs. One of them is the Protein-Protein Interaction Database for Maize (PPIM) which covers nearly 2,762,560 PPIs occurring in 14,000 proteins. The PPIM contains not only accurately predicted PPIs validated through experimental data but also those molecular interactions which are indicated through text mining. The database is freely available at http://comp-sysbio.org/ppim/ with an easily accessible search engine (Zhu et al. 2016). Similarly, many other PPIs for other organisms have been developed recently. The Human Integrated Protein-Protein Interaction rEference (HIPPIE; http://cbdm.uni-mainz.de/hippie/) is a highly reliable human PPI network database with annotated descriptions of all the possible interactions predicted and related information pertaining to human proteomes (Alanis-Lobato et al. 2017). IID is another comprehensive context-specific human PPI network available at http://iid.ophid.utoronto.ca. A recent update shows the involvement of 18 species with 4.8 million PPIs in 133 tissues. IID provides unique functionality with reduced false negatives and even supports non-human species, many of which have biological significance (Kotlyar et al. 2019). WormBase (www.wormbase.org) provides the protein interaction data from almost all of the *Caenorhabditis elegans* and has 4.51-fold more interaction annotations than BioGRID. In WormBase, PPI data include all physical, genetic, regulatory and predicted interactions information related to the gene of interest. To visualise the overall interaction map, it utilises the 'Cytoscape' program (Cho et al. 2018). The Mouse Integrated Protein-Protein Interaction rEference (MIPPIE) inherits a robust infrastructure from HIPPIE and allows for the assembly of reliable networks supported by different evidence sources and high-quality experimental techniques. In MIPPIE, users have access to 42,610 interactions between 10,886 proteins. The confidence score, together with the tools and comparison with human interactomes via HIPPIE, is one of the key characteristics that differentiate MIPPIE from other PPI resources (Alanis-Lobato et al. 2020).

REFERENCES

Alanis-Lobato G., M. A. Andrade-Navarroand and M. H. Schaefer. "HIPPIE v2.0: enhancing meaningfulness and reliability of protein-protein interaction networks." *Nucleic Acids Research* 45 (2017): 408–414.

Alanis-Lobato G., J. S. Möllmann, M. H. Schaefer and M. A. Andrade-Navarro. "MIPPIE: the mouse integrated protein-protein interaction reference." *Database* (2020): 1–10.

Alonso-López D., F. J. Campos-Laborie, M. A. Gutiérrez, L. Lambourne, M. A. Calderwood, M. Vidal and J. D. L. Rivas. "APID database: redefining protein-protein interaction experimental evidences and binary interactomes." *Database* (2019): 1–8.

Becerra A., V. A. Buchelr and P. A. Moreno. "Prediction of virus-host protein-protein interactions mediated by short linear motifs." *BMC Bioinformatics* 18 (2017): 163–174.

Chang Y. C., Z. Hu, J. Rachlin, B. A. Anton, S. Kasif, R. J. Roberts and M. Steffen. "COMBREX-DB: an experiment centered database of protein function – knowledge, predictions and knowledge gaps." *Nucleic Acids Research* 44(1) (2016): 330–335.

Cho J., C. A. Grove, K. V. Auken, J. Chan, S. Gao and P. W. Sternber. "Update on protein-protein interaction data in WormBase." (2018) https://doi.org/10.17912/MICROPUB.BIOLOGY.000074

Dayhoff, M. O., R. V. Eck, M. A. Chang and M. R. Sochard(Eds.) *Atlas of protein sequence and structure* (1965). Silver Spring, MD: National Biomedical Research Foundation.

Doshi J., R. R. Kuppili, S. Gurdasani, N. Venkatakrishnan, A. Saxena and K. Bose. "PDZscape: a comprehensive PDZ-protein database." *BMC Bioinformatics* 19 (2018): 160–166.

El-Gebali S., J. Mistry, A. Bateman, S. R. Eddy, A. Luciani, S. C. Potter, M. Qureshi et al. "The Pfam protein families database in 2019." *Nucleic Acids Research* 47 (2019): 427–432.

Fang C., Y. Shang and Xu Dong. "MUFOLD-SS: new deep inception-inside-inception networks for protein secondary structure prediction." *Proteins* 86(5) (2018): 592–598.

Gaudet P., Pierre-Andre Michel, M. Zahn-Zabal, A. Britan, I. Cusin, M. Domagalski, Paula D. Duek et al. "The neXtProt knowledgebase on human proteins: 2017 update." *Nucleic Acids Research* 45 (2017): 177–182.

Hanson J., K. Paliwal, T. Litfin, Y. Yang and Y. Zhou. "Improving prediction of protein secondary structure, backbone angles, solvent accessibility and contact numbers by using predicted contact maps and an ensemble of recurrent and residual convolutional neural networks." *Bioinformatics* 35(14) (2019): 2403–2410.

Hashemifar S., B. Neyshabur, A. A. Khan and J. Xu. "Predicting protein-protein interactions through sequence-based deep learning." *Bioinformatics* 34 (2018): 802–810.

Hayashi T., Y. Matsuzaki, K. Yanagiswaw, M. Ohue and Y. Akiyama. "MEGADOCK-Web: an integrated database of high-throughput structure-based protein-protein interaction predictions." *BMC Bioinformatics* 19(Suppl 4) (2018): 62–119.

Hayat S., C. Peters, N. Shu, K. D. Tsirigos and A. Elofsson. "Inclusion of dyad-repeat pattern improves topology prediction of transmembrane β-barrel proteins." *Bioinformatics* 32(10) (2016): 1571–1573.

Hrabe T., L. Zhanwen, M. Sedova, P. Rotkiewicz, L Jaroszewski and A. Godzik. "PDBFlex: exploring flexibility in protein structures." *Nucleic Acids Research* 44(1) (2016): 423–428.

Hu G., Z. Wu, C. J. Oldfield, C. Wang and L. Kurgan. "Quality assessment for the putative intrinsic disorder in proteins." *Bioinformatics* 35(10) (2019): 1692–1700.

Jankauskaite J., B. Jiménez-Garcia, J. Dapkūnas, J. Fernández-Recio, I. H Moal. "SKEMPI 2.0: an updated benchmark of changes in protein-protein binding energy, kinetics and thermodynamics upon mutation." *Bioinformatics* 35(3) (2019): 462–469.

Khandelwal I., P. K. Agrawal, A. Sharma and R. Srivastav (eds.). "Bioinformatics database resources." *Library and Information Services for Bioinformatics Education and Research* (2016): 45–90. IGI Global.

Kotlyar M., C. Pastrello, Z. Malik and I. Jurisica. "IID 2018 update: context-specific physical protein-protein interactions in human, model organisms and domesticated species." *Nucleic Acids Research* 47 (2019): 581–589.

Krivák R. and D. Hoksza. "P2Rank: machine learning based tool for rapid and accurate prediction of ligand binding sites from protein structure." *Journal of Cheminformatics* 10(39) (2018): 1–12.

Kuhlman B. and P. Bradley. "Advances in protein structure prediction and design." *Nature Reviews Molecular Cell Biology* 20(11) (2019): 681–697.

Kulandaisamy A., S. B. Priya, R. Sakthivel, S. Tarnovskaya, I. Bizin, P. Hönigschmid, D. Frishman and M. Michael Gromiha. "MutHTP: mutations in human transmembrane proteins." *Bioinformatics* 34(13) (2018): 2325–2326.

Liao Y., R. D. Schaeffer, J. Pei and N. V. Grishin. "A sequence family database built on ECOD structural domains." *Bioinformatics* 34(17) (2018): 2997–3003.

Li-Ping L., W. Yan-Bin, Y. Zhu-Hong, L. Yang and A. Ji-Yong. "PCLPred: a bioinformatics method for predicting protein-protein interactions by combining relevance vector machine model with low-rank matrix approximation." *International Journal of Molecular Sciences* 19(1029) (2018): 1–13.

Ludwiczak J., A. Winski, A. M. da SilvaNeto, K. Szczepaniak, V. Alva and S. Dunin-Horkawicz. "PiPred – a deep-learning method for prediction of π-helices in protein sequences." *Scientific Reports* 9 (2019): 6888–6897.

Mahlich Y., M. Steineggar, B. Rost and Y. Bromberg. "HFSP: high speed homology-driven function annotation of proteins." *Bioinformatics* 34(13) (2018): 304–312.

Mirdita M., L. V. D. Driesh, C. Galiez, M. J. Martin, J. Söding and M. Steinegger. "Uniclust databases of clustered and deeply annotated protein sequences and alignments." *Nucleic Acids Research* 45 (2017): 170–176.

Mishra D., V. R. Chaturvedi, V. P. Snijesh, N. A. Shaik and M. P. Singh. "Other biological databases." In: *Essentials of Bioinformatics*, ed. N. A. Shaik et al. Volume I (2019): 75–96. Switzerland: Springer Nature.

Mohamed I., A. El-Morsi, E. Soweha, H. Zaghloul and M. El- Hefnawi. "Phylogenetic analysis and secondary and tertiary (3D) structure prediction of the p7 protein of Hepatitis C Virus." *Journal of Agricultural Chemistry and Biotechnology* 8(9) (2017): 225–230.

Necci M., D. Piovesan, Z. Dosztanyi and S. C. E. Tosatto. "MobiDB-lite: fast and highly specific consensus prediction of intrinsic disorder in proteins." *Bioinformatics* 33(9) (2017): 1402–1404.

Necci M., D. Piovesan, Z. Dosztanyi and S. C. E. Tosatto. "A comprehensive assessment of long intrinsic protein disorder from the DisProt database." *Bioinformatics* 34(3) (2018): 445–452.

Northey T. C., A. Baresic and A. C. R. Martin. "IntPred: a structure-based predictor of protein-protein interaction sites." *Bioinformatics* 34(2) (2018): 223–229.

O'Brien K. T., C. Mooney, C. Lopez, G. Pollastri and D. C. Shields. "Prediction of polyproline II secondary structure propensity in proteins." *Royal Society of Open Science* 7 (2020): 191239–191248.

Oughtred R., C. Stark, Bobby-Joe Breitkreutz, J. Rust, L. Boucher, C. Chang, N. Kolas et al. "The BioGRID interaction database: 2019 update." *Nucleic Acids Research* 47 (2019): 529–541.

Pundir S., M. J. Martin and C. O. Donovan. "UniProtProtein Knowledgebase." *Methods in Molecular Biology* 1558 (2017): 41–55.

Rawlings N. D., A. J. Barret, P. D. Thomas, X. Huang, A. Bateman and R. D. Finn. "The MEROPS database of proteolytic enzymes, their substrates and inhibitors in 2017 and a comparison with peptidases in the PANTHER database." *Nucleic Acids Research* 46 (2018): 624–632.

Sanchez-Garcia R., C. O. S. Sorzano, J. M. Carazo and J. Segura. "BIPSPI: a method for the prediction of partner-specific protein-protein interfaces." *Bioinformatics* 35(3) (2019): 470–477.

Sillitoe I., T. E. Lewis, A. Cuff, S. Das, P. Ashford, N. L. Dawson, N. Furnham et al. "CATH: comprehensive structural and functional annotations for genome sequences." *Nucleic Acids Research* 43 (2015): 376–81.

Solomon O., E. Eyal, N. Amariglio, R. Unger and G. Rechavi. "e23D: database and visualization of A-to-I RNA editing sites mapped to 3D protein structures." *Bioinformatics* 32(14) (2016): 2213–2215.

Stephenson J. D., R. A. Laskowski, A. Nightingale, M. E. Hurles and J. M. Thornton. "VarMap: a web tool for mapping genomic coordinates to protein sequence and structure and retrieving protein structural annotations." *Bioinformatics* 35(22) (2019): 4854–4856.

Szklarczyk D., A. L. Gable, D. Lyon, A. Junge, S. Wyder, J. Huerta-Cepas, M. Simonovic et al. "STRING v11: protein-protein association networks with increased coverage, supporting functional discovery in genome-wide experimental datasets." *Nucleic Acids Research* 47(2019): 607–613.

Trevizani R., F. L. Custodio, K. B. dos Santos and L. E. Dardenne. "Critical features of fragment libraries for protein structure prediction." *PLoS One* 12(1) (2017): 1–22.

Varga J., L. Dobson and G. E. Tusnady. "TOPDOM: database of conservatively located domains and motifs in proteins." *Bioinformatics* 32(17) (2016): 2725–2726.

Varga, J. K. and G. E. Tusnady. "TMCrys: predict propensity of success for transmembrane protein crystallization." *Bioinformatics* 34(18) (2018): 3126–3130.

Xue L. C., J. P. Rodrigues, P. L. Kastritis, A. M. Bonvin and A. Vangone. "PRODIGY: a web server for predicting the binding affinity of protein-protein complexes." *Bioinformatics* 32(23) (2016): 3676–3678.

Yang Y., R. Heffernan, K. Paliwal, J. Lyons, A. Dehzangi, A. Sharma, J. Wang et al. "SPIDER2: a package to predict secondary structure, accessible surface area, and main-chain torsional angles by deep neural networks." *Methods in Molecular Biology* 1484 (2017): 55–63.

Zhu G., A. Wu, Xin-Jian Xu, Pei-Pei Xiao, Le Lu, J. Liu, Y. Cao et al. 2016. "PPIM: a protein-protein interaction database for Maize." *Plant Physiology* 170(2) (2016): 618–626.

5

CBL-CIPK: The Ca$^+$ Signals during Abiotic Stress Response

Pitambri Thakur and Neelam Prabha Negi
University Institute of Biotechnology, Chandigarh University

CONTENTS

5.1 Introduction

Environmental conditions are changing day by day and various signalling processes are adapting to the changeable environment. As we all know plants are not mobile organisms, so plants must face adverse environmental conditions. So, to adjust in external stress conditions, plants have created a special system for instant transmit signals. In the plant cell, calcium (Ca^{2+}) ions act as a secondary messenger and regulate various developmental and functional processes (Gilroy and Trewavas 2001; Sanders et al. 2002; Batistic and Kudla 2012). Ca^{2+} signals are recognised by Ca^{2+} sensor proteins and specifically transmit signals to the downstream pathway in Ca^{2+} regulatory mechanisms (Roberts and Harmon 1992). In abiotic stress-responsive mechanisms of plants, CBL-CIPK (calcineurin B-like protein-CBL-interacting protein kinase) occupies a very important position (Luan et al. 2002; Luan 2009; Weinl and Kudla 2009). The CBL was first described in *Arabidopsis thaliana* and animals. CBL family has sequence similarity with neuronal Ca^{2+} sensors. CIPK family is a protein kinase with functions coinciding adenosine monophosphate (AMP) dependent kinase (AMPK) in animals (Liu and Zhu 1998; Shi et al. 1999; Liu et al. 2000).

In plant cells, CBL proteins are a separate family of Ca^{2+} sensors. CBL-CIPK functions as a nutrient transport systems regulator in regulating sodium (Na^+) (Kolukisaoglu et al. 2004; Batistic et al. 2010; Drerup et al. 2013), magnesium (Mg^{2+}) (Kolukisaoglu et al. 2004), potassium (K^+) and nitrate (NO_3^-) (Cheong et al. 2007; Held et al. 2011) and proton regulator (Drerup et al. 2013). Abiotic stress affects photosynthesis, rate of germination, leaf expansion, total biomass, etc., hence it affects all the developmental and physiological processes of the plants. CBL-CIPK works as an essential component of the Ca^{2+} sensor to receive the signal from the environment. So, the CBL-CIPK and the abiotic stress study has become the burning area of research because it is helping plants to adjust to the adverse environmental conditions. CBL protein binds to the C-terminal region of CIPK and stimulates the kinase (Drerup et al. 2013) and N-terminal direct CBL-CIPK complex towards the target (Batistic et al. 2010).

According to bioinformatics research reports, different plants do have a different number of CBL and CIPK proteins. There are 10 CBLs and 26 CIPKs reported in *A. thaliana* (Qiu et al. 2002), while 10 CBLs and 31 CIPKs are reported in *Oryza sativa* (Zhu et al. 1998); in *Populustrichocarpa*,10 CBLs and 27 CIPKs are reported (Qiu et al. 2002); 8 CBLs and 43 CIPKs are reported in *Zea mays* (Zhu et al. 1998); 8 CBLs and 21 CIPKs in *Vitisvinifera* (Shi et al. 2002) and 6 CBLs and 32 CIPKs in *Sorghum bicolor* (Shi et al. 2002). First, the execution of CBL-CIPK was discovered in salt overly sensitive (SOS) pathways (Tang et al. 2015). SOS3 and SOS2 were also determined to up-regulate the plasma membrane (PM) activity, under high-salt concentration (Ho et al. 2009). The *Arabidopsis* mutants SOS1, SOS2 and SOS3 (called CBL4 and CIPK24) under high-salt concentration also produce the same salt-sensitive phenotype. CBL-CIPK pathways drive nutrient transport systems regulators, regulating Na^+ (Batistic et al. 2010; Drerup et al. 2013), K^+(Qiu et al. 2002)Mg^{2+} (Held et al. 2011), NO_3^- (Qiu et al. 2002) and homeostasis H^+ (proton) (Shi et al. 2002). A few previous findings also have drawn consideration towards the activity of the CBL-CIPK in different ion sensitivity groups (Ho et al. 2009; Tang et al. 2015).

This chapter elaborates on the CBL-CIPK Ca^{2+} signals during the abiotic stress response and attaches various responses important to the environmental stresses, such as K^+, NO_3^-, mMg^{2+}, and hormonal stresses. It also summarises the CBL-CIPK physiological role in discovering certain innovations in the future and certainly, this chapter provides new insight into the further study of the functional classification and characterisation in the understanding of the CBL-CIPK response with respect to various abiotic stresses. Few new directions in the research and future aspects are also discussed in this chapter.

5.2 CBL-CIPK Pathway Mechanisms

The CBL-CIPK structural characteristics provide a base for their interaction study (Sánchez-Barrena et al. 2007). They disclose the CBL-CIPK complex functioning to decode the intracellular Ca^{2+} signals produced by external stresses (Sánchez-Barrena et al. 2013). Four elongation factor hands are accommodated by the CBL protein; each elongation factor hand comprises a conserved α-helix-loop-α-helix structure (Sánchez-Barrena et al. 2013). These elongation factor hands from EF1 to EF4 are coordinated in a fixed space that is placed at 22, 25 and 32 amino acids (Nagae et al. 2003; Sánchez-Barrena et al. 2005). The amino acids placed in positions 1(X), 3 (Y), 5 (Z), 7(–7), 9(–Y) and 12(–Z) are answerable for Ca^{2+} coordination (Li et al. 2009). Ca^{2+}-binding affinity is described by various amino acid positions, and bind Ca^{2+} that is based upon the side-chain donor (Nagae et al. 2003).

The CIPK is composed of two domains, one conserved N-terminal kinase catalytic domain, and the other C-terminal regulatory domain. N-terminal domain encloses the phosphorylation site, containing the activation loop. The regulatory domain encloses NAF/FISL motif which is named because of highly conserved amino acids N (Asn), A (Ala), F (Phe), I (Ile), S (Ser) and L (Leu) (Guo et al. 2001) and

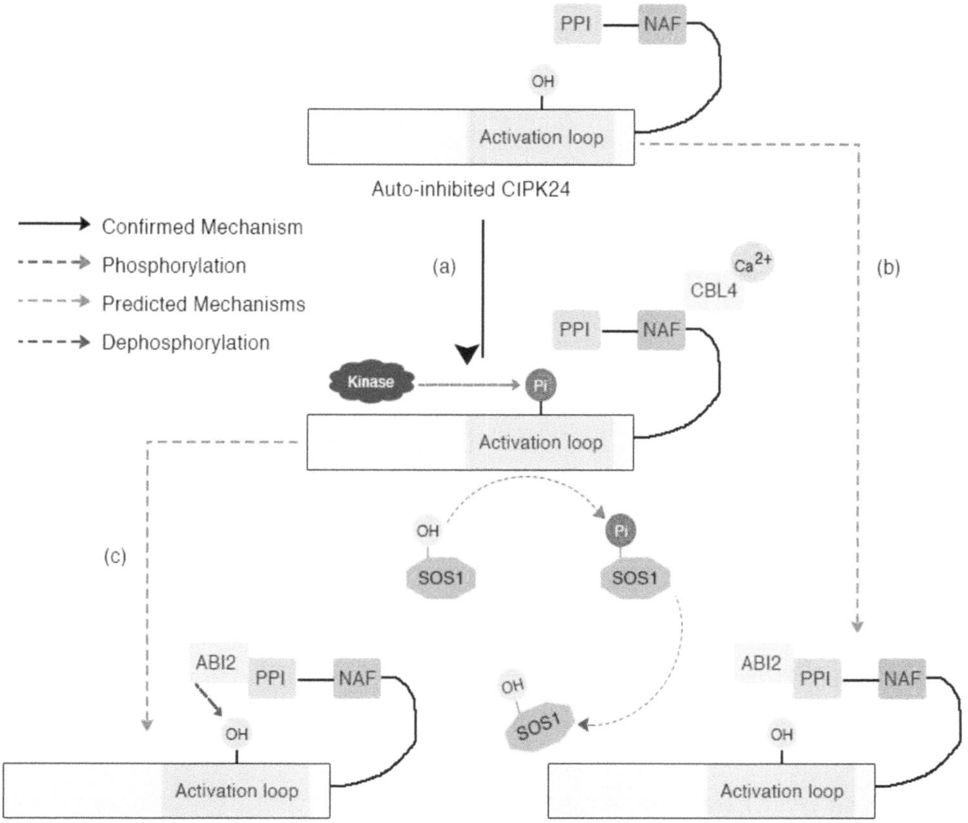

FIGURE 5.1 CBL-CIPK24 signalling mechanism. (a) CBL4-Ca^{2+}bonded attaches with the CIPK24 NAF and phosphorylates salt overly sensitive 1 (SOS1). (b) Abscisic acid-insensitive 2 (ABI2) attaches the PPI (phosphatase interaction) region of CIPK24, and dephosphorylates SOS1. (c) Activated CIPK24 is inhibited by ABI2.

a PPI (phosphatase interaction) motif (Guo et al. 2001). NAF motif is a must for assisting the CIPK24-CBL4 interaction and for keeping the C-terminal regulatory domain auto-inhibited (Figure 5.1) (Guo et al. 2001). CBL4 molecular surface properties modify by the attachment of Ca^{2+} to EF-hands (Kolukisaoglu et al. 2004) then via NAF motif CBL4 which interacts with CIPK24 and exposes its activation loop (Guo et al. 2001). (Figure 5.1a). Then CIPK24 phosphorylates the SOS1 situated on the PM for the elimination of the extra Na^+ from the cell (Zhu 2003). ABI2 (abscisic acid-insensitive 2) is believed to perform the dephosphorylation process in SOS1 (Figure 5.1b) and CIPK24 (Figure 5.1c) (Ragel et al. 2015).

5.3 Physiological Activity of CBL-CIPK Signalling

To clarify the CBL-CIPK complexes' physiological activity, various research attempts were conducted and divided these complexes into PM targeting and tonoplast targeting pathways.

5.3.1 Plasma Membrane Targeting

PM targeting CBL-CIPK generally participates in regulating the ion exchange to adapt plants to a changeable environment.

5.3.1.1 CBL1/9-CIPK23 Pathways

In *Arabidopsis*, this pathway was first discovered for the K^+ uptake systems depending upon the HAK5 (high-affinity K^+ transporter 5) for the high-affinity system and depending upon AKT1 (*Arabidopsis* K^+ transporter 1) for the low-affinity system. CBL1/9-CIPK23 complexes help to activate the K^+ channel incoming rectifier AKT1 and HAK5 to uptake K^+ (Ragel et al. 2015). CBL1/9-CIPK23 also participates in the stomatal closure with a counter abscisic acid (ABA) and increases anion efflux of the guard cells. SLAC1 (slow anion channel associated 1) and SLAH3 (slow anion channel 1 homolog 3) that reduce guard cell volume and shut down the stomata are phosphorylated by this complex (Maierhofer et al. 2014).

Under NO_3^--deficient conditions, CBL1/9-CIPK23 complexes regulate NO_3^- uptake and phosphorylate NPF6.3 (NO_3^- transporter 1/peptide transporter family 6.3) (Liu and Tsay 2003). When NO_3^- is in low concentration, NPF6.3 switches to a high-affinity transporter by phosphorylating Thr101 (Sánchez-Barrena 2007). Under high NO_3^- concentration, this activity is stopped by CBL9/CIPK23-dependent phosphorylation of Thr101 (Leran et al. 2015). Leran et al. (2013) reported that NPF6.3 is the outward transporter secreting NO_3^- into the xylem, although the molecular mechanism of the NPF6.3 is not yet clear. CBL9-CIPK23 complex inhibits the NO_3^- efflux.

5.3.1.2 CBL1/9-CIPK26 Pathways

The study of CBL1/9-CIPK26 revealed a novel role in the exchange among ROS and Ca^{2+} signalling and demonstrated the resilience of the CBL-CIPK network in plants (Marino et al. 2012). RBOHs (respiratory burst oxidase homologs), also called plant NOXs (NADPH oxidases), are very important complexes producing ROS. There are ten members named RBOHA to RBOHJ, out of that RBOHF (respiratory burst oxidase homolog protein F) responds to biotic stresses. For the activation of RBOHF binding of Ca^{2+} to the elongation factor hands and also CBL1/9-CIPK26 complexes, phosphorylation is a prerequisite (Pandey et al. 2004).

5.3.1.3 CBL9-CIPK3 Pathways

Regulation of ABA is performed by the CBL9-CIPK3 complex and participates in the salt-induced ABA-dependent pathway. However, the splitting of CBL9-CIPK3 alternated the expression of a few marker genes while responding to ABA (Kim et al. 2003; Pandey et al. 2008). Thus, CBL9 and CIPK3 may function independently under special conditions. CBL9 has high sequence similarity with CBL1 and CBL9 and can interact with CIPK15 which works as an ABA response negative regulator (Guo et al. 2002).

5.3.1.4 CBL2-CIPK11 Pathways

By the phosphorylation of the Thr947 (threonine 947) and PMH^+-ATPase (PMA), the PM proton pump is activated. CBL2-CIPK11 complex phosphorylate serine 931 of PMA and suppress attachment of the Thr947 to undiscovered kinase and leads to deterioration of PMA interaction (Fuglsang et al. 1999). Thus, the CBL2-CIPK11 complex participates negatively in the activation of the PMA process. CIPK11 mutant plants are receptive to a higher alkaline environment. Inactivated PMA cannot administer motivation for ion transport because the proton electrochemical gradient is not developed (Kanczewska et al. 2005).

5.3.1.5 CBL4-CIPK6 Pathways

This pathway along with AKT2 is a Ca^{2+}-dependent and phosphorylation-independent K^+ regulation pathway. As compared to other CBL-CIPK-K^+ pathways, CL4-CIPK6 cannot phosphorylate AKT2; the CIPK6 C-terminal regulatory domain is sufficient for CBL4 and AKT2 communication. For the deviation of the AKT2 complex from the endoplasmic reticulum (ER) to the PM, CBL4 is the key factor (Cheong et al. 2007).

5.3.1.6 CBL4-CIPK24 Pathways

CBL4-CIPK24 complex helps to improve salt tolerance of the plants by driving the Na^+/H^+ exchanger SOS1 to send Na^+ back into the soil. It essentially functions under high-salt conditions for Na^+(Qiu et al. 2002).

5.3.2 Tonoplast Targeting Pathways

Tonoplast CBL-CIPK pathways are associated with the growth and development of the plants (Eckert et al. 2014; Steinhorst et al. 2015). All the vacuolar membrane targeting pathways are helping cells to confine harmful ions into the vacuole for removing toxicity (Pandey et al. 2007; Liu et al. 2013; Pandey et al. 2015).

5.3.2.1 CBL2/3-CIPK3/9/23/26 Pathways

These pathways by interacting with V-ATPase (vacuolar H^+-ATPase) regulate intracellular ion homeostasis. CBL2 and CBL3 are present in tonoplast, and the cbl2cbl3 double mutant is unstable to metal ions such as Ca^{2+}, Fe^{3+}, K^+, Cu^{2+}, Mg^{2+} and Zn^{2+} but not to Na^+ (Vert and Chory 2009). CBL2/CBL3 are paralogous gene pairs with different functions but share 92% amino acid sequence similarity.

5.3.2.2 CBL2/3-CIPK12 Pathways

These complexes localised in tonoplast are essential regulators in pollen germination (Steinhorst et al. 2015). According to findings, overexpression of CBL2 and CBL3 may lead to impaired pollen tube growth. Overexpression of CIPK12 leads to vacuolar phenotypes, and loss of function leads to deterioration of polar growth. CIPK19 is the closest homolog of CIPK12, overexpression and disruption of CIPK19 lead to reduced pollen tube elongation and expansion (Zhou et al. 2015).

5.3.2.3 CBL2/3-CIPK21 Pathways

CIPK21 is in tonoplast with CBL2/3 under high Na^+ stress and is hypersensitive to high Na^+, but its role is not certain yet (Pandey et al. 2015). Further research is needed to find out the possible tonoplast targeting pathway of CBL2/3-CIPK21 in response to high Na^+ stress (Pandey et al. 2015).

5.3.2.4 CBL10-CIPK24 Pathways

CBL10-CIPK24 functions in response to the high salt concentration and provides a novel salt-tolerance mechanism and increases the activity of a vacuolar Na^+/H^+ exchanger (Qiu et al. 2004).

5.4 CBL-CIPK Signalling Responses to Environmental Abiotic Stresses

These complexes are controlling Ca^{2+} signals driven by diverse abiotic stresses such as low K^+, low Mg^{2+}, NO_3^-, high salt, low phosphorus, high pH, ABA and cold (Cheong et al. 2003; Quan et al. 2007; Fuglsang et al. 2007). All these observations have been determined by mutant studies of *Arabidopsis*. For abiotic stresses, crosstalk between the CBL-CIPK network and other pathways will help to improve tolerant crops in adverse conditions and can enhance the yield (Thapa et al. 2011). Till now, CBL-CIPK involved in the influx/efflux mechanisms of various ions helps in creating an adjustable environment under unfavourable conditions in the cell. All these pathways are discussed below in brief.

5.4.1 Sodium (Na^{2+}) Signalling

In plant cells, SOS pathways are the recognised pioneer CBL-CIPK pathway for ion homeostasis maintenance (Qiu et al. 2002). The biochemical and genetic analysis of the SOS mutant has shown a molecular mechanism where the CBL-CIPK complex can resolve the salt resistance (Pel et al. 2000). SOS pathway includes SOS2 (AtCIPK24), SOS3 (AtCBL4) and SOS1 (PM Na^+/K^+ antiporter). Also, it can increase the salt resistance of the plants by effluxing Na^{2+} from the cell via SOS1 leading to salt tolerance in plants (Qiu et al. 2002). SOS3-SOS2 complex is formed under the high-salt condition in the roots and allows SOS2 to activate SOS1 (Qiu et al. 2002). If SOS1 is unable to activate, then under salty conditions plant growth is inhibited (Cheong et al. 2003). AtCBL4/SOS3-AtCIPK24/SOS2 complex is triggered by external salt stress to start Na^+/H^+ exchange activity of SOS1 to prevent the cell from extra Na^+ (Zhu 2002). AtCBL10 is participating in activating tonoplast Na^+/H^+ NHX antiporters to confine extra Na^+ ions in the vacuole (Cheng et al. 2004).

CBL4/SOS3 and CBL10 perform their activity of salt tolerance in different ways because of their different expression patterns and locations. Various localisation experimental reports say CBL10 is localised in tonoplast (vacuolar membrane) (Kim et al. 2007). According to many reports, Ca^{2+} sensor, CBL1, interacts with CIPK24 for Na^+ regulation in the plant cell. Localisation assay has confirmed localisation in the PM, and its exposition in shoots and roots is confirmed by expression pattern analysis (Albrecht et al. 2003).

CIPKs are salinity sensitive too; *Arabidopsis* CIPK6 was more sensitive to salt stress in comparison to the wild. A yeast two-hybrid system was applied to prove the relationship between CIPK6 and CBL4/SOS3 and confirmed CIPK6 participation in the pathway (Held et al. 2011). Other than *Arabidopsis*, apple MdCIPK6L-OE, OsCBL4 in rice and BjSOS3 in *Brassica juncea* conferred tolerance to salts and gave little clarity to the salt pathway (Hu et al. 2012; Wang et al. 2012).

5.4.2 Magnesium (Mg^{2+}) Signalling

Mg^{2+}/Ca^{2+} homeostasis maintenance is essential for mineral nutrients supply (Yamanaka et al. 2010) and for serpentine-tolerant plants (Bradshaw 2005). Double mutant function analysis confirmed the participation of CBL2 and CBL3 (cbl2-cbl3) in regulating vacuole-mediated Mg^{2+} ion homeostasis in the cell. According to Tang et al. (2015), CIPK3/9/23/26 interacted physically with the CBL2/3 on the tonoplast. Thus, these findings suggest CIPK3/9/23/26 works in sync with the CBL2/3 to mitigate the toxic effects of the external high Mg^{2+} concentrations via vacuolar confinement, but the exact CBL-CIPK pair is not clear yet. Major Mg^{2+} transporters localised on the tonoplast are MGT2 and MGT3 (Conn et al. 2011). Furthermore, recognition of the transporters is required, which are stimulated under Mg^{2+} toxicity conditions, to understand the ion detoxification mechanism in the plants.

5.4.3 Nitrate (NO_3^-) Signalling

NO_3^- is the key nitrogen source of plants, and nitrogen is the principal limiting factor for plant growth and crop production (Angeli et al. 2009). Research on NO_3^- sensing and signalling molecular mechanisms has started to be explained in *A. thaliana*. Three NO_3^- transporter family members, such as 53 of AtNRT1, 7 of AtNRT2 and 7 of AtCLC, have been identified in *Arabidopsis* (Tsay et al. 1993; Forde 2000). AtNRT2.1 and AtNRT2.2 members are involved in high-affinity uptake and AtNRT1.2 in low-affinity uptake, whereas AtNRT1.1 is involved as a dual-affinity transporter (Wang et al. 1998; Liu et al. 1999).

The CHL1 role in NO_3^- signalling was first reported in *Arabidopsis* from the loss-of-function mutant (chl1) studies, which confirmed the role of CHL1 in the expression of AtNRT2.1 in response to the NO_3^- stress (Wang et al. 1998). CHL1 role in direct sensing of external NO_3^- via AtCIPK23-mediated phosphorylation needs to be understood. Besides, AtCIPK23 requires *Arabidopsis* CBL9 for the activation of CHL1 for high-affinity NO_3^- transportation (Vert and Chory 2009). Under stress conditions, *Arabidopsis* CIPK8 acts as a low-affinity NO_3^- response described by transcriptomic study, but its precise regulation mechanism needs further analysis (Hu et al. 2009).

5.4.4 Potassium (K^+) Signalling

K^+ is the key mineral nutrient involved in different physiological processes and controls the crop production yield. Preliminary K^+ signals are created by plants in the root cells in response to the external K^+ fluctuations. From roots, the signal is transferred to the cytoplasm and further sensed by Ca^{2+} sensors.

AKT1 is the low-affinity incoming K^+ channel and is regulated by Ca^{2+} sensor CBL-CIPK for K^+ homeostasis maintenance in the cell (Li et al. 2006; Nieves-Cordones et al. 2012). If the amount of external K^+ becomes low, CIPK23 is presented to the PM accelerated by CBL1 and CBL9 for AKT1 phosphorylation (Lee et al. 2007; Nieves-Cordones et al. 2012). AKT1 OE (overexpressed) *Arabidopsis* plants under low K^+ conditions did not show any symbolic performance in growth, while At/PeCBL1, AtCBL9 and AtCIPK23 OE *Arabidopsis* showed relative tolerance under similar conditions (Pilot et al. 2003; Zhang et al. 2013). AKTI activity is negatively regulated by a PP2C-type phosphatase AKT1-interacting PP2C1 (AIP1). Hence, AKTI is phosphorylated by the CBL1/CBL9-CIPK23 complex and dephosphorylated by AIP1 (Lee et al. 2007). According to findings, a Ca^{2+} sensor controls K^+ channel activity by boosting up the alteration to the PM (Tuskanet al. 2006). Further research is still needed to confirm the hypothesis that the unknown CBLs interact with distinct CIPKs under low K^+ stress conditions to sense Ca^{2+} signals (Amtmann and Armengaud 2007).

5.4.5 Phosphorus (PO_4^-) Signalling

In plants, phosphorus is a secondary macronutrient and in phosphorus-deficient conditions, inorganic phosphorus (Pi) is absorbed by plants (Bieleski 1973). In switching the metabolic pathway and in enzymatic reactions, Pi is the key element. Under low Pi conditions in *Brassica napus*, BnCBL1 and BnCIPK6 proteins cooperate in yeast two-hybrid screens (Chen et al. 2012). In *Arabidopsis*, under low Pi treatment, BnCBL1 or BnCIPK6 OE showed enhanced plant growth and was found in the development of the lateral roots. Hence BnCBL1 and BnCIPK6 control the Pi-deficient plant responses, but the pathways are still unknown. This area still needs further research to identify the pathways (Chen et al. 2012).

5.4.6 Abscisic Acid (ABA) Signalling

The most important phytohormone in plants is ABA, seed germination to plant growth and responses to abiotic stresses; all these activities are controlled by ABA (Nakashima and Yamaguchi-Shinozaki 2013). Ca^{2+} sensors are involved in the ABA signalling pathway, which was confirmed by the detection of a Ca^{2+} responder in an early step of the signalling (Leung and Giraudat 1998; Allen et al. 2000, 2001). Further studies on OE and mutant lines of CBL-CIPK confirmed its involvement in the ABA signalling pathway. Two mechanisms, namely ABA dependent and ABA independent, regulate ABA signalling pathways, which synchronously control stress-responsive genes.

In regulating osmotic stress-responsive genes, the ABA-dependent pathway shows a crucial performance (Chinnusamy et al. 2004). In regulating the ABA responses, AtCBL9 forms a specific complex with the AtCIPK3 (Pandey et al. 2008; Wanga et al. 2020) and clarifies that during seed germination, the AtCBL9-CIPK3 complex negatively regulates ABA signalling. Findings report that CBL is also positively regulating the GA (gibberellic acid) pathway. Hence it is confirmed from the reports that the CBL-CIPK system shows a vital role in the hormonal signalling.

5.5 CBL-CIPK Calcium Signalling in Response to Drought, Salinity and Cold

Various environmental stimuli are regulated in plants by CBL-CIPK genes. According to Wanga et al. (2020), in the tea plant 8 CBL and 25 CIPK genes were determined and they performed the expression analysis of these genes in response to cold, salinity, and drought in young shoots and mature leaves. They reported CBL1/3/5 and CIPK1/4/5/6a/7/8/10b/10c/12/14a/19/23a/24 were induced by drought and salinity, under cold stress CBL9 and CIPK4/6a/6b/7/11/14b/19/20 were induced in young shoot and mature leaves and in young shoots only CIPK5/25 was induced. Also, yeast two-hybrid analysis confirmed CBL1 interacts with CIPK1/10b/12 and CBL9 interacts with CIPK1/10b/12/14b.

According to Sun et al. (2015), for the low K^+, salt and osmotic stress responses 10 CBLs and 26 CIPKs from *Arabidopsis,* 7 CBLs and 20 CIPKs from wheat, 10 CBLs and 30 CIPKs from rice, 8 CBLs and 20 CIPKs from grape have been identified via genomic analysis. For protecting plants from Mg^{2+} toxicity, AtCBL2/3 and AtCIPK3/9/23/26 interacting networks have been identified by Tang et al.(2015).

According to Han et al. (2019), in the AtCBL1/9–AtCIPK26 complex and AtCIPK11 for the regulation of ROS production, RBOHF is phosphorylated. Ma et al. (2019) reported that the sucrose transporter MdSUT2.2 can be phosphorylated by MdCIPK13 and MdCIPK22. Liu et al. (2019) reported 7 CsCBLs and 18 CsCIPKs in the tea plant and response to the abiotic stress; only 8 genes were identified for expression pattern analysis (Liu et al. 2019). Ma et al. (2019) reported 9 CaCBL and 26 CaCIPK genes in pepper and based on chromosomal order, the genes were named. Quantitative real-time PCR (rt-qPCR) assay suggested when these genes were expressed in various tissues and exposed to different stresses such as salt, salicylic acid, osmotic stress, ABA, cold, heat and *Phytophthora capsici*. Zhao et al. (2019) conducted a systematic genome analysis of CIPK genes in foxtail millet (*Setaria italic* L.) and identified 35 CIPK members that were divided into four subgroups based on phylogenetic relationships. These CIPKs are involved in various stress signalling pathways in response to hormones, light signalling and abiotic stimuli.

5.6 Conclusions and Future Perspectives

The knowledge about the activity of a single protein in various physiological processes has been greatly enriched by various studies on CBLs and CIPKs. The target-regulated studies on CBL-CIPK complexes have advanced our understanding of the signalling system. The identification of the diverse pathways will contribute plenty of knowledge about the physiological responses under abiotic stress conditions. The CBL-CIPK families in various plant species need to be progressed to understand the complete pathways responding to abiotic stress conditions. CBL-CIPK studies are limited to the expression analyses and interaction studies of these families. More experimental work has been done on other species such as poplar, pea, rice and maize to explain CBL-CIPK's role in response to abiotic stresses. Many CBLs-CIPK members have been identified from the above species in response to salt, cold, plant hormones and drought stresses (Tang and Luan 2017; Dubeaux et al. 2018; Jiang et al. 2019; Yang et al. 2019). Gene knockout approaches and dissecting gene functions of mutants should be utilised to identify further signalling components for future research. The exact complex interaction network between Ca^{2+} sensors and kinases needs to be investigated by using available genome sequences. To understand the plant's response against abiotic stresses, the CBL-CIPK signalling model highlights the future research importance of focusing on the molecular mechanisms and provides a competent method of determining enhanced tolerance to different abiotic stresses by providing molecular targets for genetically engineered crops.

Acknowledgement

Authors acknowledge DST-SERB grant no EEQ/ 2018/001236 for financial help.

REFERENCES

Albrecht, V., S. Weinl and D. Blazevic, et al. 2003. The calcium sensor CBL1 integrates plant responses to abiotic stresses. *The Plant Journal* 36(4): 457–470.

Allen, G.J., S.P. Chu, C.L. Harrington, et al. 2001. A defined range of guard cell calcium oscillation parameters encodes stomatal movements. *Nature* 411(6841): 1053–1057.

Allen, G.J., S.P. Chu, K. Schumacher, et al. 2000. Alteration of stimulus-specific guard cell calcium oscillations and stomatal closing in *Arabidopsis* det3 mutant. *Science* 289(5488): 2338–2342.

Amtmann, A. and P. Armengaud.2007. The role of calcium sensor interacting protein kinases in plant adaptation to potassium deficiency: new answers to old questions. *Cell Research* 17(6): 483–485.

Angeli, A.D., D. Monachello, G. Ephritikhine, et al. 2009. CLC mediated anion transport in plant cells. *Philosophical Transactions of the Royal Society B: Biological Sciences* 364(1514): 195–201.

Batistic, O. and J. Kudla. 2012. Analysis of calcium signaling pathways in plants. *Biochemistry and Biophysics Acta* 1820:1283–1293.

Batistic, O., R. Waadt, L. Steinhorst, K. Held and J. Kudla.2010. CBL-mediated targeting of CIPKs facilitates the decoding of calcium signals emanating from distinct cellular stores. *The Plant Journal* 61(2): 211–222.

Bieleski, R.L.1973. Phosphate pools, phosphate transport, and phosphate availability. *Annual Review of Plant Physiology* 24(1): 225–252.

Bradshaw Jr, H.B. 2005. Mutations in CAX1 produce phenotypes characteristic of plants tolerant to serpentine soils. *New Phytologist* 167(1): 81–88.

Chen, L., F. Ren, L. Zhou, Q.Q. Wang, H. Zhong and X.B. Li. 2012. The *Brassica napus* Calcineurin B-Like 1/CBL-interacting protein kinase 6 (CBL1/CIPK6) component is involved in the plant response to abiotic stress and ABA signaling. *Journal of Experimental Botany* 63(17): 6211–6222.

Cheng, N.H., J.K. Pittman, J.K. Zhu and K.D. Hirschi. 2004. The protein kinase SOS2 activates the Arabidopsis H+/Ca2+ antiporter CAX1 to integrate calcium transport and salt tolerance. *The Journal of Biological Chemistry* 279(4): 2922–2926.

Cheong, Y.H., K.N. Kim, G.K. Pandey, R. Gupta, J.J. Grant and S. Luan. 2003. CBL1, a calcium sensor that differentially regulates salt, drought, and cold responses in *Arabidopsis. The Plant Cell* 15(8): 1833–1845.

Cheong, Y.H., G.K. Pandey and J.J. Grant. 2007. Two calcineurin B-like calcium sensors, interacting with protein kinase CIPK23, regulate leaf transpiration and root potassium uptake in *Arabidopsis. The Plant Journal* 52(2): 223–239.

Chinnusamy, V., K. Schumaker and J.K. Zhu. 2004. Molecular genetic perspectives on cross-talk and specificity in abiotic stress signalling in plants. *Journal of Experimental Botany* 55(395): 225–236.

Conn, S.J., V. Conn, S.D. Tyerman, B.N. Kaiser, R.A. Leigh and M. Gilliham. 2011. Magnesium transporters, MGT2/MRS2-1 and MGT3/MRS2-5, are important for magnesium partitioning within *Arabidopsis thaliana* mesophyll vacuoles. *New Phytologist* 190(3): 583–594.

Drerup, M.M., K. Schlucking, K. Hashimoto, et al. 2013. The calcineurin B-like calcium sensors CBL1 and CBL9 together with their interacting protein kinase CIPK26 regulate the *Arabidopsis* NADPH oxidase RBOHF. *Molecular Plant* 6(2):559–569.

Dubeaux, G., et al. 2018. Metal sensing by the IRT1 transporter receptor orchestrates its own degradation and plant metal nutrition. *Molecular Cell* 69: 953–964.

Eckert, C., J.N. Offenborn, T. Heinz, T. Armarego-Marriott, S. Schültke, C. Zhang, S. Hillmer, M. Heilmann, K. Schumacher and R. Bock. 2014. The vacuolar calcium sensors CBL2 and CBL3 affect seed size and embryonic development in *Arabidopsis thaliana. Plant Journal* 78: 146–156.

Forde, B.G. 2000. Nitrate transporters in plants: structure, function and regulation. *Biochimica et Biophysica Acta — Biomembranes* 1465(1-2): 219–235.

Fuglsang, A.T., Y. Guo, T.A. Cuin, Q. Qiu, C. Song, K.A. Kristiansen, K. Bych, A. Schulz, S. Shabala, K.S. Schumaker, et al. 2007. *Arabidopsis* protein kinase PKS5 inhibits the plasma membrane H+-ATPase by preventing interaction with 14-3-3 protein. *Plant Cell* 19:1617–1634.

Fuglsang, A.T., S. Visconti, K. Drumn, T. Jahn, A. Stensballe, B. Mattei, O.N. Jensen, P. Aducci and M.G. Palmgren. 1999. Binding of 14-3-3 protein to the plasma membrane H-ATPase AHA2 involves the three C-terminal residues Tyr946-Thr-Val and requires phosphorylation of Thr947. *Journal of Biological Chemistry* 274: 36774–36780.

Gilroy, S. and A. Trewavas. 2001. Signal processing, and transduction in plant cells: The end of the beginning? *Nature Reviews Molecular Cell Biology* 2: 307–314.

Guo, Y., U. Halfter, M. Ishitani and J.K. Zhu. 2001. Molecular characterization of functional domains in the protein kinase SOS2 that is required for plant salt tolerance. *Plant Cell* 13: 1383–1400.

Guo, Y., L. Xiong, C.P. Song, D. Gong, U. Halfter and J.K. Zhu. 2002. A calcium sensor and its interacting protein kinase are global regulators of abscisic acid signaling in Arabidopsis. *Developmental Cell* 3: 233–244.

Han, J.P., P. Köster, M.M. Drerup, M. Scholz, S. Li, K.H. Edel, K. Hashimoto, K. Kuchitsu, M. Hippler and J. Kudla. 2019. Fine-tuning of RBOHF activity is achieved by differential phosphorylation and Ca^{2+} binding. *New Phytologist* 221: 1935–1949.

Held, K., F. Pascaud and C. Eckert. 2011. Calcium-dependent modulation and plasma membrane targeting of the AKT2 potassium channel by the CBL4/CIPK6 calcium sensor/protein kinase complex. *Cell Research* 21(7): 1116–1130.

Ho, C.H., S.H. Lin, H.C. Hu and Y.F. Tsay. 2009. CHL1 functions as a nitrate sensor in plants. *Cell* 138: 1184–1194.

Hu, D.G., M. Li, H. Luo, et al. 2012. Molecular cloning and functional characterization of MdSOS2 reveals its involvement in salt tolerance in apple callus and Arabidopsis. *Plant Cell Report* 31(4): 713–722.

Hu, H.C, Y.Y. Wang and Y.F. Tsay. 2009. AtCIPK8, a CBL interacting protein kinase, regulates the low-affinity phase of the primary nitrate response. *The Plant Journal* 57(2):264–278.

Jiang, Z, et al. 2019. Plant cell-surface GIPC sphingolipids sense salt to trigger Ca^{2+} influx. *Nature* 572: 341–346.

Kanczewska, J., S. Marco, C. Vandermeeren, O. Maudoux, J.L. Rigaud and M. Boutry. 2005. Activation of the plant plasma membrane H^{+}-ATPase by phosphorylation and binding of 14-3-3 proteins converts a dimer into a hexamer. *Proceedings of National Academy of Sciences USA* 102: 11675–11680.

Kim, B.G., R. Waadt and Y.H. Cheong. 2007. The calcium sensor CBL10 mediates salt tolerance by regulating ion homeostasis in Arabidopsis. *The Plant Journal* 52(3): 473–484.

Kim, K.N., Y.H. Cheong, J.J. Grant, G.K. Pandey and S. Luan. 2003. CIPK3, a calcium sensor-associated protein kinase that regulates abscisic acid and cold signal transduction in *Arabidopsis*. *Plant Cell* 15: 411–423.

Kolukisaoglu, U., S. Weinl, D. Blazevic, O. Batistic and J. Kudla. 2004. Calcium sensors and their interacting protein kinases: genomics of the *Arabidopsis* and rice CBL-CIPK signaling networks. *Plant Physiology* 134(1): 43–58.

Lee, S.C., W.Z. Lan and B.G. Kim. 2007. A protein phosphorylation/dephosphorylation network regulates a plant potassium channel. *Proceedings of National Academy of Sciences USA* 104(40): 15959–15964.

Leran, S., K. H. Edel, M. Pervent, K. Hashimoto, C. Corratge-Faillie, J.N. Offenborn, P. Tillard, A. Gojon, J. Kudla and B. Lacombe. 2015. Nitrate sensing and uptake in Arabidopsis are enhanced by ABI2, a phosphatase inactivated by the stress hormone abscisic acid. *Science Signaling* 8: 1–7.

Leran, S., S. Munos, C. Brachet, P. Tillard, A. Gojon and B. Lacombe. 2013. Arabidopsis NRT1.1 is a bidirectional transporter involved in root-to-shoot nitrate translocation. *Molecular Plant* 6: 1984–1987.

Leung, J. and J. Giraudat. 1998. Abscisic acid signal transduction. *Annual Review of Plant Biology* 49(1):199–222.

Li, L., B.G. Kim, Y.H. Cheong, G.K. Pandey and S. Luan. 2006. A Ca^{2+} signaling pathway regulates a K^{+} channel for low-K response in Arabidopsis. *Proceedings of National Academy of Sciences USA* 103(33): 12625–12630.

Li, R., J. Zhang, J. Wei, H. Wang, Y. Wang and R. Ma. 2009. Functions and mechanisms of the CBL-CIPK signaling system in plant response to abiotic stress. *Progress in Natural Science* 19: 667–676.

Liu, H., Y. Wang, H. Li, R. Teng, Y. Wang and J. Zhuang. 2019. Genome-wide identification and expression analysis of calcineurin B-like protein and calcineurin B-like protein-interacting protein kinase family genes in tea plant. *DNA Cell Biology* 38: 824–839.

Liu, J., M. Ishitani, U. Halfter, C.S. Kim and J.K. Zhu. 2000. The *Arabidopsis thaliana* SOS2 gene encodes a protein kinase that is required for salt tolerance. *Proceedings of National Academy of Sciences USA* 97: 3730–3734.

Liu, J. and J.K. Zhu. 1998. A calcium sensor homolog required for plant salt tolerance. *Science* 280: 1943–1945.

Liu, K.H., C.Y. Huang and Y.F. Tsay. 1999. CHL1 is a dual-affinity nitrate transporter of Arabidopsis involved in multiple phases of nitrate uptake. *Plant Cell* 11(5): 865–874.

Liu, K.H. and Y.F. Tsay. 2003. Switching between the two action modes of the dual-affinity nitrate transporter CHL1by phosphorylation. *EMBO Journal* 22: 1005–1013.

Liu, L.L., H.M. Ren, L.Q. Chen, Y. Wang and W.H. Wu. 2013. A protein kinase, calcineurin B-like protein-interacting protein kinase 9, interacts with calcium sensor calcineurin B-like protein3 and regulates potassium homeostasis under low-potassium stress in Arabidopsis. *Plant Physiology* 161: 266–277.

Luan, S. 2009. The CBL-CIPK network in plant calcium signaling. *Trends in Plant Science* 14: 37–42.

Luan, S., J. Kudla, M. Rodriguez-Concepcion, S. Yalovsky and W. Gruissem. 2002. Calmodulins and calcineurin B-like proteins calcium sensors for specific signal response coupling in plants. *Plant Cell* 14: S389–S400.

Ma, Q.J., M.H. Sun, H. Kang, J. Lu, C.X. You and Y.J. Hao.2019. A CIPK protein kinase targets sucrose transporter MdSUT2.2 at Ser254 for phosphorylation to enhance salt tolerance. *Plant Cell and Environment* 42: 918–930.

Maierhofer, T., M. Diekmann, J.N. Offenborn, C. Lind, H. Bauer, K. Hashimoto, K.A. Al-Rasheid, S. Luan, J. Kudla and D. Geiger. 2014. Site-and kinase-specific phosphorylation-mediated activation of SLAC1, a guard cell anion channel stimulated by abscisic acid. *Science Signaling* 7: 1–11.

Marino, D., C. Dunand, A. Puppo and N. Pauly. 2012. A burst of plant NADPH oxidases. *Trends in Plant Science* 17: 9–15.

Nagae, M., A. Nozawa, N. Koizumi, H. Sano, H. Hashimoto, M. Sato and T. Shimizu. 2003. The crystal structure of the novel calcium-binding protein AtCBL2 from *Arabidopsis thaliana*. *Journal of Biological Chemistry* 278: 42240–42246.

Nakashima, K. and K. Yamaguchi-Shinozaki. 2013. ABA signaling in stress-response and seed development. *Plant Cell Reports* 32(7): 959–970.

Nieves-Cordones, M., F. Caballero, V. Martinez and F. Rubio. 2012. Disruption of the *Arabidopsis thaliana* inward-rectifier K+ channel AKT1 improves plant responses to water stress. *Plant and Cell Physiology* 53(2): 423–432.

Pandey, G.K., Y.H. Cheong, B.G. Kim, J.J. Grant, L. Li and S. Luan. 2007. CIPK9: a calcium sensor-interacting protein kinase required for low-potassium tolerance in *Arabidopsis*. *Cell Research* 17:411–421.

Pandey, G.K., Y.H. Cheong, K.N. Kim, J.J. Grant, L. Li, W. Hung, C. D'Angelo, S. Weinl, J. Kudla and S. Luan. 2004. The calcium sensor calcineurin B-like 9 modulates abscisic acid sensitivity and biosynthesis in *Arabidopsis*. *Plant Cell* 16:1912–1924.

Pandey, G.K., J.J. Grant, Y.H. Cheong, B.G. Kim, L.G. Li and S. Luan. 2008. Calcineurin-B-like protein CBL9 interacts with target kinase CIPK3 in the regulation of ABA response in seed germination. *Molecular Plant* 1(2): 238–248.

Pandey, G.K., P. Kanwar, A. Singh, L. Steinhorst, A. Pandey, A.K. Yadav, I. Tokas, S.K. Sanyal, B.G. Kim and S.C. Lee. 2015. Calcineurin B-Like protein-interacting protein kinase CIPK21 regulates osmotic and salt stress responses in *Arabidopsis*. *Plant Physiology* 169: 780–792.

Pel, Z.M., Y. Murata, G. Benning, et al. 2000. Calcium channels activated by hydrogen peroxide mediate abscisic acid signalling in guard cells. *Nature* 406(6797): 731–734.

Pilot, G., F. Gaymard, K. Mouline, I. Cherel and H. Sentenac. 2003. Regulated expression of Arabidopsis shaker K+ channel genes involved in K+ uptake and distribution in the plant. *Plant Molecular Biology* 51(5): 773–787.

Qiu, Q.S., Y. Guo, M.A. Dietrich, K.S. Schumaker and J.K. Zhu. 2002. Regulation of SOS1, a plasma membrane Na^+/H^+ exchanger in *Arabidopsis thaliana*, by SOS2 and SOS3. *Proceedings of National Academy of Sciences USA* 99: 8436–8441.

Qiu, Q.S., Y. Guo, F.J. Quintero, J.M. Pardo, K.S. Schumaker and J.K. Zhu. 2004. Regulation of vacuolar Na^+/H+ exchange in *Arabidopsis thaliana* by the salt-overly-sensitive (SOS) pathway. *Journal of Biological Chemistry* 279: 207–215.

Quan, R, H. Lin, I. Mendoza, et al. 2007. SCABP8/CBL10, a putative calcium sensor, interacts with the protein kinase SOS2 to protect *Arabidopsis* shoots from salt stress. *The Plant Cell* 19(4):1415–1431.

Ragel, P., R. Ródenas, E. García-Martín, Z. Andrés, I. Villalta, M. Nieves-Cordones, R.M. Rivero, V. Martínez, J.M. Pardo and F.J. Quintero. 2015. CIPK23 regulates HAK5-mediated high-affinity K+ uptake in *Arabidopsis* roots. *Plant Physiology* 169: 2863–2873.

Roberts, D.M. and A.C. Harmon. 1992. Calcium-modulated proteins: targets of intracellular calcium signals in higher plants. *Annual Review of Plant Physiology and Plant Molecular Biology* 43: 375–414.

Sánchez-Barrena, M.J., H. Fujii, I. Angulo, M. Martínez-Ripoll, J.K. Zhu and A. Albert. 2007. The structure of the C-terminal domain of the protein kinase AtSOS2 bound to the calcium sensor AtSOS3. *Molecular Cell* 26: 427–435.

Sánchez-Barrena, M.J., M. Martínez-Ripoll and A. Albert. 2013. Structural biology of a major signaling network that regulates plant abiotic stress: the CBL-CIPK mediated pathway. *International Journal of Molecular Science* 14: 5734–5749.

Sánchez-Barrena, M.J., M. Martínez-Ripoll, J.K. Zhu and A. Albert. 2005. The structure of the *Arabidopsis thaliana* SOS3: molecular mechanism of sensing calcium for salt stress response. *Journal of Molecular Biology* 345: 1253–1264.

Sanders, D., J. Pelloux, C. Brownlee and J.F. Harper. 2002. Calcium at the crossroads of signaling. *Plant Cell* 14:S401–S417.

Shi, H., F.J. Quintero, J.M. Pardo and J.K. Zhu. 2002. The putative plasma membrane Na+/H+ antiporter SOS1controls long-distance Na+ transport in plants. *Plant Cell* 14: 465–477.

Shi, J., K.N. Kim, O. Ritz, V. Albrecht, R. Gupta, K. Harter, S. Luan and J. Kudla. 1999. Novel protein kinases associated with calcineurin B-like calcium sensors in Arabidopsis. *Plant Cell* 11: 2393–2405.

Steinhorst, L., A. Mähs, T. Ischebeck, C. Zhang, X. Zhang, S. Arendt, S. Schültke, I. Heilmann and J. Kudla. 2015. Vacuolar CBL-CIPK12 Ca^{2+}-sensor-kinase complexes are required for polarized pollen tube growth. *Current Biology* 25: 1475–1482.

Sun, T., Y. Wang, M. Wang, T. Li, Y. Zhou, X. Wang, S. Wei, G. He and G. Yang. 2015. Identification and comprehensive analyses of the CBL and CIPK gene families in wheat (*Triticum aestivum* L.). *BMC Plant Biology* 15: 269.

Tang, R.J. and S. Luan. 2017. Regulation of calcium and magnesium homeostasis in plants: from transporters to signaling network. *Current Opinion in Plant Biology* 39: 97–105.

Tang, R.J., F.G. Zhao, V.J. Garcia, T.J. Kleist, L. Yang, H.X. Zhang and S. Luan. 2015. Tonoplast CBL-CIPK calcium signaling network regulates magnesium homeostasis in Arabidopsis. *Proceedings of National Academy of Sciences USA* 112: 3134–3139.

Thapa, G., M. Dey, S. Sahoo and S.K. Panda. 2011. An insight into the drought stress induced alterations in plants. *Biologia Plantarum* 55(4): 603–613.

Tsay, Y.F, J.I. Schroeder, K.A. Feldmann and N.M. Crawford. 1993. The herbicide sensitivity gene CHL1 of *Arabidopsis* encodes a nitrate-inducible nitrate transporter. *Cell* 72(5): 705–713.

Tuskan, G.A., S. DiFazio, S. Jansson, et al. 2006. The genome of black cottonwood, *Populus trichocarpa* (Torr. & Gray). *Science* 313(5793): 1596–1604.

Vert, G. and J. Chory. 2009. A toggle switch in plant nitrate uptake. *Cell* 138(6): 1064–1066.

Wang, R., D. Liu and N.M. Crawford. 1998. The Arabidopsis CHL1 protein plays a major role in high-affinity nitrate uptake. *Proceedings of National Academy of Sciences USA* 95(25): 15134–15139.

Wang, R.K., L.L. Li, Z.H. Cao, et al. 2012. Molecular cloning and functional characterization of a novel apple MdCIPK6L gene reveals its involvement in multiple abiotic stress tolerance in transgenic plants. *Plant Molecular Biology* 79(1–2): 123–135.

Wanga, L., X. Fenga, L. Yaoa, C. Dinga, L. Leia, X. Haoa, N. Lia, J. Zenga, Y. Yanga and X. Wanga. 2020. Characterization of CBL–CIPK signaling complexes and their involvement in cold response in tea plant. *Plant Physiology and Biochemistry* 154: 195–203.

Weinl, S. and J. Kudla. 2009. The CBL-CIPK Ca^{2+}-decoding signaling network: function and perspectives. *New Phytology* 184: 517–528.

Yamanaka, T., Y. Nakagawa, K. Mori, et al. 2010. MCA1 and MCA2 that mediate Ca^{2+} uptake have distinct and overlapping roles in *Arabidopsis*. *Plant Physiology* 152(3): 1284–1296.

Yang, Y., et al. 2019. Calcineurin B-like proteins CBL4 andCBL10 mediate two independent salt tolerance pathways in Arabidopsis. *International Journal of Molecular Science* 20: 2421.

Zhang, H., F. Lv, X. Han, X. Xia and W. Yin. 2013. The calcium sensor PeCBL1, interacting with PeCIPK24/25 and PeCIPK26, regulates Na+/K+ homeostasis in *Populuseuphratica*. *Plant Cell Reports* 32(5): 611–621.

Zhao, J., A. Yu, Y. Du, G. Wang, Y. Li, G. Zhao, et al. 2019. Foxtail millet (*Setariaitalica* L.) P. Beauv) CIPKs are responsive to ABA and abiotic stresses. *PLoS One* 14(11): e0225091. https://doi. org/10.1371/journal. pone.0225091.

Zhou, L., W. Lan, B. Chen, W. Fang and S. Luan. 2015. A calcium sensor-regulated protein kinase, calcineurin B-like protein-interacting protein kinase 19, is required for pollen tube growth and polarity. *Plant Physiology* 167: 1351–1360.

Zhu, J.K. 2002. Salt and drought stress signal transduction in plants. *Annual Review of Plant Biology* 53: 247–273.

Zhu, J.K. 2003. Regulation of ion homeostasis under salt stress. *Current Opinion in Plant Biology* 6: 441–445.

Zhu, J.K., J. Liu and L. Xiong. 1998. Genetic analysis of salt tolerance in Arabidopsis: Evidence for a critical role of potassium nutrition. *Plant Cell* 10: 1181–1191.

6

Transgenic Technology and Its Progressive Implications

Aradhana L. Hans and Sangeeta Saxena
Babhasaheb Bhimrao Ambedkar University

Ritesh Mishra
The Hebrew University of Jerusalem

CONTENTS

6.1 Introduction

The population globally is on the rise, and there is a need to feed the growing population. Equipping agricultural practices to yield more to secure the ever-increasing food supply is very important. The need to increase crop production also becomes more challenging with the decline in the cultivable area, especially in developing countries. With the limited land resource, the need to elevate production becomes more prominent. Applied plant science research towards agriculture aims to increase crop yield, increasing nutritional levels in crops with better management and protection strategies for crop and environment. Merging integrated pest management (IPM) to manage the pest situation along with sustainable agriculture and environmental protection is promising better protection for crops. Genetic engineering (GE) of crops provides a wide array of benefits that complement a wider IPM program. Better performing crops against all kinds of biotic and abiotic stresses, improved nutrition and better shelf life of crops are some of the prominent advantages of GE crops. Chemical pesticides and fertilizers also protect crops from damaging pests and enhance production, but they leave residues that linger on in the environment. This causes a huge health risk to the farmers and the beneficial insects and environment.

There is an increase in urban population due to the shift from rural to urban areas in the search of a better livelihood. Farmland is being replaced by housing developments and urban development plans, resulting in an increase in urban terrace gardening or kitchen gardening amidst rapid urban expansion. Therefore, a new generation of crops that would fit into a newer environment having limited sun exposure and different nutrient requirement is needed. Finding genes that would enable plants to survive and flourish in limited energy and space is required. Plants tolerant to urban pollution with better photosynthetic and nutrient uptake ability need to be developed. Also, there is more awareness among people and a preference towards low-fat, nutrient-dense, cleaner crop requirements due to changes in diet habits. For such specific requirements looking for genes specific to such requirements and raising such plants with such traits is the upcoming need of the new generation. Developing such transgenic plants expressing various genes and utilising them accordingly is the need of the hour. The change of perspective is to accept this technology as a crop improvement tool to accomplish the requirements. This chapter discusses such effective use of transgenics and the application of this technology.

6.2 Status of GM Crops and Applications

Advents in agricultural and plant biotechnology have led to the development of genetically modified (GM) plants and crops. GM plants or transgenic plants are developed by the transfer of certain genes between unrelated species artificially through various biotechnological methods and tools. From one gene to the stacking of multiple genes to target multiple stress at a time, GM technology has moved to an advanced level, providing a wider range of protection to the crops coping with broader stress levels. Recently, another pioneering technology named clustered regularly interspaced short palindromic repeats (CRISPR) technology has immense potential to augment biotechnological application. It is characteristic of a bacterial defence system and is used as a genome-editing technology (namely CRISPR-Cas9 gene-editing technology). Such technologies can bring about change in the technological and economic scenario of the agriculture of the country, progressing from GE crops from pest protection mainly involving *Bacillus thuringiensis* (*Bt*) to new traits based on RNA interference (RNAi) of more novel proteins from non-*Bt* sources.

Transgenic crops displaying resistance against many biotic and abiotic stresses are developed and cultivated in the field. The development of GM plants having fungal and bacterial resistance is in progress. GM plants are acting as edible vaccines that serve as antibodies enhancing the immunity. They have the gene encoding bacterial or viral disease causing agents which can be introduced in plants without losing its immunogenic property (Gunasekaran and Gothandam 2020). Phytoremediation by growing transgenic plants capable of growing in degraded lands helps to utilise unused land (Mosa et al. 2016). Apart from all these *Bt* technologies remains the most successfully applied and adopted transgenic technology globally which has resulted in a reduction in the application of hazardous chemical pesticides all over the world.

Globally GM crops have been rapidly growing and adopted by 18 million farmers cultivating over 2 billion hectares of land (Arshad et al. 2018). In India, the first GM crop to be approved for commercial cultivation was *Bt*-cotton in 2002. *Bt*-cotton has the endotoxin-producing Cry genes (Cryl Ac and Cry2Ab) from the bacteria *B. thuringiensis*. These genes are incorporated in the cotton genome using recombinant DNA technology, producing δ-endotoxins conferring resistance against lepidopteran pests. The application of *Bt*-cotton had been growing immensely and now almost 90% of cotton cultivated in India is *Bt*-cotton. It reduced the repeated application of pesticides and resulted in a higher yield of cotton that benefited the farmers. More than twofold increase in cotton yield was evident from the adoption of *Bt* technology in a decade. India now stands as the highest producer of cotton in the world. Although, recently there has been perspective advocating that the increase in yield is due to the increase in the use of fertilizers (Kranthi and Stone 2020). The other consequence of the application of *Bt*-cotton is the resurgence of other pests that were regarded as minor pests. These pests sprouted up and became major pests, e.g., bollworms, which were taken care of by the expression of δ-endotoxins. As a result, a new range of pest – hemipteran pests – started damaging the crops leading to major yield loss. In 2015, Punjab suffered a significant loss of more than 70% of cotton crops due to the infestation of whiteflies (Sahu and Samal 2020). These are sap-sucking pests that feeds on the phloem sap of the plant, damaging the plant. They multiply fast and soon the population reaches an unmanageable number. These sap-sucking pests also secrete honey due to which sooty mould grows on the cotton bolls compromising its quality. Thus, a more efficient approach in which there is a range of transgenic cotton with better and wider insect-resistant technology needs to be adopted. The new range of novel proteins should be explored from natural sources protecting against such emerging pests. Also, a wide range of new proteins will help to keep the insects from developing quick resistance. Plants expressing such proteins are being reported as showing good protection against whiteflies (Shukla et al. 2016). An array of non-*Bt* proteins is required, and a strategic application field needs to be adapted for better protection against a wide range of pests. Apart from cotton, there are over 20 crops that are undergoing research in different institutes and few are under contained field trails. Because it is an edible crop, *Bt*-brinjal has been developed but its release was withheld due to health risk issues raised by various environmental activists. There is a need to readdress the transgenic-related perspective with proper scientific reasoning and impart the right knowledge in public.

Some transgenic plants also display herbicide resistance, such as glyphosate-resistant crops, which are resistant to the broad-spectrum herbicide glyphosate. It controls a broad range of weeds by inhibiting the 5-enolpyruvylshikimate-3-phosphate (EPSP) synthase which is a key enzyme in the shikimate pathway of aromatic amino acids. Herbicide-resistant transgenic plants have made farming easier as no tilling is required to control weeds growing in the field. Even though there is a decrease in pesticide usage as a result of *Bt*-crops, herbicide for control of weeds still needs to be applied. The herbicide application not only kills weeds but has a detrimental effect on the susceptible plants nearby. Reportedly, there is resistance development against glyphosate in more than 15 weed species, globally posing a significant problem. This in turn affects the environment which is inhabited by many ecologically important flora and fauna. The optimisation of the application of an adequate amount of herbicides with proper measures ensuring minimum damage to the surrounding environment should be ensured. The aim is to program user-friendly and convenient cost-effective strategies that can be easily implemented with the existing farming practices. Especially in crops like cotton and maize, post-emergence weed management might be a better strategy to control weeds in the future.

In front of advances in genetic improvement of crops, RNAi technology comes as a breakthrough. In plants, RNAi technology has been used in engineering metabolic pathways against different targets and traits. It is also used in increasing nutritional values in plants, in conferring resistance against pathogens, pests, and diseases and improving several crop plants. RNAi is a gene silencing mechanism that is sequence specific, instigated by dsRNA (where ds is double-stranded) leading to degradation of mRNA. It helps in switching on and off the targeted gene, resulting in a significant impact in developmental biology. RNAi is used in the development of tomato with enhanced flavonoid and carotenoid both possessing similarly carotenoid health benefits. Similarly, the carotenoid content of *Brassica napus* was also augmented by using RNAi technology to down-regulate lycopene epsilon expression enhancing nutritional value. In sweet potato, the amylose content was also increased, genetically modifying the fatty acid composition of cottonseed oil with the application of RNAi. RNAi technology also has been successfully applied in protecting papaya plants against papaya ring spot virus (PRSV). The RNA-mediated resistance in papaya is based on posttranscriptional gene silencing (PTGS), host-mediated defence response to a foreign RNA sequence. It depends on sequence homology between the transgene and the corresponding viral genome. Transgenic papaya has resistance against various isolates found in different geographical areas. Management of PRSV is tricky as it is transmitted in a non-persistent manner by aphids and there are significant diverse PRSV strains in different geographical locations (Jia et al. 2017). Also, transgenic papaya expressing the PRSV coat protein gene is extremely successful in providing resistance against the virus. It is cultivated in large areas in Hawaii providing resistance to the non-transgenic papaya plants (Lindbo et al. 2017).

6.3 Challenges and Change in Perspective

Despite the global application of transgenic crops due to the advantages they confer, there are challenges in their unanimous successful implementation worldwide. IPM strategies converging more efficient and sustainable application are needed for wider global acceptance. IPM programs should be more versatile rather than uniform for better applicability due to varied application strategies based on environment and different regions. There should be no confusion in considering crop improvement safe to fight against hunger due to the increasing population. Proper flow of knowledge considering all the advantages and possible limitations of transgenic crops should be provided. Non-scientific views based on lack of scientific data should be restricted. This has unnecessarily created much confusion without any rational debating or discussion related to GM crops. Therefore, decades of research and field trials of many GM crops have not seen the light of day because of the dilemma of unknown uncertainty.

The concern related to the cultivations and possible implications in the farming system is genuine. Scientists should make an effort to explain the possible effects and suitable utilisation of transgenic technology to the farmers and general public clarifying doubts. This would help in making people aware of the actual scenario rather than creating unnecessary chaos. Without knowledge of the subject, giving non-scientific views causes the most trouble for the release and application of GM crops. The mind should be clear, and only then will the actions prevail. Thus, the right amount of knowledge should be imparted regarding the advantages and limitations attached to GM crops. For application and development, public opinion is an extremely important factor. Therefore, a proper risk assessment, prediction and prevention must be done to overrule or check for the negative impacts, if any, that are caused due to GM crops. Also, instead of generalised prediction, a case-by-case study can help to resolve any concerns. The negative propaganda jeopardising the end users is a serious concern and should never be overlooked. Unfortunately, most of the generalised information that the public has and the concerns regarding GM crops are sceptical as transgenic technology has tremendous potential in addressing the present-day crop-related issues.

6.4 Proper Risk Assessment of Transgenic Crops

To rule out the possible negative consequences or uncertainty underlying GM crops, they should go through proper risk assessment (Figure 6.1). To move towards the approval process, evaluation of potential hazards or limitations is essential. Risk can be determined by the probability or consequence of a possible hazard. Risk assessment is an iterative process that can be applied in the field of transgenic technology. It includes identification of a possible hazard, dose-response assessment, exposure characterisation and risk conclusion.

6.4.1 Identification of a Possible Hazard

The identification of possible hazards in dietary risk assessment begins with toxicity as a result of consumption. To check the potential of toxicity, an animal model is used for predicting human health risks. Especially in the case of edible crops like *Bt*-maize, the target pest is generally concerned with the leaf, but the protein is expressed in other parts of the plant that are consumed by humans and animals. Toxicity trials are extremely important to see the effects on non-target organisms. Ecological risk assessment is equally important to see the impact on beneficial and non-target insects, micro-organisms and so forth. Pollen dispersal of transgenic crops and their possible consequence seems to be checked in contained field trials before open field trails.

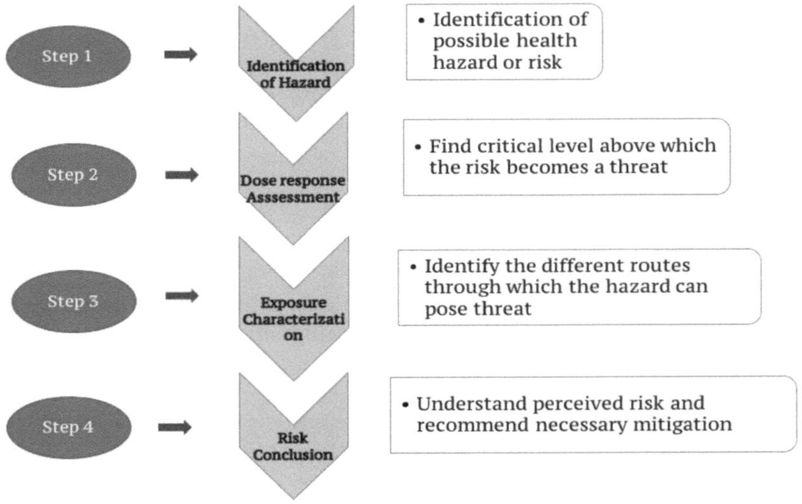

FIGURE 6.1 Risk assessment steps for transgenic crops and possible action and mitigation needed. (Adapted from Carzoli et al. 2018; licence number 4882960806285.)

6.4.2 Dose-Response Assessment

To ensure that no adverse or negative reactions occur, repeated dose-response assessment is done. When no such unwanted effect is observed, the value is termed as no observed adverse effect level (NOAEL). Such dietary dose analysis follows the mathematical model to determine NOAEL. Such dose assessments demonstrate toxic effects resulting from various amounts of toxins consumed by instars of insects. It depicts that the first instar larvae lose more weight or show early mortality after consuming candidate gene-encoded proteins used in GM crops. Based on the NOAEL value, the potential risk on humans consuming GM products can be calculated.

6.4.3 Exposure Characterisation

Exposure characterisation deals with the time the target is exposed to the toxin. It can be seen in the case of pollens that the amount of pollen present in the environment depends on the proximity of the transgenic crop and pollen release. The pollen counts usually drop at farther distances, therefore, creating buffer zones around transgenic crops which will give sensible protection from the possible escape of pollen in the nearby ecological niches.

6.4.4 Risk Conclusion

The overall risk can be summarised by comparing the results of all the three aforementioned parameters. Identification of possible hazards, dose risk assessment and exposure characterisation would collectively show a clear picture of actual risk, if any, posed by the transgenic crop. It should be taken into account that *Bt*-corn and its products have been consumed in many countries for several decades and have not shown any adverse effect on humans. The effect of *Bt*-maize on the monarch butterfly and risk assessment was carried out and some other non-target pest and was found to not affect them (Carzoli et al. 2018; Shu et al. 2018). Therefore, many other ecological, dietary risk assessments can be done strategically to find the impact of transgenic plants on the environment and other areas. Apart from these concerns regarding resistance development in insect-resistant plants, horizontal gene transfer, the effect on the overall nutrition composition of the plant, increased the use of chemicals for other minor pests and loss in biodiversity should also be looked into.

6.5 Future Prospects of Transgenic Crops

Because of the increase in population and decrease in arable land, new improved crop varieties will have to be utilised instead of sticking to conventional agriculture systems. The modified plants exhibit properties that are not possible to attain through regular breeding programs. GE makes breeding programs more effective through the transfer of foreign genes and incorporated genes of unrelated species. Transgenic plants are developed and cultivated all over the world except for a few European countries that are still unsure about their implications. The applicability of transgenic technology is immense from providing pest resistance, disease resistance, herbicide tolerance, abiotic stress tolerance, nutritional enhancement, phytoremediation and so forth and is increasing with every passing year. The changing environment and limited land resources demand improved varieties with better performing capabilities. Various transgenic crops like tomato, potato, corn, soya bean, mustard, brinjal, cabbage, lettuce and many more are developed with various improved traits. There can be few challenges like the distribution of GM seeds which need to be purchased by the farmer every year. This can be challenging for developing countries that have feeble legislation policies causing the real price of GM crops to sprawl up. The proper implication of refugia plants during the application of transgenic crops in the fields should be taken care of very strictly. Planned application procedures and adequate insecticide applications with proper safety measures should always be ensured. Sanitising farmers with the proper application of GM crops, possible risk and mitigation should be ensured from time to time. Such measures will lead to the ease of dealing with a topic which otherwise is blown out of proportion. On the other hand, the policymakers should assess all the risk analysis reports and make decisions to improve the contained and field trails. The biosafety analysis should be done on a larger scale to see the possible impacts. New technology cannot be blindly applied, but this does not mean it should never be applied. With thorough trials, risk assessments and biosafety trials, GM technology can see a new dawn and lead to fruitful yields in various spheres of agriculture.

6.6 Conclusion

Agriculture is an important aspect of any developing nation and so are the innovations associated with it. The myriad applications of GE for desirable traits in plants cannot just be overruled. There might be controversies surrounding transgenic technology largely because of a lack of logical and scientific awareness. GM regulation policies should be more standardised so that the GM crops can undergo proper assessment. Once they are through the trails and safety risk assessments, the availability of seeds to common farmers should be a hassle-free process. GM technology has the potential to mitigate the challenges of the present situation and can help in boosting agricultural growth and the economy. GM seed technology is one of the fastest-growing sectors, giving valuable returns to the farmers and society. Proper channelising of seeds to the end users will greatly help to implement the application of transgenic technology.

Other alternatives can be used to develop crops with improved traits, such as genome editing, cisgenesis and intragenesis. Since crops developed using gene-editing techniques would be mostly parallel to crops developed through conventional breeding, it can be said that such crops could possibly move faster through the regulatory channel leading to field application. Apparently, the agreeable conclusion that can be made is that the benefits of GM are such that it can neither be completely rejected nor fully accepted amidst sceptical opinions but is a compatible option. The farmers would undoubtedly benefit from GM seeds providing them better crop options resistant to various pests and herbicides and reap the benefits of high yields. The most appropriate approach would be educating farmers as well as non-farmers about the applicability and mitigation of any possible risk.

REFERENCES

Arshad, Muhammad, Rashad Rasool Khan, Asad Aslam, and Waseem Akbar. "Transgenic Bt cotton: effects on target and non-target insect diversity." *Past, Present and Future Trends in Cotton Breeding* (2018). London: IntechOpen.

Carzoli, Andrea K., Siddique I. Aboobucker, Leah L. Sandall, Thomas T. Lübberstedt, and Walter P. Suza. "Risks and opportunities of GM crops: Bt maize example." *Global Food Security* 19 (2018): 84–91.

Gunasekaran, B., and K. M. Gothandam. "A review on edible vaccines and their prospects." *Brazilian Journal of Medical and Biological Research* 53, no. 2 (2020): e87498.

Jia, Ruizong, Hui Zhao, Jing Huang, Hua Kong, Yuliang Zhang, Jingyuan Guo, Qixing Huang, et al. "Use of RNAi technology to develop a PRSV-resistant transgenic papaya." *Scientific Reports* 7, no. 1 (2017): 1–9.

Kranthi, K. R., and Glenn Davis Stone. "Long-term impacts of Bt cotton in India." *Nature Plants* 6, no. 3 (2020): 188–196.

Lindbo, John A., and Bryce W. Falk. "The impact of "coat protein-mediated virus resistance" in applied plant pathology and basic research." *Phytopathology* 107, no. 6 (2017): 624–634.

Mosa, Kareem A., Ismail Saadoun, Kundan Kumar, Mohamed Helmy, and Om Parkash Dhankher. "Potential biotechnological strategies for the cleanup of heavy metals and metalloids." *Frontiers in Plant Science* 7 (2016): 303.

Sahu, Bhupen Kumar, and Ipsita Samal. "Sucking pest complex of cotton and their management: A review." *The Pharma Innovation Journal* 9, no. 5 (2020): 29–32.

Shu, Yinghua, Jörg Romeis, and Michael Meissle. "No interactions of stacked Bt maize with the non-target aphid *Rhopalosiphum padi* and the spider mite *Tetranychus urticae*." *Frontiers in Plant Science* 9 (2018): 39.

Shukla, Anoop Kumar, Santosh Kumar Upadhyay, Manisha Mishra, Sharad Saurabh, Rahul Singh, Harpal Singh, Nidhi Thakur et al. "Expression of an insecticidal fern protein in cotton protects against whitefly." *Nature Biotechnology* 34, no. 10 (2016): 1046–1051.

7

PR Proteins: Key Genes for Engineering Disease Resistance in Plants

Aditi Sharma
Dr. Yashwant Singh Parmar University of Horticulture and Forestry

Ashutosh Sharma and Rahul Kumar
DAV University

Indu Sharma
Sant Baba Bhag Singh University

Akshay Kumar Vats
Maharishi Markandeshwar University

CONTENTS

7.1 Introduction

In recent years, the rapid increase in the world population had a huge impact on agriculture (mainly crop protection practices) because of the rapid decrease in the area under cultivation. Further, there has been a gradual shift in consumer behaviour towards processed foodstuffs (Stranieri et al. 2017). Plants are constantly challenged by various types of plant pathogens (Agrios 2005), viz., fungi, bacteria, phytoplasma, viruses, viroids and nematodes etc., which are evolving as better pathogens than ever, and which affect both the food production and food quality. One of the greatest challenges in the food supply chain has been seen in the current COVID-19 pandemic, as global food supply chains are disrupted, so there has been a need to be vocal to utilise and improve local agricultural resources (Chakraborty 2020).To ensure the local production and to strengthen local food supply chains, engineering local plants for better resistance against plant pathogens is pivotal. To overcome the invading plant pathogens, plants utilise an array of defence mechanisms for their survival. Plant immune responses against the pathogens fall under two different modes, i.e., pathogen-associated molecular pattern-triggered immunity (PTI or PAMP-triggered immunity) and effector-triggered immunity (ETI) (Dangl and Jones 2001; Andersen et al. 2018). Pattern recognition receptor molecules (PRRs) within the host cell membrane detect pathogen-associated molecular patterns (PAMPs), and wall-associated kinases (WAKs) recognise damage-associated molecular patterns (DAMPs) during infection (Zipfel 2014). The pathogen effectors after recognition induce many structural and biochemical defence responses in plants such as formation and deposition of callose, abscission layer, tyloses, gum deposition and programmed cell death, followed by the generation of reactive oxygen species (ROS) and the production of hydroxyproline-rich proteins, phytoalexins, tannins and other specific groups of proteins called pathogenesis-related (PR) proteins (van Loon et al. 1994; Hammond-Kosack and Jones 1996; van Loon et al. 2006).

Out of these, PR proteins are among one of the well-studied defence molecules that are produced after the induction of systemic acquired resistance (SAR) in plants. They were first reported from tobacco plants infected by tobacco mosaic virus (TMV) and initially, only five major classes of PR proteins, viz., PR-1 to PR-5, were reported from tobacco (Breen et al. 2017). Later, different PR proteins were identified from various plants. At present PR proteins are grouped into 17 major families based on their sequence similarities and the biological functions. They mainly have been classified as thaumatin/osmotin-like proteins, chitinases, glucanases, defensins and thionins, oxalate oxidase, oxalate oxidase like proteins/germin-like proteins, lipid transfer proteins (LTPs) and so forth, based on their mode of action. Initial studies suggested that PRs are capable of conferring resistance against invading plant pathogens, and the recent studies are suggesting that they have another major role in conferring abiotic stress tolerance (van Loon et al. 2006; Wu et al. 2016; Kamle et al. 2020; Sinha et al. 2020). There are 17 major families of PR proteins (Table 7.1) which are structurally diverse and may differ in their mode of action.

During the last few years, functional characterisations and downstream signalling partners of these proteins have been the areas of interest for plant pathologists and plant molecular biologists. With the advances in plant transformation technologies, high-throughput reverse genetic studies were initiated for establishing gene-phenotype relationships (Pereira 2000). The PR proteins have recently also been described as potential food preservatives (Thery et al. 2019, 2020), thereby suggesting the possible implication of PR protein overexpressing genetically modified (GM) crops in processed food items. Further, they are also found to reduce mycotoxin production (Sundaresha et al. 2010, 2016). Some other beneficial roles of PR proteins other than the defence against plant pathogens are described in Figure 7.1. However, the allergenic potential of some PR proteins has also been reported (Jain and Khurana 2018; which may be elevated in foodstuffs derived from PR protein overexpressing transgenics. Therefore, a case-by-case safety evaluation of such GM crops may be required. Further, the regulation of expression of PR proteins can help in understanding the other associated interacting partners like transcription factors, mitogen-activated protein kinases, G proteins, ubiquitin, calcium, plant growth regulators (PGRs) and so forth for getting a stable and desired level of their expression. The present chapter provides an overall overview on the PR proteins, their role in plant defence (anti-bacterial, anti-fungal, anti-viral, anti-nematodal, anti-insect properties), the relation of PR proteins, transgenic plant overexpressing PR proteins, role in abiotic stress tolerance, interaction with PGRs and the possible implication of PR overexpressing transgenics on human health. The success and snags of transgenic plants expressing PR proteins and peptides are also highlighted in this chapter.

7.2 Anti-Fungal Activity

Plants are continuously being challenged by the fungal pathogens that have evolved alongside the evolution of the defence mechanisms of the plant hosts. Many PR proteins have been shown to possess anti-fungal activities (Table 7.2). Further, the constitutive overexpression of PR proteins has been proven excellent for the development of disease-resistant varieties (Dai et al. 2016; Moosa et al. 2017; Ali et al. 2018c; Jain and Khurana 2018). PR-1 proteins can bind with the sterols, thereby resulting in pathogen growth inhibition (Gamir et al. 2017). Some PR proteins target fungal cell walls by hydrolysing them,

TABLE 7.1

Representative Members of the Major PR Proteins Families

Families	Plant Source	Type Member	Gene Accession Number	Gene Symbol	Size (kDa)	Properties	References
PR-1	*Nicotiana tabacum*	Tobacco PR-1a	YOO707	*Ypr1*	15	Anti-fungal	Antoniw et al. (1980)
PR-2	*N. tabacum*	Tobacco PR-2	M59443.1	*Ypr2, [Gns2 ('Glb')]*	30	ß-1,3-Glucanase	Antoniw et al. (1980)
PR-3	*N. tabacum*	Tobacco P, Q	X77111.1	*Ypr3, Chia*	25–30	Chitinase types I, II and IV–VII	van Loon (1982)
PR-4	*N. tabacum*	Tobacco 'R'	NW_015888419.1	*Ypr4, Chid*	15–20	Chitinase types I and II	van Loon (1982)
PR-5	*N. tabacum*	Tobacco S	NW_015793016	*Ypr5*	25	Thaumatin-like	van Loon (1982)
PR-6	*Solanum lycopersicum*	Tomato inhibitor I	NW_004196001.1	*Ypr6, Pis ('Pin')*	8	Proteinase-inhibitor	Green and Ryan (1972)
PR-7	*S. lycopersicum*	Tomato P69	NC_015445.2	*Ypr7*	75	Endo-proteinase	Vera and Conejero (1988)
PR-8	*Cucumis sativus*	Cucumber chitinase	NC_026660.1	*Ypr8, Chib*	28	Chitinase type III	Metraux et al. (1988)
PR-9	*S. tuberosum*	Potato peroxidase	AJ401150	*Prx2*	35	Peroxidase	Lagrimini et al. (1987)
PR-10	*Petroselinum crispum*	Parsley 'PR1'	NC_026940.1	*Ypr10*	17	'Ribonuclease-like'	Somssich et al. (1986)
PR-11	*N. tabacum*	Tobacco 'class V' chitinase	gi|899342	*Ypr11, Chic*	40	Chitinase, type I	Melchers et al. (1994)
PR-12	*Raphanus raphanistrum*	Radish Rs-AFP3	NC_025209.1	*Ypr12*	5	Defensin	Terras et al. (1995)
PR-13	*Arabidopsis thaliana*	*Arabidopsis* THI2.1	gi|1181531	*Ypr13, Thi*	5	Thionin	Epple et al. (1995)
PR-14	*Hordeum vulgare*	Barley LTP4	gi|1045201	*Ypr14, Ltp*	9	Lipid-transfer protein	García-Olmedo et al. (1995)
PR-15	*H. vulgare*	Barley OxOa (germin)	gi|2266668	*Ypr15*	20	Oxalate oxidase	Zhang et al. (1995)
PR-16	*H. vulgare*	Barley OxOLP	gi|1070358	*Yrp16*	20	'Oxalate oxidase-like"	Wei et al. (1998)
PR-17	*N. tabacum*	Tobacco PRp27		*Yrp17*	27	Unknown	Okushima et al. (2000)

leading to the death of plant cells. The major anti-fungal families of PR proteins are PR-2 to PR-5, PR-8 and PR-12 to PR-14. They mainly include osmotin-like proteins, thaumatin-like proteins, chitinases, glucanases, defensins, thionins, oxalate oxidase or oxalate oxidase like proteins and LTPs (Ali et al. 2018b).

Several studies have confirmed that many PR proteins provide resistance against most devastating fungal pathogens belonging to genera such as *Phytopthora, Pyricularia, Puccinia, Neovossia, Botrytis, Pseudoperonospora, Colletotrichum, Fusarium, Rhizoctonia, Sclerotinia, Cercospora, Sphaerotheca, Blumeria, Uncinula* and so forth (Fujimori et al. 2016; Misra et al. 2016; Moosa et al. 2017; Wang et al. 2019). In another experiment, it was found that GAPEP1, a novel small peptide derived from PR proteins of *Gossypium arboretum*, enhanced the resistance against *Verticillium dahlia* (Yuan et al. 2019). Similarly, β-1,3-glucanase has been shown to cause significant inhibition of hyphal growth, spore formation and mycelial morphology of fungi associated with wheat kernel such as *Alternaria* sp., *Fusarium graminearum, Aspergillus flavus, A. niger*, and *Penicillium* sp. (Cools et al. 2017; Zhang et al. 2019). Recently, it was recorded that a wheat PR protein, *TaLr35PR2* gene, is required for *Lr35*-mediated adult plant resistance against *Puccinia triticina* (Liang et al. 2019). The osmotin protein belonging to the PR-5

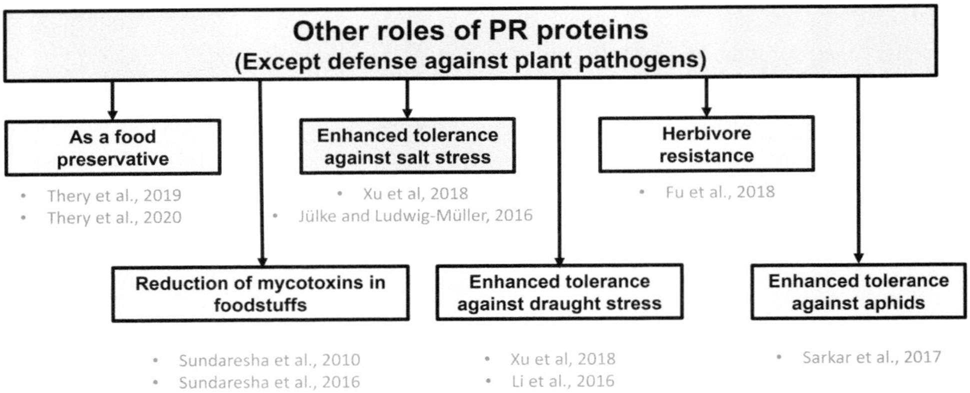

FIGURE 7.1 Some other beneficial roles of PR proteins, other than the defense against plant pathogens.

TABLE 7.2

Examples of Major Classes of Anti-Fungal PR Proteins with Their Target Fungal Pathogens

Anti-Fungal Proteins	Family	Genes	Mode of Action	Target Fungi
Thaumatin-like or osmotin-like proteins	PR-5	*P23, Mal d2, TLP-D34, thaumatin-II, tlp-D34, GbTLP1, PpTLP, ObTLP1, OsOSM1, CsTLP*	Inhibits fungal enzymes such as xylanase, trypsin, a-amylase and β-glucanase, produces transmembrane pores, blocks the function of plasma membrane receptor molecules involved in cAMP/RAS2 signalling pathways, osmotin activates mitogen-activated protein kinase pathway	*Phytopthora citrophthora, Rhizoctonia, Botrytis, Sclerotinia, Pseudoperonospora cubensis, Alternaria alternate, Fusarium pallidoroseum, Colletotrichum gloeosporioides* and *C. dematium*
Chitinases and glucanases	Chitinases – PR-3, PR-4, PR-8 and PR-11 Glucanases – PR-2	*RCC2, pcht28, BjCHI1, RCC2, TcChi1, chit cDNA, chi194, rcc2 and rcg3, RCH10, VvGHF17*	Hydrolytic enzymes cause lysis of fungal hyphae, inhibit fungal growth, degrade chitin	*Uncinula nectar, Pyricularia oryzae, Verticillium dahlia, Puccinia striiformis* f.sp. *tritici, Rhizoctonia solani, Puccinia coronate, Cercospora arachidicola, C, gloeosporioides, Aspergillus flavus, F. graminearum, C. higginsianum, Mycosphaerella fijiensis* and *Botrytis cinerea*
Defensins and thionins	Defensins – PR-12 Thionins – PR-13	*Wasabi defensin, RsAFP2, MsDef1, RsAFP2, thionin, PhDef1 and PhDef2, Drr23oa, MtDef4.2*	Disrupts fungal membrane by pore formation that ultimately leads to cell death and modifies membrane permeability of fungi	*Magnaporthe grisea, Cercospora arachidicola, Pheaoisariopsis personata, Phytophthora parasitica* var. *nicotianae, F. oxysporum* f.sp. *cubense, F. moniliforme, R. solani, F. tucumaniae, Puccinia triticina* and *C. gossypii* var. *cephalosporioides*
Lipid transfer proteins	PR-14	*Ace-AMP1*	Interacts with biological membranes leading to altered membrane permeability	*Sphaerotheca pannosa* var. *rosae, Blumeria graminis* f.sp. *tritici and Neovossia indica, M. grisea and R. solani*

Source: After Ali et al. (2018b).

protein family has shown resistance against many pathogens and overexpression of the osmotin gene, *OsmWS*, of *Withania somnifera* has shown a high level of resistance against *A. solani* in potato cultivar 'Kufri Chipsona 1' (Kaur et al. 2020).

7.3 Anti-Bacterial Activity

Bacterial pathogens enter into the host through wounds created by insects, natural openings and with the help of vectors. The initial penetration process is subsequently followed by pathogen recognition and activation of defence responses in the resistant plants. PR proteins are also well-characterised weapons against bacterial pathogens (Zaynab et al. 2019). Different PR protein families such as PR-10 and PR-12 to PR-14 have shown broad-spectrum anti-bacterial properties against most devastating bacterial pathogens (van der Weerden and Anderson 2013; Kraszewska et al. 2016). PR proteins generally consist of a short sequence of amino acids known as anti-microbial peptides (AMPs). Among them, defensins is a class of the best-characterised AMPs that are synthesised in diverse plant species after pathogen attack (Tam et al. 2015; Iqbal et al. 2019). Recent studies have enlightened the role of defensins as anti-bacterial proteins. Defensins like fabatin-2 from broad bean, SOD1-7 from spinach, ZmESR-6 from maize, Cp-thionin from cowpea, DmAMP1 from dahlia, CtAMP1 from *Clitoria* and AhAMP1 from *Aesculus* have been reported to show potential anti-bacterial activity against gram-positive and gram-negative bacteria (Balandin et al. 2005; Franco et al. 2006; Kraszewska et al. 2016; Velivelli et al. 2018).

Further, the potent anti-bacterial action of defensin OsDef7 and OsDef8 has been reported against *Xanthomonas oryzae* pv. *oryzae*, *X. oryzae* pv. *oryzicola* and *Erwinia carotovora* subsp. *Atroseptica* (Tantong et al. 2016). These defensins may form dimers and are also known to exhibit resistance against *Xanthomonas campestris* pv. *glycines* (Guillén-Chable et al. 2017; Weerawanich et al. 2018). Another bi-domain defensin peptide MtDef5 has been isolated from *Medicago truncatula* that displayed inhibitory activity against *X. campestris* pv. *campestris* (Velivelli et al. 2018; Sathoff and Samac 2019). PR-1 gene (*MuPR1*) extracted from mulberry plants is known to express resistance against *Pseudomonas syringae* pv. *tomato* (Li-Jiang et al. 2019). PR-1 protein (NtPR1a) from tobacco has also enhanced disease resistance against *Ralstonia solanacearum* by activating the defence-related genes (Liu et al. 2019).

7.4 Anti-Viral Activity

Viruses being obligate plant pathogens are able to hijack plant machinery and utilise hosts to synthesise their own proteins and nucleic acid. Compared with the plant defence against fungal and bacterial pathogens, plant anti-viral defence could be more complex. Although this direct interaction allows the plant to evolve new defence mechanisms targeting viral factors, plant viruses can also take advantage of the interaction to explore the weak points of plant anti-viral barriers and exploit their error-prone polymerases and multiple functional proteins to quickly escape host defences and multiply (Wu et al. 2019). PR proteins have proven a viable host response to the attacking viruses and accumulate in the plant tissues after infection. Several anti-viral PR proteins and peptides such as PR-2a, PR-3, PR-9 (peroxidase), PR-10 and PR-12 to PR-14, RNA-binding proteins (RBPs) and ribosome-inactivating proteins (RIPs) are known to suppress viruses and blocks virus propagation (Sindelarova and Sindelar 2005; Musidlak et al. 2017).

Further, PR proteins, viz., PR-2a and PR-3 from tobacco, were found to possess anti-viral properties against TMV (Sindelarova and Sindelar 2005). Up-regulation of PR-1, PR-4 and PR-5 proteins in tobacco was encountered by inoculation with medicinal plant extracts resulting in the resistance against TMV and black shank disease (Wang et al. 2018). Sabokkhiz and co-workers (2019) also reported the anti-viral activity of such anti-microbial proteins against TMV.

7.5 Anti-Nematodal and Anti-Insect Properties

Plant pathogenic nematodes and insect pests are also types of biotic agents that pose a potential threat to agriculture. Besides their direct harm, many nematodes and insect pests are also well-known vectors of plant viruses. In response to the parasitic nematode and insect pests, the plants have evolved various mechanisms to defend themselves. The soybean PR protein, GmPR08-Bet VI provides resistance against the soybean cyst nematode *Heterodera glycines* via a molecular interaction between the proteins GmSHMT08 and GmSNAP18 (Lakhssassi et al. 2020). It has been recently shown that the *Arabidopsis* thionin-like genes are involved in providing resistance against the beet cyst nematode, *H. schachtii,* since the transgenic lines overexpressing thionin-like proteins displayed a lower number of nematodes, which developed on the roots after inoculation (Almaghrabi et al. 2019).

Production of PR proteins is among the most crucial events of inducible defence responses in plants against leaf-chewing and phloem-feeding insects (Carr and Klessig 1989; Antony and Palaniswami 2006; Rashid and Chung 2017). The newly synthesised plant proteins after insect feeding or wounding were categorised into three types (Ryan 2000). The first group includes defensive proteins such as protease inhibitors (PIs), the second includes proteins involved in signalling and the third group includes the proteins involved in metabolic activities. Chitinases belonging to PR proteins 3, 4, 8 and 11 families can digest chitin that is a component of insect exoskeletons and peritrophic membranes (Kramer and Muthukrishnan 1997). In tomato, PR protein genes are known to be expressed against *Trialeurodes vaporariorum* and *Bemisia tabaci* (Puthoff et al. 2010).

PIs and amylase inhibitors, i.e., α-AIs belonging to PR protein families, occur naturally in plants and after insect attack, their abundance simultaneously increases in response to wounding (Sharma 2015; Wielkopolan and Obręsalska-Stęplowska 2016). PR-4 proteins are other important proteins that were originally referred to as wound-inducible proteins and are less extensively studied defensive proteins. They are known to show anti-fungal, RNase-, DNase- and insect-inhibiting activity against *Helicoverpa armigera* (Singh et al. 2018a). Some PR proteins like β-1, 3-glucanase, chitinase, polyphenol oxidase and peroxidases are also involved in defence against cotton whitefly (Soliman et al. 2019).

7.6 Regulation of PR Proteins

The regulation of PR proteins is quite a complicated process involving a number of elements/factors like treatment with elicitors and/or PGRs, induction of abiotic stress and the nature of the promoter region. Besides the regulation by the elicitors/PGRs, the transcriptional regulation is studied in both monocots and dicots and the role of several transcription factors and other signalling intermediates has also been recently found to play a key role in it. In an attempt to analyse promoter regions of several families of PR proteins of *Arabidopsis thaliana* and *Oryza sativa*, the marked differences were found. Analysis of the distribution of *cis*-acting regulatory elements indicated their functional multiplicity. The CpG islands are observed in the *O. sativa* PRs but not in *A. thaliana*, indicating that the monocot genome consisted of more GC-rich motifs than the dicot genome. Further, the tandem repeats were also observed in 5′ UTR of the PR genes of both (Kaur et al. 2017).

Several transcription factors have also been found to be involved in the regulation of PR proteins. The transgenic *Nicotiana benthamiana* plant overexpressing a potato NAC family transcription factor, i.e., *StNACb4,* was also found to be capable of activating *StPR10* gene expression thereby increasing its tolerance towards the pathogen *R. solanacearum* causing bacterial wilt (Chang et al. 2020). It has also been suggested that the rice CaM-binding transcription factor (OsCBT) regulates the plant defence by modulating the expression (differential expression) of several genes, including 21 genes encoding PR proteins, based on the studies on the mutant allele of *oscbt-1* which exhibited a broad-spectrum resistance against various rice pathogens (Chung et al. 2020). Another rice ERF transcription factor, OsERF83, provides resistance towards *Magnaporthe oryzae*, the rice blast pathogen by the regulation of the PR proteins. Several PR genes belonging to their diverse classes, viz., PR-1 to PR-3, PR-5 and PR-10, were found up-regulated in transgenic plants overexpressing *OsERF83* of defence-related genes in rice (Tezuka et al. 2019). Recently, Sathoff et al.(2020), found that the promoters of *M. truncatula* and *M. sativa* PR proteins, viz., *MtPR5*, *MtPR10* and *MsPR10,* were found to be pathogen inducible and their several putative transcriptional regulatory elements were predicted which included binding sites for selected WRKY, ERF, MYB and GBF3 transcription factors.

Moreover, the role of signalling intermediates like MAP kinases (MAPKs) has also come into light in understanding the defence-providing mechanism of PR proteins. Recently, it was concluded using a yeast two-hybrid assay that an LTP from tobacco, NtLTP4, interacts with the MAPK wound-induced protein kinase (WIPK), thereby providing abiotic stress tolerance (Xu et al. 2018). The application of elicitors like chitosan and its nanoparticles were found to induce the expression of PR proteins; however, chitosan nanoparticles were more efficient than the chitosan in the induction of PR proteins. Up to 1.15- and 1.48-fold up-regulation was recorded in chitinases and β-1,3-glucanases genes, respectively (Chun and Chandrasekaran 2019). The application of glycerol is also found to be effective in wheat against powdery mildew via the induction of three types of PR genes, i.e., PR-1, PR-3 and PR-10 (Li et al. 2020).

7.7 Transgenic Expression of PR Proteins

With the development of genetic transformation techniques in plants (Keshava Reddy et al. 2018), the transgenic plants became a powerful tool to study various aspects including the study of gene function (Pereira 2000; Michielse et al. 2005). Recently the transgenic expression of various groups of PR proteins (thaumatin and osmotin-like proteins, chitinases, glucanases, defensins and thionins, defensins, thionins, oxalate oxidase, oxalate oxidase like proteins/germin-like proteins and LTPs) has remarkably enhanced the resistance of the transformed plant against a number of plant pathogens (Table 7.3).

PR-5 proteins fall under the category of thaumatin-like or osmotin-like proteins. The thaumatin-like proteins are known to produce transmembrane pores in fungal cells, whereas the osmotin proteins are known to maintain the osmolarity in cellular compartments of the compatible solutes (Moosa et al. 2017; Kaur et al. 2020). Transgenic poplar and tobacco plants overexpressing thaumatin genes showed enhanced resistance towards *Marssonina brunnea* and *A. panax,* respectively (Li et al. 2020; Sun et al. 2020). Similarly, the overexpression of osmotin gene in potato was responsible for imparting resistance towards the early blight pathogen *A. solani* (Kaur et al. 2020). PR proteins belonging to chitinase and glucanases (mainly PR-2 to PR-4, PR-8 and PR-11 proteins) are the enzymes that help in the degradation of fungal cell walls as they are made up by the cross-linking of chitin and glucan units as the structural units of the fungal cell walls (Moosa et al. 2017). The transgenic overexpression of various chitinase genes has been recently shown to enhance the disease resistance against the phytopathogens like *Rhizoctonia solani, Pseudomonas tabaci, A. alternata, Exserohilum turcicum, Curvularia lunata, Ganoderma boninense* and so forth (Kang et al. 2017; Liu et al. 2020; Hanin et al. 2020). Further, the transgenic overexpression of a potato glucanase (a PR-2 protein), *endo-1,3-β-D-glucanase* gene showed enhanced protection towards blister blight pathogen, *Exobasidium vexans,* in tea (Singh et al. 2018b). Recently, the transgenic overexpression of *M. sativa glucanase AGLU1* showed enhanced resistance against *G. boninense* in *Elaeis guineensis* (Hanin et al. 2020).

Defensins and thionins belong to PR-12 and PR-13 families of PR proteins, respectively, and are small cysteine-rich AMPs (Moosa et al. 2017). Transgenic expression of defensin and defensin-like proteins have been shown to provide resistance against a number of plant pathogens like *Typhula ishikariensis, F. graminearum, P. triticina, F. solani* and so forth (Sasaki et al. 2016; Kaur et al. 2017; Wang et al. 2019). Similarly, the transgenic overexpression of *A. thaliana* thionin genes, *AT1G12660* and *AT1G12663,* have been shown to enhance resistance against *F. solani* and *F. oxysporum* in potato by inhibiting spore germination (Hammad et al. 2017). Oxalate oxidase, oxalate oxidase like and germin-like proteins belong to the cupin and germin superfamily of proteins (belonging to PR-15 and PR-16) and break down oxalic acid that is secreted by some phytopathogens during host-pathogen interaction. Further this oxalic acid acts as a non-host-specific toxin, leading to disease development (Moosa et al. 2017). Transgenic overexpression of oxalate oxidase genes has shown recently to enhance disease resistance towards the pathogens *Sclerotinia sclerotiorum* and *Phytophthora infestans* (Ghosh et al. 2016; Yang et al. 2019). *GhABP19* and *GhGLP2*, two *Gossypium hirsutum* genes encoding a germin-like protein overexpressing transgenic *A. thaliana,* showed increased resistance against *Verticillium dahliae* and *F. oxysporum* (Pei et al. 2019, 2020). LTPs (mainly belonging to PR-14) are small cystine-rich cationic peptides that have been produced by plants in response to several pathogens (Moosa et al. 2017). Recently, *A. thaliana* LTP

TABLE 7.3

Some Examples of Transgenic Plants Expressing Different Groups of PR Proteins to Enhance the Resistance towards Plant Pathogens (2016 Onwards)

Source Organism	Nature of Protein	Gene Encoding	Transgenic Expression System	Target Pathogen(s)	Response(s)	References
A. Thaumatin/Osmotin-Like Proteins						
Poplar cultivar 'Nanlin895' (*Populus deltoides* × *P. euramericana*)	Thaumatin-like protein	*PeTLP*	Poplar cultivar 'Nanlin895' (*P. deltoides* × *P. euramericana*)	*Marssonina brunnea*	Enhanced resistance	Sun et al. (2020)
Panax notoginseng	Thaumatin-like protein	*PnTLP2*	*Nicotiana tabacum*	*Alternaria panax*	Enhanced resistance	Li et al. (2020)
Withania somnifera	Osmotin	*OsmWS*	*Solanum tuberosum* cultivar 'kufrichipsona 1'	*A. solani*	Enhanced resistance	Kaur et al. (2020)
B. Chitinases and Glucanases						
Zoysia japonica	Chitinase class II	*Zjchi2*	*Z. japonica*	*Rhizoctonia solani* AG2-2	Enhanced resistance	Kang et al. (2017)
Leymus chinensis	Class II chitinase	*LcCHI2*	*N. tabacum*; *Zea mays*	*Pseudomonas tabaci, A. alternata, Exserohilum turcicum, Curvularia lunata*	Enhanced resistance	Liu et al. (2020)
Oryza sativa	Chitinase	*RCH10*	*Elaeis guineensis*	*Ganoderma boninense*	Enhanced resistance	Hanin et al. 2020
Medicago sativa	Glucanase	*AGLU1*	*E. guineensis*	*G. boninense*	Enhanced resistance	Hanin et al. (2020)
S. tuberosum	Glucanase	*endo-1,3-β-D-glucanase*	*Camellia sinensis*	*Exobasidium vexans*	Enhanced resistance	Singh et al. (2018b)
C. Defensins and Thionins						
Triticum aestivum	Defensin	*TAD1*	*T. aestivum*	*Typhula ishikariensis, Fusarium graminearum*	Enhanced resistance	Sasaki et al. (2016)
Medicago truncatula	Defensin	*MtDef4.2*	*T. aestivum*	*Puccinia triticina*	Enhanced resistance	Kaur et al. (2017)
P. notoginseng	Defensin-like protein	*PnDEFL1*	*N. tabacum*	*Fusarium solani*	Enhanced resistance	Wang et al. (2019)
Arabidopsis thaliana	Thionin	*AT1G12660* and *AT1G12663*	*S. tuberosum*	*F. solani and F. oxysporum*	Enhanced resistance	Hammad et al. (2017)
D. Oxalate Oxidase, Oxalate Oxidase Like Proteins and Germin-Like Proteins						
T. aestivum	Oxalate oxidase	*OXO*	*Glycine max*	*Sclerotinia sclerotiorum*	Enhanced resistance	Yang et al. (2019)
O. sativa	Oxalate oxidase	*Osoxo4*	*S. tuberosum*	*Phytophthora infestans*	Enhanced resistance	Ghosh et al. (2016)
Gossypium hirsutum	Germin-like protein	*GhABP19*	*A. thaliana*	*Verticillium dahliae* and *F. oxysporum*	Enhanced resistance	Pei et al. (2019)
G. hirsutum	Germin-like protein	*GhGLP2*	*A. thaliana*	*V. dahliae* and *F. oxysporum*	Enhanced resistance	Pei et al. (2020)
E. Lipid Transfer Proteins						
A. thaliana	Lipid transfer protein	LTP1, LTP3, LTP4 and LTP8 Clubroot susceptibility	*A. thaliana* roots	*Plasmodiophora brassicae*	Enhanced resistance	Jülke and Ludwig-Müller (2016)
A. thaliana	Lipid transfer protein	*AtLTP4.4*	*T. aestivum*	*F. graminearum*	Enhanced resistance	McLaughlin et al. (2020)

overexpressing transgenics has been shown to increase resistance towards pathogens *Plasmodiophora brassicae* and *F. graminearum* (Jülke and Ludwig-Müller 2016; McLaughlin et al. 2020).

7.8 Multiple Stress Tolerance (Abiotic Stresses) Induced by PR Proteins

Besides providing resistance against the plant pathogens, some PR proteins are also known to enhance the resistance towards different abiotic stresses. Some examples of PR protein overexpressing transgenics displaying the tolerance towards different abiotic stresses have been listed in Table 7.4. The up-regulation in the expression of jasmonic acid inducible PR-10 gene *JIOsPR10* under salt and drought stress conditions is displayed too. The constitutive overexpression of *JIOsPR10* in transgenic rice plants showed enhanced salt and drought stress tolerance besides the reduced susceptibility against rice blast fungi compared with the wild type (Wu et al. 2016). In particular, PR-10 family containing ribonuclease-like proteins are also known to be involved in providing enhanced tolerance towards various kinds of abiotic stresses (Sinha et al. 2020).

In cucumber, *Trichoderma atroviride* induced resistance to the disease caused by *R. solani* (Nawrocka et al. 2018). *T. atroviride*-associated disease protection in cucumber was observed to regulate the expression of PR proteins. The expression of SAR-related PR genes, namely *PR-1* and *PR-5* genes, were up-regulated by *T. atroviride*; whereas the expression of induced systemic resistance (ISR) related genes, i.e., *PR-4* gene, was up-regulated in the presence of both *T. atroviride* and *R. solani*. However, the expression of the *PR-4* gene was not induced in *T. atroviride* treated plants. It was also reported that the up-regulation of the *PR-4* gene alone (without any change in expression of *PR-1* and *PR-5* genes) did not affect either susceptibility or disease resistance of the cucumber plants to *R. solani*. It was further concluded by Nawrocka et al. (2018) that volatile compounds such as ethylhexyl salicylate (EHS), methyl salicylate (MeSA) and β-cyclocitral were also key participants in overexpression (up-regulation) of SAR-associated PR genes (*PR-1* and *PR-5* genes) during *T. atroviride*-associated disease (*R. solani*) resistance in cucumber plants.

An LTP gene, *NtLTP4*, from *Nicotiana tabacum*, when overexpressed in *N. tabacum* enhanced resistance towards both salt and drought stresses. During salt stress, the *NtLTP4* overexpressing plants accumulated less Na^+ with the higher expression of the salt-responsive genes *NHX1* (Na^+/H^+ exchangers) and *HKT1* (high-affinity K^+ transporter 1) than the wild type. Whereas during the drought stress, the *NtLTP4* overexpressing lines also enhanced their tolerance towards drought stress by reducing the rate of transpiration rate (Xu et al. 2018). Using *in silico* approaches, 13 novel *PR-1* genes in tomato were identified. Besides their role in providing the resistance against the plant pathogens, it was also suggested that all of these genes are up-regulated under drought conditions and were therefore considered as good

TABLE 7.4

Some Examples of Transgenic Plants Expressing Various PR Proteins to Enhance the Resistance against Various Abiotic Stresses (2016 Onwards)

Source Organism	Nature of Protein	Gene(s) Encoding	Transgenic Expression System(s)	Abiotic Stress(es)	Response(s)	References
Leymus chinensis	Class II chitinase	*LcCHI2*	*Nicotiana tabacum*; *Zea mays*	Saline-alkali stress	Improved the stress tolerance	Liu et al. (2020)
N. tabacum	Lipid transfer protein	*NtLTP4*	*N. tabacum*	Salt and drought stress	Improved the stress tolerance	Xu et al. (2018)
Arabidopsis thaliana	Lipid transfer protein	*LTP1* and *LTP3*	*A. thaliana* roots	Salt stress	Improved the stress tolerance	Jülke and Ludwig-Müller (2016)
Glycine max	Germin-like protein	*GmGLP7*	*A. thaliana*	Salt, drought and oxidative	Improved the stress tolerance	Li et al. (2016)

candidates for the molecular markers for developing abiotic stress-resistant/tolerant varieties in tomato (Akbudak et al. 2020).

7.9 Interaction of PR Proteins with PGRs

Any invading pathogen, when perceived by the host plant, leads to the cascade of plant stress-responsive defence mechanisms (Naidoo et al. 2013). Among these defence responses, various types of PGRs such as salicylic acid (SA), jasmonic acid (JA/jasmonates), ethylene, brassinosteroids (BRs), abscisic acid (ABA) and so forth have been activated (Naidoo et al. 2013; Sharma et al. 2018). The pathogen-induced stimulation of SA, JA and/or ethylene has been reported to activate several SAR signature defence genes such as PR genes (Naidoo et al. 2013; Taif et al. 2020). Therefore, the PR genes have also been employed as the molecular indicators of the signalling pathways of these PGRs (SA, JA and ethylene). In cucumber plants, the volatile compounds like EHS, β-cyclocitral and MeSA up-regulated both SA-dependent response and expression of *PR-1* and *PR-5* genes in the presence of *T. atroviride* (Nawrocka et al. 2018). Whereas unsaturated fatty acids (derivatives of linoleic and linolenic acid), namely E-2-hexenal, Z-3-hexanal and Z-3-hexenol, were reported to be associated with the JA-dependent responses and up-regulation of *PR-4* which induced *T. atroviride*-mediated diseased resistance in cucumber plants against *R. solani.*

It has also been reported by Ali et al. (2017; 2018a) that phytopathogen-associated activation of SA/JA/ethylene pathways results in the accumulation of PR genes to reduce the pathogen load or to minimise the onset of disease in the uninfected plant organs. The biotrophic phytopathogens induce the activation of the SA signalling cascade, which further activates the transcription of *NPR1* (non-expressor of pathogen-related gene 1). The stimulation of *NPR1* transcripts subsequently activates the expression of three SA signature genes, viz., *PR-1, PR-2* (α-1, 3-glucanase) and *PR-5* (thaumatin like) which ensure SAR (Ali et al. 2017). The necrotrophic phytopathogens activate the JA signalling cascade which further stimulates the expression of three JA signature PR genes, viz., *PR-3* (chitinases), *PR-12* (plant defensins) and *PR-13.* The activation of *PR-3*, *PR-12* and *PR-13* genes results in the local accumulation of their products and thus leads to local acquired resistance (LAR) only. Recent transcript profiling studies have revealed that in *Arabidopsis* the enhanced expressions of three *PR* defence genes, i.e., *PR-1, PR-2* and *PR-5,* revealed the stimulation of the SA signalling pathway; whereas in *Eucalyptus*, declined expression of two *PR* defence genes, i.e., *EgrPR3* and *EgrLOX,* when the levels of SA treatments were enhanced (Naidoo et al. 2013; Taif et al. 2020). Further, the transcript levels of the *Eucalyptus* PR gene, i.e., *EgrPR2*, were reported to be declined when the concentrations of methyl jasmonate (MeJA: derivative of JA) were increased (Naidoo et al. 2013).

The efficacy of a plant activator (prepared from the extract of yeast cell wall), viz., Housaku Monogatari (HM), was observed against fungal and bacterial pathogen-induced infections in *Brassica rapa* and *A. thaliana* leaves (Narusaka et al. 2015). HM treatments were reported to induce disease resistance in these plants via increasing the expressions of PR genes and inducing late activation of SA signalling pathways. Also, the early stimulation of JA and ethylene signalling cascades. The enhanced expressions of *PR-1* and *PR-2* (*glucanase*) were related to the stimulation of the SA signalling pathway. However, *PR-3* (basic-chitinase) and *PR-4* (endo-chitinases) transcripts were reported to be potentiated when ethylene and JA signalling pathways were activated (Narusaka et al. 2015).

In *Lotus japonicus*, an ABA-responsive defence gene (*LjGlu1*: β-1,3-glucanase) had been identified (Osuki et al. 2016). In *B. juncea*, treatment of ABA significantly reduced the transcript levels of PR defence genes (*PR-1, PR-2* and *PR-5*), which are SA signature genes, hence indicating the antagonistic relationship between ABA and SA (Ali et al. 2017). However, the transcript levels of JA signature genes (*PR-3, PR-12* and *PR-13*) were enhanced by ABA treatments, emphasising the synergistic interactions between ABA and JA. Exogenous treatments of SA induced SA responsive genes, i.e., *PR-1, PR-2* and *PR-5*; whereas SA treatments did not induce JA responsive defence genes (*PR-3, PR-12* and *PR-13*). However, exogenous treatments of JA induced JA responsive genes (*PR-3, PR-12* and *PR-13*) and it could not induce SA responsive defence genes (*PR-1, PR-2* and *PR-5*). Recently, it has been recorded that the proteins encoded by *Manihot esculenta* genes *MeAux/IAA* are the positive regulators of plant disease resistance against cassava bacterial blight via the induction of PR protein genes at the transcriptional level (Fan et al. 2020).

7.10 PR Proteins and Human Health

Recently, some PR proteins were documented to cause allergic reactions in humans. It has been reported that some families of PR proteins are homologous to the proteins present in fruit and vegetables (Smeekens et al. 2018). Such PR proteins often cause IgE (an allergen-specific immunoglobulin E antibody in humans) cross-reactivity in humans resulting in oral allergy syndrome. PR proteins, namely PR-5 (thaumatin-like proteins), PR-10 (ribosome inactivating proteins) and PR-14 (LTPs), have been reported to be the most common PR proteins responsible for oral allergy syndrome. Kleine-Tebbe et al. (2002) reported that the ingestion of food items containing soy protein caused immediate-type allergic symptoms in patients who had birch (*Betula verrucosa*) pollen allergy. This allergy in patients was reported due to cross-reactivity of birch pollen allergen (Bet v 1)-specific IgE to the PR-10 protein (homologous PR proteins). Another study revealed the presence of an allergen PR-10 protein, i.e., Jug r5 protein, which is similar to the Bet v1-like protein in English walnut (*Juglans regia*) through immunoblot experiments (Wangorsch et al. 2017). However, the amount of Jug r5 in the walnut extract was reported to be less. In patients, the walnut allergy was displayed through IgE cross-reactivity with Jug r5.

The PR protein-induced allergenicity may also be associated with other factors such as environmental pollutants, agrochemicals and so forth (Jain and Khurana 2018). A PR-5 protein, i.e., Cup a 3, was derived and cloned from *Cupressus arizonica* (cypress) pollen protein and its expression was investigated under a polluted environment (Cortegano et al. 2004). It was analysed through radioallergosorbent test inhibition and skin tests that increased Cup a 3 protein expression under polluted conditions was directly associated with the occurrence of pollen allergenicity in cypress allergic patients. In about 63% of patients, specific IgE antibodies against Cup a 3 protein had been reported and the major allergen was identified to be homologous to PR-5 proteins, which contributed to the overall allergenicity due to cypress pollens. Cypress allergy was higher and the expression of Cup a 3 protein was enhanced when the pollution was high in the area from where the pollen had been collected. It has been further revealed that the PR-5-like protein, Cup a 3, is present in the pollen cytoplasm and walls of pollen grains of cypress that cause winter allergic respiratory disease (Suárez-Cervera et al. 2008). Cup a 3 was reported to be present during both the air dispersion and hydration stages. This PR-5 protein (Cup a 3) in the cypress pollen grains had been observed to be more abundant in polluted air conditions compared with unpolluted areas, thereby strengthening the cypress pollen allergenicity.

Since air pollutants were reported to enhance the allergenicity, a study was focused to investigate the effect of cadmium (Cd)-polluted soil on the pollen allergenicity of the annual bluegrass (*Poa annua* L.) (Aina et al. 2010). It was reported that pollens from the plants grown in Cd-contaminated soils had more pollen allergic severity in grass pollen-allergic patients compared with the pollens derived from control plants. Further, it was observed that Cd-exposed pollen was observed to be released easily into the environment and the released allergens contained a PR-3 (class I chitinase like) protein. These PR-3 proteins were associated with the enhanced tendency to bind specific IgE causing increased allergenicity in grass pollen-allergic patients. Several PR proteins had been reported to cause birch-mugwort celery-spice syndromes, pollen-related food syndrome and latex-fruit syndrome in humans (Gupta et al. 2019). The inherent allergic potential of a PR-protein, i.e., α-dioxygenase fragment, isolated from chickpea was reported to cause the degranulation of mast cells, thereby leading to the secretion of various allergic mediators resulting in the allergic manifestation in BALB/c mice. Finkina et al. (2017) documented that members of PR-10 and PR-14 protein families are the most ubiquitous allergens derived from plants (plant foods and pollen grains) and have the potential to sensitise the human immune system and cause cross-reactive allergic reactions.

Burgeoning studies revealed that about eight classes of PR proteins (viz., PR-1 to PR-5, PR-8, PR-10 and PR-14) have caused allergy in humans (Jain and Khurana 2018). These studies have emphasised the employment of PR proteins as preservative agents in the food industry as well as for the production of disease-resistant plants through plant genetic engineering techniques. Several investigations reported that the phytopathogenic fungi-induced diseases can be protected through overexpression of PR genes and transgenic plants that overexpress PR genes induce host plant resistance to such fungi. However, such GM crops that overexpressed PR proteins may result in enhanced allergenicity/toxicity.

Furthermore, the recent studies based on *in silico* approaches have suggested that all the remaining classes of PR proteins also possess the allergenicity potential. Therefore, the inclusion of PR proteins in the food may be linked to the possible inclusion of more potential allergens in the foodstuffs. However, in recent years, the positive roles of PR proteins on their inclusion in foodstuffs have also been listed. PR proteins, because of their anti-microbial properties, are described as food preservatives (Thery et al. 2019, 2020). For example, Younas et al. (2016) also suggested the use of a chitinase isolated from peanut as a food preservative in bread might be useful against bread moulds. However, the role of PR proteins in the foodstuffs has also been suggested by several workers in the past (Thery et al. 2019). Therefore, the inclusion of PR proteins may lead to the safer consumption of those foodstuffs that are more prone to mycotoxin consumption.

7.11 Conclusions and Future Prospects

PR proteins were primarily known as the proteins that were induced and accumulated in the plant tissue in response to the pathogen attack. Several families of PR proteins (from PR-1 to PR-17) have been identified. In recent years, numerous attempts have been made to produce PR proteins overexpressing transgenics, which have helped their functional annotation and have advanced our understanding towards their mode of action, the underlying mechanism and their other interacting partners. The transgenic plants expressing more than one PR protein simultaneously have also been used to provide a broad-spectrum resistance against the plant pathogens. Further, beyond the role of PR proteins in conferring the plant defence against the invading pathogen, now they also have been linked to the enhanced resistance against multiple biotic stresses and even the abiotic stresses. They also have been known to be regulated by wounding and the application of chemical inducers. However, some PR proteins have been identified to cause food allergy in humans, thereby limiting the role of PR overexpressing transgenics. Further, the PR proteins from grapes/other fruits are also known for haze formation in white wine or fruit juice, thereby reducing their commercial value. However, on the other hand, their role as a food-preserving agent also has been documented recently, which may be highly beneficial to the food industry. In the past few years the approach of the scientific community working on the transgene expression of various PR proteins has been shifted from their functional characterisation to making them future-ready and safe GM crops overexpressing desired PR proteins; therefore, the careful evaluation of PR overexpressing transgenics in terms of their toxicity or allergenicity may be evaluated before commercialising them as future GM crops.

REFERENCES

Agrios, G. N. 2005. *Plant pathology.* 5th ed. Elsevier Academic Press, Burlington, MA, pp. 79–103.

Aina, R., Asero, R., Ghiani, A., Marconi, G., Albertini, E. and Citterio, S. 2010. Exposure to cadmium-contaminated soils increases allergenicity of *Poa annua* L. pollen. *Allergy* 65(10): 1313–1321.

Akbudak, M. A., Yildiz, S. and Filiz, E. 2020. Pathogenesis related protein-1 (PR-1) genes in tomato (*Solanum lycopersicum* L.): Bioinformatics analyses and expression profiles in response to drought stress. *Genomics* 112: 4089–4099.

Ali, S., Ganai, B. A., Kamili, A. N., Bhat, A. A., Mir, Z. A., Bhat, J. A., Tyagi, A., Islam, S.T., Mushtaq, M., Yadav, P. and Rawat, S. 2018a. Pathogenesis-related proteins and peptides as promising tools for engineering plants with multiple stress tolerance. *Microbiological Research* 212: 29–37.

Ali, S., Ganai, B.A., Kamili, A.N., et al. 2018b. Pathogenesis-related proteins and peptides as promising tools for engineering plants with multiple stress tolerance. *Microbiological Research* 212–213: 29–37.

Ali, S., Mir, Z. A., Bhat, J. A., Tyagi, A., Chandrashekar, N., Yadav, P., Rawat, S., Sultana, M. and Grover, A. 2018c. Isolation and characterization of systemic acquired resistance marker gene PR1 and its promoter from *Brassica juncea*. *Biotechnology* 8: 10–23.

Ali, S., Mir, Z. A., Tyagi, A., Bhat, J. A., Chandrashekar, N., Papolu, P. K., Rawat, S. and Grover, A. 2017. Identification and comparative analysis of *Brassica juncea* pathogenesis-related genes in response to hormonal, biotic and abiotic stresses. *Acta Physiologiae Plantarum* 39(12): 268.

Almaghrabi, B., Ali, M. A., Zahoor, A., Shah, K. H. and Bohlmann, H. 2019. *Arabidopsis* thionin-like genes are involved in resistance against the beet-cyst nematode (*Heterodera schachtii*). *Plant Physiology and Biochemistry* 140: 55–67.

Andersen, E. J., Ali, S., Byamukama, E., Yen, Y. and Nepal, M. P. 2018. Disease resistance mechanisms in plants. *Genes* 9(7): 339.

Antoniw, J. F., Ritter, C. E., Pierpoint, W. S. and van Loon, L. C. 1980. Comparison of three pathogenesis-related proteins from plants of two cultivars of tobacco infected with TMV. *Journal of General Virology* 47: 79–87.

Antony, B. and Palaniswami, M. S. 2006. *Bemisia tabaci* feeding induces pathogenesis-related proteins in cassava (*Manihot esculenta* Crantz). *Indian Journal of Biochemistry and Biophysics* 43(3): 182–185.

Balandin, M., Royo, J., Gomez, E., Muniz, L. M., Molina, A. and Hueros, G. 2005. A protective role for the embryo surrounding region of the maize endosperm, as evidenced by the characterisation of ZmESR-6, a defensin gene specifically expressed in this region. *Plant Molecular Biology* 58: 269–282.

Breen, S., Williams, S. J., Outram, M., Kobe, B. and Solomon, P. S. 2017. Emerging insights into the functions of pathogenesis-related protein 1. *Trends in Plant Science* 22: 871–879.

Carr, J. P. and Klessig, D. F. 1989. The pathogenesis-related proteins of plants. In: Setlow, J. K. (ed.), *Genetic Engineering: Principle and Methods*. Plenum Press, New York, pp. 65–109.

Chakraborty, U. 2020. Vocal for local: reviewing global experience with an Indian insight. *Online International Interdisciplinary Research Journal* 10: 115–121.

Chang, Y., Yu, R., Feng, J., Chen, H., Eri, H. and Gao, G. 2020. NAC transcription factor involves in regulating bacterial wilt resistance in potato. *Functional Plant Biology* 47: 925–936. doi: 10.1071/FP19331.

Chun, S. C. and Chandrasekaran, M. 2019. Chitosan and chitosan nanoparticles induced expression of pathogenesis-related proteins genes enhance biotic stress tolerance in tomato. *International Journal of Biological Macromolecules* 125: 948–954.

Chung, J. S., Koo, S. C., Jin, B. J., Baek, D., Yeom, S. I., Chun, H. J., Choi, M. S., Cho, H. M., Lee, S. H., Jung, W. H., Choi, C. W., Chandran, A. K. N., Shim, S. I., Chung, J., Jung, K. H. and Choi, C. W. 2020. Rice CaM-binding transcription factor (OsCBT) mediates defence signalling via transcriptional reprogramming. *Plant Biotechnology Reports* 14: 309–321.

Cools, T. L., Struyfs, C., Cammue, B. P. and Thevissen, K. 2017. Antifungal plant defensins: increased insight in their mode of action as a basis for their use to combat fungal infections. *Future Microbiology* 12: 441–454.

Cortegano, I., Civantos, E., Aceituno, E., Del Moral, A., Lopez, E., Lombardero, M., Del Pozo, V. and Lahoz, C. 2004. Cloning and expression of a major allergen from *Cupressus arizonica* pollen, Cup a 3, a PR-5 protein expressed under polluted environment. *Allergy* 59(5): 485–490.

Dai, L., Wang Xie, D., Zhang, X. C., Wang, X., Xu, Y., Wang, Y. and Zhang, J. 2016. The novel gene vpPR4-1 from *Vitis pseudoreticulata* increases powdery mildew resistance in transgenic *Vitis vinifera* L. *Frontiers in Plant Sciences* 7: 695.

Dangl, J. L. and Jones, J. D. G. 2001. Plant pathogens and integrated defence responses to infection. *Nature* 411: 826–833.

Epple, P., Apel, K. and Bohlmann, H. 1995. An *Arabidopsis thaliana* thionin gene is inducible via a signal transduction pathway different from that for pathogenesis-related proteins. *Plant Physiology* 109: 813–820.

Fan, S., Chang, Y., Liu, G., Shang, S., Tian, L. and Shi, H. 2020. Molecular functional analysis of auxin/indole-3-acetic acid proteins (Aux/IAAs) in plant disease resistance in cassava. *Physiologia Plantarum* 168(1): 88–97.

Finkina, E. I., Melnikova, D. N., Bogdanov, I. V. and Ovchinnikova, T. V. 2017. Plant pathogenesis-related proteins PR-10 and PR-14 as components of innate immunity system and ubiquitous allergens. *Current Medicinal Chemistry* 24(17): 1772.

Franco, O. L., Murad, A. M., Leite, J. R., Mendes, P. A., Prates, M. V., Bloch, C. Jr., et al. 2006. Identification of a cowpea gamma-thionin with bactericidal activity. *FEBS Journal* 273: 3489–3497.

Fujimori, N., Enoki, S., Suzuki, A., Naznin, H. A., Shimizu, M. and Suzuki, S. 2016. Grape apoplasmic b-1, 3-glucanase confers fungal disease resistance in Arabidopsis. *Scientia Horticulturae* 200: 105–110.

Gamir J., Darwiche R., Van'T Hof, P., Choudhary, V., Stumpe, M., Schneiter, R. and Mauch, F. 2017. The sterol-binding activity of pathogenesis-related protein 1 reveals the mode of action of an antimicrobial protein. *The Plant Journal* 89: 502–509.

García-Olmedo, F., Molina, A., Segura, A. and Moreno, M. 1995. The defensive role of nonspecific lipid-transfer proteins in plants. *Trends in Microbiology* 3: 72–74.

Ghosh, S., Molla, K. A., Karmakar, S., Datta, S. K. and Datta, K. 2016. Enhanced resistance to late blight pathogen conferred by expression of rice oxalate oxidase 4 gene in transgenic potato. *Plant Cell, Tissue and Organ Culture* 126(3): 429–437.

Green, T. R. and Ryan, C. A. 1972. Wound-induced proteinase inhibitor in plant leaves: a possible defense mechanism against insects. *Science* 175: 776–777.

Guillén-Chable, F., Arenas-Sosa, I., Islas-Flores, I., Corzo, G., Martinez-Liu, C. and Estrada, G. 2017. Antibacterial activity and phospholipid recognition of the recombinant defensin J1-1 from *Capsicum* genus. *Protein Expression and Purification* 136: 45–51.

Gupta, R. K., Sharma, A., Verma, A., Ansari, I. A. and Dwivedi, P. D. 2019. Inherent allergic potential of α-dioxygenase fragment: A pathogenesis related protein. *Immunobiology* 224(2): 207–219.

Hammad, I. A., Abdel-Razik, A. B., Soliman, E. R. and Tawfik, E. 2017. Transgenic potato (*Solanum tuberosum*) expressing two antifungal thionin genes confer resistance to *Fusarium* spp. *Journal of Pharmaceutical and Biological Sciences* 12: 69–79.

Hammond-Kosack, K. E. and Jones, J. D. G. 1996. Inducible plant defense mechanisms and resistance gene function. *Plant Cell* 8: 1773–1791.

Hanin, A. N., Parveez, G. K. A., Rasid, O. A. and Masani, M. Y. A. 2020. Biolistic-mediated oil palm transformation with alfalfa glucanase (AGLU1) and rice chitinase (RCH10) genes for increasing oil palm resistance towards *Ganoderma boninense*. *Industrial Crops and Products* 144: 112008.

Iqbal, A., Raham, S. K., Kashmala, S., Anum, I., Faryal, A., Syeda, A., Shahen, S. and Masahiro, M. 2019. Antimicrobial peptides as effective tools for enhanced disease resistance in plants. *Plant Cell, Tissue and Organ Culture* 1–15.

Jain, D. and Khurana, J. P. 2018. Role of pathogenesis-related (PR) proteins in plant defense mechanism. In: Sing A. and Singh I. (eds.), *Molecular Aspects of Plant-Pathogen Interaction*. Springer, Singapore, pp. 265–281.

Jülke, S. and Ludwig-Müller, J. 2016. Response of *Arabidopsis thaliana* roots with altered lipid transfer protein (ltp) gene expression to the clubroot disease and salt stress. *Plants* 5(1): 2.

Kamle, M., Borah, R., Bora, H., Jaiswal, A. K., Singh, R. K. and Kumar, P. 2020. Systemic acquired resistance (SAR) and induced systemic resistance (ISR): Role and mechanism of action against phytopathogens. In: Hesham, A. L., Upadhyay, R., Sharma, G., Manoharachary, C. and Gupta V. (eds.), *Fungal Biotechnology and Bioengineering. Fungal Biology*. Springer, Cham, Switzerland, pp 457–470.

Kang, J. N., Park, M. Y., Kim, W. N., Kang, H. G., Sun, H. J., Yang, D. H., Ko, S. M. and Lee, H. Y. 2017. Resistance of transgenic zoysia grass overexpressing the zoysia grass class II chitinase gene Zjchi2 against *Rhizoctonia solani* AG2-2 (IV). *Plant Biotechnology Reports* 11(4): 229–238.

Kaur, A., Pati, P. K., Pati, A. M. and Nagpal, A. K. 2017. *In-silico* analysis of cis-acting regulatory elements of pathogenesis-related proteins of *Arabidopsis thaliana* and *Oryza sativa*. *PLoS One*. 12(9): e0184523.

Kaur, A., Reddy, M. S., Pati, P. K. and Kumar, A. 2020. Over-expression of osmotin (OsmWS) gene of *Withania somnifera* in potato cultivar 'Kufri Chipsona 1'imparts resistance to *Alternaria solani*. *Plant Cell Tissue and Organ Culture* 142: 131–142.

Kaur, J., Fellers, J., Adholeya, A., Velivelli, S. L., El-Mounadi, K., Nersesian, N. and Shah, D. 2017. Expression of apoplast-targeted plant defensin MtDef4. 2 confers resistance to leaf rust pathogen *Puccinia triticina* but does not affect mycorrhizal symbiosis in transgenic wheat. *Transgenic Research* 26(1): 37–49.

Keshava Reddy, G., Kumar, A. R. V. and Ramu, V. S. 2018. Methods of plant transformation-a review. *International Journal of Current Microbiology and Applied Sciences* 7(7): 2656–2668.

Kleine-Tebbe, J., Wangorsch, A., Vogel, L., Crowell, D. N., Haustein, U. F. and Vieths, S. 2002. Severe oral allergy syndrome and anaphylactic reactions caused by a Bet v 1–related PR-10 protein in soybean, SAM22. *Journal of Allergy and Clinical Immunology* 110(5): 797–804.

Kramer, K. J. and Muthukrishnan, S. 1997. Insect chitinases: molecular biology and potential use as biopesticides. *Insect Biochemistry and Molecular Biology* 27: 887–900.

Kraszewska, J., Beckett, M. C., James, T. C. and Bond, U. 2016. Comparative analysis of the antimicrobial activities of plant defensin-like and ultrashort peptides against food-spoiling bacteria. *Applied Environmental Microbiology* 82: 4288–4298.

Lagrimini, L. M., Burkhart, W., Moyer, M. and Rothstein, S. 1987. Molecular cloning of complementary DNA encoding the lignin forming peroxidase from tobacco: molecular analysis and tissue-specific expression. *Proceedings of National Academy of Sciences USA* 84: 7542–7546.

Lakhssassi, N., Piya, S., Bekal, S., Liu, S., Zhou, Z., Bergounioux, C., Miao, L., Meksem, J., Lakhssassi, A., Jones, K. and Kassem, M. A. 2020. A pathogenesis-related protein GmPR08-Bet VI promotes a molecular interaction between the GmSHMT08 and GmSNAP18 in resistance to *Heterodera glycines*. *Plant Biotechnology Journal* 18: 1–20.

Li, Y., Qiu, L., Liu, X., Zhang, Q., Zhuansun, X., Fahima, T., Krugman, T., Sun, Q. and Xie, C. 2020. Glycerol-induced powdery mildew resistance in wheat by regulating plant fatty acid metabolism, plant hormones cross-talk, and pathogenesis-related genes. *International Journal of Molecular Sciences* 21(2): 673.

Li, Y., Zhang, D., Li, W., Mallano, A. I., Zhang, Y., Wang, T. and Li, W. 2016. Expression study of soybean germin-like gene family reveals a role of GLP7 gene in various abiotic stress tolerances. *Canadian Journal of Plant Science* 96(2): 296–304.

Liang, F., Du, X., Zhang, J., Li, X., Wang, F., Wang, H. and Liu, D. 2019. Wheat TaLr35PR2 gene is required for Lr35-mediated adult plant resistance against leaf rust fungus. *Functional Plant Biology* 47(1): 26–37.

Li-Jing, Fang, Rong-Li, Q., Zhuang, L., Chao-R, L., Ying-P, G. and Xian, L. J. 2019. Expression and functional analysis of a PR-1 gene, *MuPR1*, involved in disease resistance response in mulberry (*Morus multicaulis*). *Journal of Plant Interactions* 14(1): 376–385.

Liu, X., Yu, Y., Liu, Q., Deng, S., Jin, X., Yin, Y., Guo, J., Li, N., Liu, Y., Han, S., Wang, C. and Hao, D. 2020. A Na_2CO_3-responsive chitinase gene from *Leymus chinensis* improves pathogen resistance and saline-alkali stress tolerance in transgenic tobacco and maize. *Frontiers in Plant Science* 11: 504.

Liu, Y., Liu, Q., Tang, Y. and Ding, W. 2019. NtPR1a regulates resistance to *Ralstonia solanacearum* in *Nicotiana tabacum* via activating the defense-related genes. *Biochemical and Biophysical Research Communications* 508(3): 940–945.

McLaughlin, J. E., Al Darwish, N., Garcia-Sanchez, J., Tyagi, N., Trick, H. N., McCormick, S. and Tumer, N. E. 2020. A lipid transfer protein has antifungal and antioxidant activity and suppresses *Fusarium* head blight disease and DON accumulation in transgenic wheat. *Phytopathology* doi: 10.1094/PHYTO-04-20-0153-R.

Melchers, L. S., Potheker-de Groot, M. A., Van der Knaap, J. A., Ponstein, A. S., Sela- Buurlage, M. B., Bol, J. F., Cornelissen, B. J. C., Van den Elzen, P. J. M. and Linthorst, H. J. M. 1994. A new class of tobacco chitinases homologous to bacterial exo-chitinases displays antifungal activity. *Plant Journal* 5: 469–480.

Meng, X. and Zhang, S. 2013. MAPK cascades in plant disease resistance signalling. *Annual Review of Phytopathology* 51: 245–266.

Metraux, J. P., Streit, L. and Staub, T. 1988. A pathogenesis-related protein in cucumber is a chitinase. *Physiology and Molecular Plant Pathology* 33: 1–9.

Michielse, C. B., Hooykaas, P. J., van den Hondel, C. A. and Ram, A. F. 2005. *Agrobacterium*-mediated transformation as a tool for functional genomics in fungi. *Current Genetics* 48(1): 1–17.

Misra, R. C., Kamthan, M., Kumar, S. and Ghosh, S. 2016. A thaumatin-like protein of *Ocimum basilicum* confers tolerance to fungal pathogen and abiotic stress in transgenic Arabidopsis. *Scientific Reports* 6(1): 1–14.

Moosa, A., Farzand, A., Sahi, S. T. and Khan, S. A. 2017. Transgenic expression of antifungal pathogenesis-related proteins against phytopathogenic fungi–15 years of success. *Israel Journal of Plant Sciences* 1: 1–7.

Musidlak, O., Robert Nawrot, I. D. and Gózdzicka-Józefiak, A. 2017. Which plant proteins are involved in antiviral defense? Review on *in vivo* and *in vitro* activities of selected plant proteins against viruses. *International Journal of Molecular Science* 18: 2300.

Naidoo, R., Ferreira, L., Berger, D. K., Myburg, A. A. and Naidoo, S. 2013. The identification and differential expression of *Eucalyptus grandis* pathogenesis-related genes in response to salicylic acid and methyl jasmonate. *Frontiers in Plant Science* 4: 43.

Narusaka, M., Minami, T., Iwabuchi, C., Hamasaki, T., Takasaki, S., Kawamura, K. and Narusaka, Y. 2015. Yeast cell wall extract induces disease resistance against bacterial and fungal pathogens in *Arabidopsis thaliana* and *Brassica* crop. *PLoS One* 10(1): e0115864.

Nawrocka, J., Małolepsza, U., Szymczak, K. and Szczech, M. 2018. Involvement of metabolic components, volatile compounds, PR proteins, and mechanical strengthening in multilayer protection of cucumber plants against *Rhizoctonia solani* activated by *Trichoderma atroviride* TRS25. *Protoplasma* 255(1): 359–373.

Okushima, Y., Koizumi, N., Kusano, T. and Sano, H. 2000. Secreted proteins of tobacco cultured BY2 cells: identification of a new member of pathogenesis-related proteins. *Plant Molecular Biology* 42: 479–488.

Osuki, K. I., Hashimoto, S., Suzuki, A., Araragi, M., Takahara, A., Kurosawa, M., Kucho, K. I., Higashi, S., Abe, M. and Uchiumi, T. 2016. Gene expression and localization of a β-1, 3-glucanase of *Lotus japonicus*. *Journal of Plant Research* 129(4): 749–758.

Pei, Y., Li, X., Zhu, Y., Ge, X., Sun, Y., Liu, N. and Hou, Y. 2019. GhABP19, a novel germin-like protein from *Gossypium hirsutum*, plays an important role in the regulation of resistance to verticillium and fusarium wilt pathogens. *Frontiers in Plant Science* 10: 583.

Pei, Y., Zhu, Y., Jia, Y., Ge, X., Li, X., Li, F. and Hou, Y. 2020. Molecular evidence for the involvement of cotton GhGLP2, in enhanced resistance to *Verticillium* and *Fusarium* Wilts and oxidative stress. *Scientific Reports* 10(1): 1–15.

Pereira, A. 2000. A transgenic perspective on plant functional genomics. *Transgenic Research* 9(4–5): 245–260.

Puthoff, D. P., Holzer, F. M., Perring, T. M. and Walling, L. L. 2010. Tomato pathogenesis-related protein genes are expressed in response to *Trialeurodes vaporariorum* and *Bemisia tabaci* Biotype B feeding. *Journal of Chemical Ecology* 36 (11): 1271–1285.

Rashid, M. H. and Chung, Y.R. 2017. Induction of systemic resistance against insect herbivores in plants by beneficial soil microbes. *Frontiers in Plant Science* 8: 1816.

Ryan, C. A. 2000. The systemin signaling pathway: differential activation of plant defensive genes. *Biochemica et Biophysica Acta* 1477: 112–121.

Sabokkhiz, M. A., Tanhaeian, A. and Mamarabadi, M. 2019. Study on antiviral activity of two recombinant antimicrobial peptides against tobacco mosaic virus. *Probiotics and Antimicrobial Proteins* 11(4): 1370–1378.

Sarkar, P., Jana, K. and Sikdar, S.R. 2017. Overexpression of biologically safe Rorippa indica defensin enhances aphid tolerance in Brassica juncea. *Planta* 246(5): 1029–1044.

Sasaki, K., Kuwabara, C., Umeki, N., Fujioka, M., Saburi, W., Matsui, H. and Imai, R. 2016. The cold-induced defensin TAD1 confers resistance against snow mold and *Fusarium* head blight in transgenic wheat. *Journal of Biotechnology* 228: 3–7.

Sathoff, A. E. and Samac, D. A. 2019. Antibacterial activity of plant defensins. *Molecular Plant-Microbe Interactions* 32(5): 507–514.

Sathoff, A. E., Dornbusch, M. R., Miller, S. S. and Samac, D. A. 2020. Functional analysis of Medicago-derived pathogen-induced gene promoters for usage in transgenic alfalfa. *Molecular Breeding* 40(7): 1–12.

Sharma, I., Bhardwaj, R., Gautam, V., Kaur, R. and Sharma, A. 2018. Brassinosteroids: occurrence, structure and stress protective activities. *Frontiers in Natural Product Chemistry* 4: 204–239.

Sharma, K. 2015. Protease inhibitors in crop protection from insects. *International Journal of Current Research Aca Review* 3(2): 55–70.

Sindelarova, M. and Sindelar, L. 2005. Isolation of pathogenesis-related proteins from TMV-infected tobacco and their influence on infectivity of TMV. *Plant Protection Science* 41: 52–57.

Singh, A., Jain, D., Tyagi, C., Singh, S., Kumar, S. and Singh, I. K. 2018a. *In silico* prediction of active site and *in vitro* DNase and RNase activities of Helicoverpa-inducible pathogenesis related-4 protein from *Cicer arietinum*. *International Journal of Biological Macromolecules* 113: 869–880.

Singh, H. R., Hazarika, P., Agarwala, N., Bhattacharyya, N., Bhagawati, P., Gohain, B., Bandyopadhyay, T., Bharalee, R., Gupta, S., Deka, M. and Das, S. 2018b. Transgenic tea over-expressing *Solanum tuberosum* endo-1, 3-beta-D-glucanase gene conferred resistance against blister blight disease. *Plant Molecular Biology Reporter* 36(1): 107–122.

Sinha, R. K., Verma, S. S. and Rastogi, A. 2020. Role of pathogen-related protein 10 (PR 10) under abiotic and biotic stresses in plants. *Phyton* 89(2): 167.

Smeekens, J. M., Bagley, K. and Kulis, M. 2018. Tree nut allergies: Allergen homology, cross-reactivity, and implications for therapy. *Clinical & Experimental Allergy* 48(7): 762–772.

Soliman, A., Idriss, M., El-Meniawi, F. and Rawash, I. 2019. Induction of pathogenesis-related (PR) proteins as a plant defense mechanism for controlling the cotton whitefly *Bemisia tabaci*. *Alexandria Journal of Agricultural Sciences* 64(2): 107–122.

Somssich, I. E., Schmelzer, E., Bollmann, J. and Hahlbrock, K. 1986. Rapid activation by fungal elicitor of genes encoding "pathogenesis- related" proteins in cultured parsley cells. *Proceedings of National Academy of Sciences USA* 83: 2427–2430.

Stranieri, S., Ricci, E. C. and Banterle, A. 2017. Convenience food with environmentally sustainable attributes: A consumer perspective. *Appetite* 116: 11–20.

Suárez-Cervera, M., Castells, T., Vega-Maray, A., Civantos, E., del Pozo, V., Fernández-González, D., Moreno-Grau, S., Moral, A., López-Iglesias, C., Lahoz, C. and Seoane-Camba, J. A. 2008. Effects of air pollution on Cup a 3 allergen in *Cupressus arizonica* pollen grains. *Annals of Allergy, Asthma & Immunology* 101(1): 57–66.

Sun, W., Zhou, Y., Movahedi, A., Wei, H. and Zhuge, Q. 2020. Thaumatin-like protein (Pe-TLP) acts as a positive factor in transgenic poplars enhanced resistance to spots disease. *Physiological and Molecular Plant Pathology* 112: 101512.

Sundaresha, S., Kumar, A.M., Rohini, S., Math, S.A., Keshamma, E., Chandrashekar, S.C. and Udayakumar, M. 2010. Enhanced protection against two major fungal pathogens of groundnut, Cercospora arachidicola and Aspergillus flavus in transgenic groundnut over-expressing a tobacco β 1 -3 glucanase. *European Journal of Plant Pathology* 126(4): 497–508.

Sundaresha, S., Rohini, S., Appanna, V.K., Arthikala, M.K., Shanmugam, N.B., Shashibhushan, N.B., Kishore, C.H., Pannerselvam, R., Kirti, P.B. and Udayakumar, M. 2016. Co-overexpression of Brassica juncea NPR1 (BjNPR1) and Trigonella foenum-graecum defensin (Tfgd) in transgenic peanut provides comprehensive but varied protection against Aspergillus flavus and Cercospora arachidicola. *Plant Cell Reports* 35(5): 1189–1203.

Taif, S., Zhao, Q., Pu, L., Li, X., Liu, D. and Cui, X. 2020. A β-1, 3-glucanase gene from *Panax notoginseng* confers resistance in tobacco to *Fusarium solani*. *Industrial Crops and Products* 143: 111947.

Tam, J. P., Wang, S., Wong, K. H. and Tan, W. L. 2015. Antimicrobial peptides from plants. *Pharmaceuticals* 8: 711–757.

Tantong, S., Pringsulaka, O., Weerawanich, K., Meeprasert, A., Rungrotmongkol, T., Sarnthima, R., Roytrakul, S. and Sirikantaramas, S. 2016. Two novel antimicrobial defensins from rice identified by gene coexpression network analyses. *Peptides* 84: 7–16.

Terras, F. R., Eggermont, K., Kovaleva, V., Raikhel, N. V., Osborn, R. W., Kester, A., Rees, S. B., Vanderleyden, J., Cammue, B. P. and Broekaert, W. F. 1995. Small cysteine-rich antifungal proteins from radish: their role in host defense. *Plant Cell* 7: 573–588.

Tezuka, D., Kawamata, A., Kato, H., Saburi, W., Mori, H. and Imai, R. 2019. The rice ethylene response factor OsERF83 positively regulates disease resistance to *Magnaporthe oryzae*. *Plant Physiology and Biochemistry* 135: 263–271.

Thery, T., Lynch, K. M. and Arendt, E. K. 2019. Natural antifungal peptides/proteins as model for novel food preservatives. *Comprehensive Reviews in Food Science and Food Safety* 18(5): 1327–1360.

Thery, T., Lynch, K. M., Zannini, E. and Arendt, E. K. 2020. Isolation, characterisation and application of a new antifungal protein from broccoli seeds–New food preservative with great potential. *Food Control* 107356.

Van der Weerden, N. L. and Anderson, M. A. 2013. Plant defensins: common fold, multiple functions. *Fungal Biology Reviews* 26: 121–131.

Van Loon, L. C. 1982. Regulation of changes in proteins and enzymes associated with active defense against virus infection. In: Wood, R.K.S. (ed.), *Active Defense Mechanisms in Plants*. Plenum Press, New York, pp. 247–273.

Van Loon, L. C., Pierpont, W.S., Boller, T. and Conejero, V. 1994. Recommendations for naming plant pathogenesis-related proteins. *Plant Molecular Biology Reporter* 12: 245–264.

Van Loon, L. C, Rep, M. and Pieterse, C. M. 2006. Significance of inducible defense-related proteins in infected plants. *Annual Reviews of Phytopathology* 44: 135–162.

Velivelli, S. L. S., Islam, K. T., Hobson, E. and Shah, D. M. 2018. Modes of action of a bi-domain plant defensin MtDef5 against a bacterial pathogen *Xanthomonas campestris*. *Frontiers in Microbiology* 9: 934.

Vera, P. and Conejero, V. 1988. Pathogenesis-related proteins of tomato: p-69 as an alkaline endoproteinase. *Plant Physiology* 87: 58–63.

Wang, Q., Qiu, B. L., Li, S., Zhang, Y. P., Cui, X. M., Ge, F. and Liu, D. Q. 2019. A methyl jasmonate induced defensin like protein from *Panax notoginseng* confers resistance against *Fusarium solani* in transgenic tobacco. *Biologia Plantarum* 63: 797–807.

Wang, R., Wang, S., Pan, W., Li, Q., Xia, Z., Guan, E., Zheng, M., Pang, G., Yang, Y. and Yi, Z. 2018. Strategy of tobacco plant against black shank and tobacco mosaic virus infection via induction of PR-1, PR-4 and PR-5 proteins assisted by medicinal plant extracts. *Physiological and Molecular Plant Pathology* 101: 127–145.

Wangorsch, A., Jamin, A., Lidholm, J., Gräni, N., Lang, C., Ballmer-Weber, B., Vieths, S. and Scheurer, S. 2017. Identification and implication of an allergenic PR-10 protein from walnut in birch pollen associated walnut allergy. *Molecular Nutrition & Food Research* 61(4): 1600902.

Weerawanich, K., Webster, G., Ma, J. K. C., Phoolcharoen, W. and Sirikantaramas, S. 2018. Gene expression analysis, subcellular localization, and *in planta* antimicrobial activity of rice (*Oryza sativa* L.) defensin 7 and 8. *Plant Physiology and Biochemistry* 124: 160–166.

Wei, Y. D., Zhang, Z. G., Andersen, C. H., Schmelzer, E., Gregersen, P. L., Collinge, D. B., Smedegaard-Petersen, V. and Thordal-Christensen, H. 1998. An epidermis/papilla-specific oxidase-like protein in the defence response of barley attacked by the powdery mildew fungus. *Plant Molecular Biology* 36: 101–112.

Wielkopolan, B. and Obrępalska-Stęplowska, A. 2016. Three-way interaction among plants, bacteria, and coleopteran insects. *Planta* 244(2): 313–332.

Wu, J., Kim, S. G., Kang, K. Y., Kim, J. G., Park, S. R., Gupta, R., Kim, Y. H., Wang, Y. and Kim, S. T. 2016. Overexpression of a pathogenesis-related protein 10 enhances biotic and abiotic stress tolerance in rice. *The Plant Pathology Journal* 32(6): 552.

Wu, X., Valli, A., García, J. A., Zhou, X. and Cheng, X. 2019. The tug-of-war between plants and viruses: Great progress and many remaining questions. *Viruses* 11(3): 203.

Xu, Y., Zheng, X., Song, Y., Zhu, L., Yu, Z., Gan, L. and Zhu, C. 2018. NtLTP4, a lipid transfer protein that enhances salt and drought stresses tolerance in *Nicotiana tabacum*. *Scientific Reports* 8(1): 1–14.

Yang, X., Yang, J., Wang, Y., He, H., Niu, L., Guo, D. and Li, Q. 2019. Enhanced resistance to sclerotinia stem rot in transgenic soybean that overexpresses a wheat oxalate oxidase. *Transgenic Research* 28(1): 103–114.

Younas, A., Saleem, M., Tariq, H., Arooj, B., Akhthar, M. S. and Tahira, R. 2016. Purification and biochemical characterization of a pathogenesis related endochitinase *Arachis hypogaea*. *International Journal of Agriculture & Biology* 18: 780–788.

Yuan, N., Dai, C., Ling, X., Zhang, B. and Du, J. 2019. Peptidomics-based study reveals that GAPEP1, a novel small peptide derived from pathogenesis-related (PR) protein of cotton, enhances fungal disease resistance. *Molecular Breeding* 39(11): 156.

Zaynab, M., Fatima, M., Sharif, Y., Zafar, M. H., Ali, H. and Khan, K. A. 2019. Role of primary metabolites in plant defense against pathogens. *Microbial Pathogenesis* 137: 103728.

Zhang, S. B., Zhang, W. J., Zhai, H. C., Lv, Y. Y., Cai, J. P., Jia, F., Wang, J. S. and Hu, Y. S. 2019. Expression of a wheat β-1, 3-glucanase in *Pichia pastoris* and its inhibitory effect on fungi commonly associated with wheat kernel. *Protein Expression and Purification* 154: 134–139.

Zhang, Z., Collinge, D. B. and Thordal-Christensen, H. 1995. Germin-like oxalate oxidase, a H_2O_2-producing enzyme, accumulates in barley attacked by the powdery mildew fungus. *Plant Journal* 8: 139–145.

Zipfel, C. 2014. Plant pattern-recognition receptors. *Trends in Immunology* 35: 345–351.

8

Approaches and Techniques in Plant Metabolic Engineering

Usha Kiran
Vanercia Institute of Technical Education

CONTENTS

8.1 Introduction

The world population is growing at an unprecedented pace. Feeding such a large population is a big challenge and requires a multidimensional approach to find an answer. Apart from this, crop production is itself challenged by the climatic, agronomic, biotic, societal and economic factors as well as reduced land coverage. Identifying the traits for good vigour and nutritional content itself is very perplexing and incorporating them into the cultivated varieties is an uphill task; therefore, the role of plant breeders has become much more challenging than before. Apart from satisfying the growing hunger, a major part of the world population is also dependent on plants for their medical needs. Plants are rich and sometimes the only source of bioactive compounds. These compounds are used as drugs and precursors of semisynthetic drugs. Also, plant extracts have been and are still used to prevent and treat a number of diseases by many ethnic groups. Further, plants could also be important in providing valuable leads for novel drugs.

In the current scenario, the world is witnessing a global shift towards economically attractive greener technologies aimed to deliver environmentally sustainable products. The unprecedented use of chemical fertilisers and pesticides used during the "Green Revolution" has altered the soil nutrients. Plummeting land coverage under agriculture and forest, limited nutrient contents in soil and reduced interest of farmers in this time-consuming process are also reducing the productivity. New innovations in plant-breeding techniques are required to attain food security and ecological and societal resilience (Yousaf et al. 2017).

The intervention of modern biotechnology principles using metabolic engineering and synthetic biology may provide an effective solution towards the development of economical and sustainable food and medicine sources (Long et al. 2015). Metabolic engineering aims to modulate the expression of one or more genes or gene networks to improve the production of a target compound, which would increase the economic or nutritional value of the plant (Garg et. al. 2018). Metabolic engineering is also a preferred tool to decrease the content of unwanted or toxic compounds in a plant to make it usable for human and animal consumption. Synthetic biology aims at restructuring the endogenous metabolic pathways to produce new compounds in the host plant. The completed genome sequence project had given plant scientists an insight into the metabolic pathways and their regulation and had enabled them to extrapolate novel pathways into economically important plants. This chapter reviews the trends and techniques used for engineering metabolic pathways in plants to produce desired results.

8.2 Trends in Metabolic Engineering in Plants

The challenge of providing food, feed, and medicine to meet global demands is a demanding task. Food, not only to kill hunger but to provide appropriate nutrition, is the need of the hour especially in developing and underdeveloped countries. The main problem is that the cultivated strains with the desired nutrient for the proper growth of human and cattle either may not exist or the given strain with appropriate qualities may not be suitable for cultivation and/or require expensive economical processing. Metabolic engineering allows reconstructing pathways which can selectively divert the carbon flux towards the production of economically important metabolites.

8.2.1 Metabolic Engineering of Medicinal Plants for Phytonutrients

Plants produce structurally diverse chemical compounds that they use to protect themselves from bacteria, fungus, insects, pests and other adverse environmental threats. These compounds are called phytonutrients or secondary metabolites. These bioactive phytonutrients are terpenoids, flavonoids, carotenoids, phenolics, phytosterols and glucosinolates. Many of these phytonutrients are reported as being used as nutraceutical, antimicrobial, antifungal, antioxidant, immunomodulatory, anticancerous, anti-osteoporotic and antidiabetic for human applications (Kinney 2006). Despite the importance of these metabolites, the biosynthetic process of only a few are known. Thus, the limited knowledge of the complexity of structure and pathways, specific stereochemistry and economic unfeasibility has restricted the development of synthetic production of these valuable compounds. Also, most medicinal plants are found in the wild and contain these compounds in very low amounts. Species-specific expression of these phytonutrients further makes these plants invaluable. Overexploitation of wild plants for medicinal purposes had led to extinction or near extinction of some of these species. Metabolic engineering has been proved to be an effective tool in enhancing the concentration of these high-value metabolites.

Metabolic manipulations are being used to exploit the immense diversity of plant metabolism for the production of medicinally important secondary metabolites. Taxol or paclitaxel is a complex diterpene with antitumor and antileukemic activity. It is extracted from *Taxus brevifolia* (Pacific yew); however, the amount obtained is very low. Geranylgeranyl diphosphate is used as a precursor for the synthesis of taxol following a complex pathway constituting 20 distinct enzymatic steps. The carbon flux, however, is divided after the synthesis of 5α-hydroxytaxa-4(20),11(12)-diene for the synthesis of 14β-hydroxy toxoids and taxol. Both 14β- hydroxylase (14OH) and 13α-hydroxylase (13OH) utilise 5α-hydroxytaxa-4(20),11(12)-diene as a substrate, thus forming the branching point in the taxol biosynthesis. Li et al. (2013) used antisense RNA inhibition to suppress 14β-hydroxylase gene (14OH) expression in the Taxus × media TM3 cell lines to increase the concentration of taxol.

Podophyllotoxin (PTOX) and its derivative are important lignans, which are used as a precursor for anticancer drugs (Liu et al. 2015). Isolated first from *Podophyllum peltatum* L., PTOX is also found in *Linum*, *Forsythia*, *Anthriscus*, *Catharanthus*, *Cassia*, *Commiphora*, *Thuja*, *Juniperus*, *Callitris*, *Diphylleia*, and *Dysosma* (Kusari et al. 2011); however, *Podophyllum* species is used as a source for isolation of PTOX. Since not many drugs have reached the clinical application in the case of cancer and Podophyllum species are becoming endangered, research on alternative plants to enhance the indigenous concentration of PTOX is important. *Anthriscus sylvestris* (L.) produces high concentrations of deoxypodophyllotoxin, which can be used as a precursor for the PTOX production (Hendrawati et al. 2011). Vasilev et al. (2006) showed that deoxypodophyllotoxin can be converted into epipodophyllotoxin using human liver cytochrome P450 3A4 (CYP3A4) and human NADPH P450 reductase, and hence it can be used commercially.

8.2.2 Crops with Altered Nutrients

Food security is becoming a rising concern with the current world population of 7.8 billion and still rising. With shrinking cultivable land and drastic climate changes, this is further aggravated. Apart from evident hunger, scientists and plant breeders are facing the challenge of fighting "hidden huger" which is a result of inadequate intake of essential micronutrients in the daily diet (Hodge 2016). Biofortification of different crop varieties offers a cost-effective, sustainable and long-term solution in delivering micronutrients to masses especially in developing and poor countries where access to diverse diets and nutrient supplements is a costly affair (Perez-Massot et al. 2013). The staggering increase in the occurrence of obesity and associated aliments is being reported from developed countries. The solution to both problems requires a reformed approach towards food consumption and supply. The new biotechnology interventions like metabolic engineering and system biology will have a dramatic impact on enriching diet to address problems associated with micronutrient deficiency. Thus the aim of obtaining a cost-effective and sustainable method of delivering maximum nutritional value food from the crops grown on shrinking landmasses could be achieved (de Steur et al. 2017).

Population essentially depending on staple crops such as wheat, rice, maize and cassava for sustenance tend to develop vitamin A, iron, calcium, iodine, zinc or selenium deficiencies. Vitamin A deficiency (VAD) is reported to be prevalent in people living in rice-based societies in developing and poor countries. Biofortification of rice with β-carotene, the 'golden rice', has been developed with the aim to enrich diet with β-carotene, a precursor of vitamin A. Genes encoding phytoene synthase (*psy*) from daffodil (*Narcissus pseudonarcissus*) and phytoene desaturase (*crtl*) from soil bacterium *Erwinia uredovora* were expressed in rice endosperm, resulting in the production of β-carotene (Burkhardt et al. 1997; Ye et al. 2000). Substantial increase of up to 23-fold in total carotenoids compared with the original golden rice was reported in rice plants transformed with phytoene synthase (*psy*) from maize (Paine et al. 2005).

Wheat, another highly consumed cereal all over the world, is a convenient source of energy, proteins and vitamins (Shewry 2009). Fruits and vegetables contain carotenoids from the carotene class (α- and β-carotene and lycopene) whereas wheat contains carotenoids from the xanthophylls class (lutein, zeaxanthin and β-cryptoxanthin) (Abdel-Aal el-Sayed 2007; Hentschel et al. 2002). Wang et al. (2014) transformed the common wheat cultivar Bobwhite with the bacterial phytoene synthase gene (*CrtB*) and carotene desaturase gene (*CrtI*) in an attempt to augment the provitamin A content in wheat grains. The results suggested a small increase in the total carotenoid content in the grains of transgenic wheat when *CrtB* or *CrtI* genes were expressed alone. However, on co-expression of *CrtB* and *CrtI* genes, a marked increase in total carotenoid content with darker red/yellow grain phenotype was observed.

Zinc, a micronutrient, plays an important role in cell growth and differentiation and efficient protein and lipid metabolism in the human body. Zinc deficiency results in stunted growth, hypogonadism, skin disorders, compromised immune function and cognitive dysfunction (Prasad 2013). Zinc contents in barley (*Hordeum vulgare*), another important cereal crop, were improved by overexpression of zinc transporters. Plants absorb micronutrients from their surroundings and improvement in absorption and utilisation efficiency of these micronutrients would increase their content in the plants. Ramesh et al. (2004) overexpressed zinc transporters from *Arabidopsis* in barley under the ubiquitin promoter. The transformed barley showed higher zinc and iron contents than the untransformed plants. The rice endosperm showed a significant increase in Fe and Zn concentrations with the constitutive overexpression of genes from the OsNAS family (Johnson et al. 2011)

8.2.3 Enhancing Photosynthetic Efficiency

Photosynthesis is of paramount importance for the conversion of solar energy into chemical energy and could be passed on from one trophic level to another (Orr et al. 2017). The enhancement of the photosynthetic efficiency of plants could, therefore, be an effective tool to meet the increased productivity demands under the uncertainties of global climate change (Nowicka et al. 2018; Simkin et al. 2019).

Rubisco (D-ribulose-1,5-bisphosphate carboxylase/oxygenase) is an important enzyme in assimilating atmospheric CO_2 into the biosphere. It, however, has a slow turnover rate (a few catalytic events per second) and has a wasteful oxygenase activity (oxygenation of RuBP) which recycles unproductive reaction products by the process of photorespiration (Timm et al. 2016). It is estimated that Rubisco fixes O_2 in up to one-third of reactions instead of CO_2 (Tcherkez et al. 2006). The competing oxygenation activity uses previously fixed CO_2 and energy and hence decreases the photosynthetic efficiency. To counterbalance this, the plants contain a large amount of Rubisco, which requires large investments in nitrogen. Efforts are, therefore, being made towards the molecular engineering of Rubisco to improve the photosynthetic efficiency of plants (Betti et al. 2016; Carmo-Silva et al. 2015; Lin et al. 2014).

Rubico is a large 550-kDa hexadecameric complex with eight small and eight large subunits. The complexity of the structure and function of Rubisco's assembly has made molecular engineering of the enzyme exceptionally challenging. A recent report suggests crop yield could be enhanced by introducing the CO_2-concentrating mechanism (CCM) from cyanobacteria into plants. Lin et al. (2014) transformed tobacco with Rubisco from the cyanobacterium *Synechococcus elongatus* PCC7942 (Se7942) after knocking out the native Rubisco large subunit gene. The Se7942 small and large subunit genes in combination with either the Se7942 assembly chaperone (RbcX) or an internal CO_2-concentrating carboxysomal protein (CcmM35) were inserted in native Rubisco to make knockouts. Both transformed lines supported autotrophic growth and had higher CO_2 fixation rates per unit of enzyme than the untransformed tobacco control plants. The cyanobacteria Rubisco exhibits higher catalytic rates than higher plant Rubisco but has lower CO_2 affinity and specificity for CO_2 over O_2. The co-engineering of a functional CCM showed enhanced efficiency of cyanobacterial Rubisco in tobacco. These photosynthetically competent transplastomic tobacco lines are an important achievement and suggest that other components of the cyanobacterial CCM could be used to further enhance plant photosynthetic efficiency (Price and Howitt 2014).

Most of the terrestrial plants use the Calvin-Benson-Bassham cycle (CBB cycle/C3 photosynthesis) to fix CO_2 into organic matter. It is estimated that 100 billion tons of carbon (15% of the carbon in the atmosphere) is fixed in a year by C3 photosynthesis (Raines 2011). The rate of CO_2 assimilation is limited by the initiation of photorespiration under high CO_2 conditions. Thus, the CBB cycle has been seen as a possible target to improve carbon fixation. Sedoheptulose-1,7-bisphosphatase (SBPase), plastid transketolase and transaldolase are the key enzymes controlling the carbon flux in the CBB cycle (Raines 2011; Singh et al. 2014). Rosenthal et al. (2011) overexpressed *Arabidopsis thaliana* SBPase in tobacco plants. The transgenic plants showed higher photosynthetic rates, greater carbon assimilation and electron transport rates and better yield compared with untransformed control plants, when grown under free-air CO_2 enrichment (FACE) conditions. The efforts have now expanded to simultaneous modulation of other Calvin-Benson enzymes to achieve cumulative impact on photosynthetic efficiency and biomass (Simkin et al. 2017).

8.2.4 Engineering Crops for Biofuel

For decades, the primary focus of plant metabolic engineering was to improve productivity and nutritional quality of crop plants. In the last few years, however, the technology has also been used to develop transgenic plants for the industrial production of biofuels (Davies et al. 2010). The sustainable yield of biofuel requires cautious selection of biomass production systems with high biomass and efficient technologies for transformation of the biomass to biofuel (Henry 2010). The bioethanol is produced by sugars (sugarcane and corn) and lignocellulose biomass from wheat, corn, sugarcane, poplar and switchgrass residual biomass. The cellulose and lignocellulose content in the plant, however, influence the enzyme accessibility to the cellular polysaccharides (Blanch et al. 2011; Himmel et al. 2007). The genetic manipulations are, therefore, required to develop plant varieties with reduced interfering substances such as lignin and enhanced desirable substances such carbohydrate. Such varieties would facilitate maximum recovery of sugars for conversions to biofuel.

Triacylglycerols (TAGs), the storage lipids in plants, have also been used as a precursor for clean, environmentally friendly and renewable biofuel production in the last few years. TAGs are an energy-rich reduced form of carbon available in nature (Durrett et al. 2008; Thelen and Ohlrogge 2002) which can be easily transformed into biodiesel by the formation of fatty acid methyl esters (Ohlrogge and

Chapman 2011). Attempts have been made towards the genetic manipulations of plants to achieve higher concentrations of TAGs in seeds and vegetative tissues (stems and leaves). The complete elucidation of the TAG biochemical pathway and the understanding of enzymes involved in biosynthesis and metabolism (Vanhercke et al. 2014, 2017; Xu et al. 2018) enabled shifting carbon-partitioning balance towards TAG production in leaves and stems for increased energy yields. Plants like *Nicotiana tabacum*, harbouring a functional TAG biosynthesis pathway, have been efficiently exploited for biofuel production. TAGs are regarded as by-products of starch production. The shared carbon flux, therefore, must be diverted to TAG biosynthesis for an increase in TAG accumulation. The ADP-glucose pyrophosphorylase gene (AGPase) is the rate-limiting enzyme in starch biosynthesis. Sanjaya et al. (2011) silenced AGPase, overexpressed WRINKLED1 (WRI1) and suppressed the AGPase gene along with overexpression of WRI1 simultaneously in three different experiments to check the effect on the diversion of carbon flux away from starch and towards TAG biosynthesis. WRI1, a positive regulator of fatty acid biosynthesis in *Arabidopsis* seeds, belongs to the APETELA2/ethylene response element-binding protein (AP2/EREBP) family of transcription factors (Cernac and Benning 2004). The AGPase suppression using RNA interference (RNAi) showed a 1.4-fold increase in TAG accumulation, whereas overexpression of WRI1 showed a 2.8-fold increase in TAG accumulation in *Arabidopsis* seedlings compared with untransformed plants. The suppression of AGPase along with the overexpression of WRI1 showed a 5.8-fold increase in TAG accumulation in *Arabidopsis* seedlings compared with untransformed plants. The approach was thus able to divert carbon flux away from starch and towards TAG biosynthesis (Sanjaya et al. 2011). The final and only committed step in converting diacylglycerol into TAG is catalysed by diacylglycerol acyltransferase (DGAT1-2) (Lehner and Kuksis 1996). The overexpression of *Arabidopsis* DGAT1 shows a sevenfold increased TAG concentration in tobacco (Wu et al. 2013). The reduction of lipid turnover is another approach for increasing TAG accumulation. Oleosins (OLEs) are important structural proteins that influence oil body formation and their longevity by protecting them from coalescence. Zale et al. (2016) constitutively overexpressed WRI1, DGAT1-2 and OLE1 genes along with the suppression of AGPase and PXA in sugarcane. The results showed a 95- and 43-fold higher accumulation of TAGs in leaves and stems, respectively, compared with untransformed control plants. Additional refinements and optimisations for higher TAG accumulation can be achieved by manipulating the TAG biosynthetic pathway and its competing pathways for carbon flux.

8.3 Strategies for Metabolic Engineering in Plants

The objective of the metabolic engineering of plants is to enhance the production of desired high-value products to meet the growing demands. The prerequisite for enhanced production of the desired compounds in any plant is a comprehensive understanding of metabolic pathways and their genetic control. The efficiency of manipulating techniques has been greatly enhanced due to data generated by high-throughput techniques such as large-scale functional genome analysis, next-generation gene sequencing (NSG), big data analysis and storage, microarray-based comparative genomic hybridisation (aCGH), high-throughput RNA sequencing and advance proteomic technologies. The applicability of this approach has been successfully extended to agriculture, environmental remediation and the production of medicines.

Biosynthesis of a compound is a multistep process involving activation and deactivation of a cascade of enzymes (Zhu et al. 2008). The goal of metabolic engineering is to enrich plants with the desired compound by increasing the precursor flux through the target pathway, up-regulating the biosynthetic pathway enzymes or by decreasing the precursor flux in the competing pathways. Another approach for the production of desired compounds is by the introduction of the exogenous biosynthetic pathways across the genetic barriers or by a completely novel pathway, unknown to nature, into easily manageable host plants (Zhu et al. 2013).

8.3.1 Enhance Metabolite by Up-Regulation of the Metabolic Pathway(s)

Metabolism is an elaborate network of highly interconnected pathways enabling cells to synthesise diverse biomolecules from much simpler precursors. Thus, modulating any metabolic pathway requires knowledge of pathway architecture, enzyme kinetics, precursors and intermediate metabolite pools. With an exponential increase in the detailed information of metabolic maps, annotation of enzymes, substrates, cofactors and products, coordinated diverse metabolic responses could be achieved. While these metabolic pathways provide a selective advantage to an organism, not all pathways are required always to be functional at the same levels. The flux of metabolites through an individual reaction or metabolic pathway is controlled to enhance the desired outcome. A metabolic pathway comprises of one or more enzyme catalysed reactions, and these metabolic pathways are connected in a network. The pathways in the network are in competition for a common precursor at the specific branching points, putting constraints on the flux inflow to the network. The flow of metabolites through a metabolic pathway is further modulated by the bottleneck (rate-limiting) enzyme at the committed step. The efficiency of the rate-limiting enzyme could be enhanced by ensuring the adequate supply of precursors. The overexpression of one or more enzymes upstream of rate-limiting enzymes increases in the effective metabolite flux in the target pathway. The enhanced production of the desired product can also be achieved by the overexpression of rate-limiting enzymes. The functional annotation of these rate-limiting enzymes and understanding their regulation can greatly help in defining the strategies for genetic manipulations to achieve the desired biological outcome (Botella-Pavía et al. 2004; Sato et al. 2007). Transcription factors control and coordinate multiple pathways. The overexpression of transcription factors, alone or together, with key enzyme gene(s) can simultaneously activate multiple pathways, resulting in enhanced biosynthesis and accumulation of targeted endogenous metabolite. In addition, clustered regularly interspersed short palindromic repeat (CRISPR)/Cas9-based activation can also be used in the stimulation of metabolic pathways (Li et al. 2017).

Panax ginseng is a rich source of pharmaceutically active ginsenosides, protopanaxadiols (PPD; Rb1, Rb2, Rc and Rd) and protopanaxatriols (PPT; Re, Rf and Rg1). Squalene synthase (SQS) is a key enzyme in ginsenoside biosynthesis in *P. ginseng*. Overexpression of PgSQS1 (*P. ginseng* SQS) enhanced the expression of all downstream genes including squalene epoxidase (SQE), dammarenediol-II synthase (DS) and β-amyrin synthase (β-AS). Transformed adventitious root cultures of *P. ginseng* showed a twofold increase in phytosterols and 1.6- to 3-fold increase in total ginsenosides compared with untransformed adventitious root cultures. Cytochrome P450s (CYP450s) are a monooxygenase protein superfamily involved in the oxidation and hydroxylation reactions in ginsenoside biosynthesis. PgCYP716A52v2 encodes a β-amyrin 28-oxidase, which catalyses the modification of β-amyrin into oleanolic acid, a precursor of an oleanane-type saponin (mainly ginsenoside Ro). Overexpression of CYP716A52v2 showed an increase in oleanane-type ginsenoside contents while the other dammarene-type ginsenoside contents were similar to the untransformed *P. ginseng* plants (Han et al. 2013).

8.3.2 Enhance Metabolite by Down-Regulation of Competitive Metabolic Pathway(s)

The key enzyme gene(s) involved in the competing pathway at the branch point divert the metabolite flux towards the undesired products and hence reduces the bioproduction of the desired metabolite. Thus to redirect the intermediates towards the target pathway, reduction in the expression of or knocking out the branch point enzyme catalysing the committed step in the undesired products could be an effective strategy. The reduction in steady-state levels of gene(s) expression could be achieved by using antisense oligonucleotide, small interfering RNA (siRNA) and RNAi techniques. The inhibition of the catabolic pathway of the desired product is another strategy for enhancing the concentration of the desired product. In addition, CRISPR/dCas9 repression systems can also be used in reducing the undesired product (Zalatan et al., 2015).

Artemisia annua is an important source of antimalarial compound; artemisinin is used in combination therapy for malaria. The carbon flux through artemisinin biosynthesis was increased by suppressing the expression of the sterol pathway which competes for the precursor in the plant. SQS is a key gene in the sterol pathway, which dedicates carbon flux to sterol biosynthesis in *A. annua*. Zhang et al. (2009) suppressed AaSQS by RNAi and observed a significant increase of artemisinin content in transgenic plants compared with untransformed plants. People living in tropical countries use cassava (*Manihot esculenta*) as a major energy source. The tuber, however, contains toxic cyanogenic glucosides. Siritunga and Sayre (2003) down-regulated cytochrome P450 enzymes CYP79D1 and

CYP79D2, which catalyse the first committed step in the biosynthesis of linamarin and lotaustralin. The transgenic plants showed more than 90% reduction in cyanogenic glucoside amounts compared with untransformed plants.

8.3.3 Production of a Novel Compound (Synthetic Biology)

Genetic engineering enables scientists to systematically engineer plants with a promise to produce novel metabolite(s). The creation of a part of or whole new metabolic pathway, from native plants to the target host plant, to produce novel metabolites is known as system biology (Hanson and Jez 2018). Enhanced knowledge about specific metabolic pathways producing high-value metabolite in the natural host, tissue-specific promoters and transporters are important prerequisites for construction of a new pathway in the host plant. Technology to develop multigene expression constructs and biosynthetic enzymes, and effective ways to test the effects of reconstructed pathways in the target host plant has helped to achieved the goals of synthetic biology (Liu and Stewart 2015; Pouvreau et al. 2018). Golden rice has been successfully developed using principles of system biology. The whole cassette of genes required to synthesise carotenoids was constructed *de novo* in rice. Carotenoids, natural pigment and the precursor of vitamin A are important phytonutrients in the fight against malnutrition and nutrient deficiency. They have significant antioxidant activities and play a preventive role in many chronic diseases (Rao and Rao 2007). The cassette comprised of daffodil (*Narcissus pseudonarcissus*) phytoene synthase (*psy*) and lycopene β-cyclase (β-*lcy*) genes cloned under the control of the endosperm-specific glutelin promoter and *Erwinia uredovora* phytoene desaturase gene (*crtI*) under the control of the constitutive 35S promoter. The overexpression of the mobilised pathway showed the expression of β-carotene synthesis in transgenic rice endosperm (Beyer et al. 2002).

8.4 Challenges for Metabolic Engineering

Plants are infinite resources for bioactive molecules. Engineering of plant metabolic pathways and networks for the production of essential minerals and vitamins, medicinal compounds, biofuel and other economically important compounds have been achieved by scientists and plant breeders in the last two decades. There are, however, significant challenges that still need to be addressed. Plants are complex systems that have a sessile life but are still able to override the effects of hostile conditions. The incomplete knowledge about the expression and regulation of genomic complexities and metabolic redundancies are restricting scientists from achieving the full benefit of bioengineering of plants. The plants have a slow growth rate and their development is affected by prevailing environmental and agronomical conditions. Thus the biosynthesis of the desired metabolite would also be affected by the environmental, physiological and developmental stage of the plants (Akula and Ravishankar 2011). Further, the manipulation of the central chassis for the desired metabolites may involve a number of genetic manipulation events together, at a precise compartment and at particular stage, which can have limited success due to constraints of the technology (Li et al. 2016). Sometimes the whole cassette of transgenes associated with the trait needs to be transferred, which may not be effective for all plants (Külahoglu et al. 2014). Further, due to the composite nature of genetic and metabolic networks, the unpredicted results such as gene silencing, no appreciable increase in target compound or enhanced expression of undesired intermediates are seen. Thus, metabolic engineering strategies for plants hold great potential to accomplish the goal of the enhanced bioactive products; however, intricate knowledge about the genetic networks and metabolic pathways is still required.

8.5 Conclusions and Future Prospects

Meeting the demands of the growing world population is a very crucial and multidimensional problem. Finding a solution for a sustainable ecosystem where produce from plants is limited by ever-changing climate, depleted soil nutrients, salinity, drought and infestation by pests and insects is an uphill task.

The plant scientists and breeders are challenged to produce cultivars that are robust, survive ecological resilience and enable food security. Long breeding cycles and limited desired content availability restrict the applicability of traditional breeding for quality and quantity improvement of high-value metabolites. Metabolic engineering is an enabling tool to improve the production of target compounds in endogenous plants by modulating one or more metabolic pathways or networks. Synthetic biology enables the transfer of diverse metabolic pathways, part of or whole, across the genetic barriers, to biosynthesise new compounds in the target host plant. Thus, techniques of metabolic engineering together would empower plant scientists and breeders to manipulate plant metabolic pathways and networks at an unprecedented level and translate the basic knowledge into tangible benefits for agriculture and medicinal purposes.

REFERENCES

Abdel-Aal El-Sayed, M., J Christopher Young, Iwona Rabalski, et al. "Identification and quantification of seed carotenoids in selected wheat species." *Journal of Agricultural Food Chemistry* 55(3), (2007): 787–794.

Akula, R, and Ravishankar GA) "Influence of abiotic stress signals on secondary metabolites in plants." *Plant Signal Behav* 6, (2011): 1720–1731.

Betti, Marco, Hermann Bauwe, Florian A Busch, et al. "Manipulating photorespiration to increase plant productivity: recent advances and perspectives for crop improvement." *Journal of Experimental Botany* 10, (2016): 2977–2988.

Beyer, P, S Al-Babili, X Ye, et al. "Golden rice: Introducing the β-carotene biosynthesis pathway into rice endosperm by genetic engineering to defeat vitamin A deficiency." *The Journal of Nutrition* 132, (2002): 506S–510S.

Blanch, Harvey W, Blake A Simmons, and Daniel Klein-Marcuschamer. "Biomass deconstruction to sugars." *Biotechnology Journal* 6, (2011): 1086–1102.

Botella-Pavía, P, O Besumbes, MA Phillips, et al. "Regulation of carotenoid biosynthesis in plants: evidence for a key role of hydroxymethylbutenyl diphosphate reductase in controlling the supply of plastidial isoprenoid precursors." *Plant Journal* 40, (2004): 188–199.

Burkhardt, P K, P Beyer, J Wünn, et al. "Transgenic rice (*Oryza sativa*) endosperm expressing daffodil (*Narcissus pseudonarcissus*) phytoene synthase accumulates phytoene, a key intermediate of provitamin A biosynthesis." *Plant Journal* 11(5), (1997): 1071–1078.

Carmo-Silva, Elizabete, Joanna C Scales, Pippa J Madgwick, et al. "Optimizing Rubisco and its regulation for greater resource use efficiency." *Plant Cell Environment* 38(9), (2015): 1817–1832.

Cernac, Alex, and Christoph Benning. "WRINKLED1 encodes an AP2/EREB domain protein involved in the control of storage compound biosynthesis in Arabidopsis." *Plant Journal* 40, (2004): 575–585.

Davies, Maelor, Malcolm Campbell, Robert Henry Campbell, et al. "The role of plant biotechnology in bioenergy production." *Plant Biotechnology Journal* 8, (2010): 243.

De Steur, Hans, Saurabh Mehta, Xavier Gellynck, et al. "GM biofortified crops: potential effects on targeting the micronutrient intake gap in human populations." *Current Opinion in Biotechnology* 44, (2017): 181–188.

Durrett, Timothy P., Christoph Benning, and John Ohlrogge. "Plant triacylglycerols as feedstocks for the production of biofuels." *Plant Journal* 54, (2008): 593–607.

Garg, Monika, Natasha Sharma, Saloni Sharma, et al. "Biofortified crops generated by breeding, agronomy, and transgenic approaches are improving lives of millions of people around the world." *Frontiers in Nutrition* 5, (2018): 12.

Han, Jung-Yeon, Min-Jun Kim, Yong-Wook Ban, et al. "The involvement of b-Amyrin 28-oxidase (CYP716A52v2) in oleanane-type ginsenoside biosynthesis in *Panax ginseng. Plant Cell Physiol.* 54, (2013): 2034–2046.

Hanson, Andrew D., and Joseph M. Jez, et al. "Synthetic biology meets plant metabolism." *Plant Science* 273 (2018): 1–2.

Hendrawati, Oktavia, Herman J Woerdenbag, Jacques Hille, et al. "Seasonal variations in the deoxypodophyllotoxin content and yield of *Anthriscus sylvestris* L. (Hoffm.) grown in the field and under controlled conditions." *Journal of Agricultural and Food Chemistry* 59(15), (2011): 8132–8139.

Henry, Robert J. "Evaluation of plant biomass resources available for replacement of fossil oil." *Plant Biotechnology Journal* 8, (2010): 288–293.

Hentschel, Verena, Katja Kranl, Jürgen Hollmann, et al. "Spectrophotometric determination of yellow pigment content and evaluation of carotenoids by high-performance liquid chromatography in durum wheat grain." *Journal of Agricultural Food Chemistry* 50(23), (2002): 6663–6668.

Himmel, Michael E, Shi-You Ding, David K Johnson, et al. "Biomass recalcitrance: engineering plants and enzymes for biofuels production." *Science* 315, (2007): 804–807.

Hodge, Judith. "Hidden hunger: approaches to tackling micronutrient deficiencies." In *Nourishing Millions: Stories of Change in Nutrition*, 35–43. Washington, DC: International Food Policy Research Institute (IFPRI) (2016).

Johnson, Alexander AT, Bianca Kyriacou, Damien L Callahan, et al. "Constitutive overexpression of the *OsNAS* gene family reveals single-gene strategies for effective iron- and zinc-biofortification of rice endosperm." *PLoS ONE* 6(9), (2011): e24476.

Kinney, Anthony J. "Metabolic engineering in plants for human health and nutrition." *Current Opinion. Biotechnology* 17(2), (2006): 130–138.

Külahoglu, C., A. K. Denton, M. Sommer, et al. "Comparative transcriptome atlases reveal altered gene expression modules between two Cleomaceae C3 and C4 plant species." *Plant Cell* 26, (2014): 3243–3260.

Kusari Souvik, Sebastian Zühlke, and Michael Spiteller. "Evaluation of the anti-cancer pro-drug podophyllotoxin and potential therapeutic analogues in Juniperus and Podophyllum species." *Phytochemical Analysis* 22, (2011): 128–143.

Lehner, R., and A. Kuksis. "Biosynthesis of triacylglycerols." *Progress in Lipid Research* 35, (1996): 169–201.

Li F-L, Ma X-J, Hu X-L, et al. "Antisense-induced suppression of taxoid 14β-hydroxylase gene expression in transgenic Taxus× media cells." *African Journal of Biotechnology* 10, (2013): 8720–8728.

Li, J., X. Meng, Y. Zong, K. Chen, H. Zhang, J. Liu, J. Li, and C. Gao. "Gene replacements and insertions in rice by intron targeting using CRISPR-Cas9. " *Nat. Plants* 2, (2016): 16139.

Li, Zhenxiang, Dandan Zhang, Xiangyu Xiong, et al. "A potent Cas9-derived gene activator for plant and mammalian cells." *Natural Plants* 3, (2017): 930–936.

Liu, Wusheng, and C. Neal Jr. Stewart. "Plant synthetic biology." *Trends Plant Science* 20, (2015): 309–317.

Liu, Ying-Qian, Jing Tian, Keduo Qian, et al. "Recent progress on C-4-modified podophyllotoxin analogs as potent antitumor agents." *Medicinal Research Reviews* 35(1), (2015): 1–62.

Lin, Myat T., Alessandro Occhialini, P John Andralojc, et al. "A faster Rubisco with potential to increase photosynthesis in crops." *Nature* 513(7519) (2014): 547–550.

Long, Stephen P., Amy Marshall-Colon, and Xin-Guang Zhu. "Meeting the global food demand of the future by engineering crop photosynthesis and yield potential." *Cell* 161, (2015): 56–66.

Muthayya, Sumithra, Jee Hyun Rah, Jonathan D Sugimoto, et al. "The global hidden hunger indices and maps: an advocacy tool for action." *PLoS One* 8(6), (2013): e67860.

Nowicka, Beatrycze, Joanna Ciura, Renata Szymańska, et al. "Improving photosynthesis, plant productivity and abiotic stress tolerance—current trends and future perspectives." *Journal of Plant Physiology* 231, (2018): 415–433.

Ohlrogge, John, and Kent Chapman. "The seeds of green energy: Expanding the contribution of plant oils as biofuels." *Biochemistry* 33(2), (2011): 34–38.

Orr Douglas, J, Auderlan M Pereira, Paula da Fonseca Pereira, et al. "Engineering photosynthesis: progress and perspectives." *F1000 Research* 6, (2017): 1891.

Paine, Jacqueline A, Catherine A Shipton, Sunandha Chaggar, et al. "Improving the nutritional value of golden rice through increased pro-vitamin A content." *Nature Biotechnology* 23, (2005): 482–487.

Perez-Massot, Eduard, Raviraj Banakar, Sonia Gomez-Galera, et al. "The contribution of transgenic plants to better health through improved nutrition: opportunities and constraints." *Genes Nutrition* 8(1), (2013): 29–41.

Prasad, Ananda S. "Discovery of human zinc deficiency: its impact on human health and disease." *Advances in Nutrition* 4, (2013): 176–190.

Price, G Dean, and Susan M Howitt. "Plant science: towards turbocharged photosynthesis." *Nature* 513(7519), (2014): 497–498.

Pouvreau, Benjamin, Thomas Vanhercke, and Surinder Singh. "From plant metabolic engineering to plant synthetic biology: the evolution of the design/build/test/learn cycle." *Plant Science* 273, (2018): 3–12.

Qi, L. W., C. Z. Wang, and C. S. Yuan. "Isolation and analysis of ginseng: advances and challenges." *Nat. Prod. Rep.* 28, (2011): 467–495.

Raines, Christine A. "Increasing photosynthetic carbon assimilation in C3 plants to improve crop yield: current and future strategies." *Plant Physiology* 155, (2011): 36–42.

Ramesh, Sunita A, Steve Choimes, and Daniel P Schachtman. "Over-expression of an *Arabidopsis* zinc transporter in *Hordeum vulgare* increases short-term zinc uptake after zinc deprivation and seed zinc content." *Plant Molecular Biology* 54(3), (2004): 373–385.

Rao, A, and L G Rao "Carotenoids and human health." *Pharmacological Research* 55, (2007): 207–216.

Rosenthal, David M, Anna M Locke, Mahdi Khozaei, et al. "Over-expressing the C(3) photosynthesis cycle enzyme Sedoheptulose-1-7 bisphosphatase improves photosynthetic carbon gain and yield under fully open air CO(2) fumigation (FACE)." *BMC Plant Biology* 31 Aug 11, (2011): 123.

Sanjaya, Timothy P Durrett, Sean E Weise, et al. "Increasing the energy density of vegetative tissues by diverting carbon from starch to oil biosynthesis in transgenic Arabidopsis." *Plant Biotechnology Journal* 9, (2011): 874–883.

Sato F, T Inui and T Takemura, "Metabolic engineering in isoquinoline alkaloid biosynthesis." *Current Pharmceutical Biotechnology* 8, (2007): 211–218.

Shewry, P R. "Wheat." *Journal of Experimental Botany* 60(6), (2009): 1537–1553.

Simkin, Andrew J, Patricia E Lopez-Calcagno, Philip A Davey, et al. "Simultaneous stimulation of sedoheptulose 1,7-bisphosphatase, fructose 1,6-bisphophate aldolase and the photorespiratory glycine decarboxylase-H protein increases CO_2 assimilation, vegetative biomass and seed yield in Arabidopsis." *Plant Biotechnology Journal* 15(7), (2017): 805–816.

Simkin, Andrew J, Patricia E López-Calcagno, and Christine A Raines. "Feeding the world: improving photosynthetic efficiency for sustainable crop production." *Journal of Experimental Botany* 70, (2019): 1119–1140.

Singh, Jitender, Prachi Pandey, Donald James, et al. "Enhancing C3 photosynthesis: an outlook on feasible interventions for crop improvement." *Plant Biotechnology Journal* 12(9), (2014): 1217–1230.

Siritunga, D, and R T Sayre "Generation of cyanogen-free transgenic cassava." *Planta* 217(3), (2003): 367–373.

Tcherkez, Guillaume G B., Graham D Farquhar, and T John Andrews. "Despite slow catalysis and confused substrate specificity, all ribulose bisphosphate carboxylases may be nearly perfectly optimized." *Proceedings of National Academy of Sciences U S A* 19, (2006): 7246–7251.

Thelen, Jay J., and John B. Ohlrogge. "Metabolic engineering of fatty acid biosynthesis in plants." *Metabolic Engineering* 4, (2002): 12–21.

Timm, Stefan, Alexandra Florian, Alisdair R Fernie, et al. "The regulatory interplay between photorespiration and photosynthesis." *Journal of Experimental Botany* 67(10), (2016): 2923–299.

Vanhercke, Thomas, Uday K Divi, Anna El Tahchy, et al. "Step changes in leaf oil accumulation via iterative metabolic engineering." *Metabolic Engineering* 39, (2017): 237–246.

Vanhercke, Thomas, Anna El Tahchy, Qing Liu, et al. "Metabolic engineering of biomass for high energy density: oilseed-like triacylglycerol yields from plant leaves." *Plant Biotechnology Journal* 12(2), (2014): 231–239.

Vasilev, Nikolay P., Mattijs K Julsing, Albert Koulman, et al. "Bioconversion of deoxypodophyllotoxin into epipodophyllotoxin in *E. coli* using human cytochrome P450 3A4." *Journal of Biotechnology* 126(3), (2006): 383–393.

Wang, Cheng, Jian Zeng, Yin Li, et al. "Enrichment of provitamin A content in wheat (*Triticum aestivum* L.) by introduction of the bacterial carotenoid biosynthetic genes CrtB and CrtI." *Journal of Experimental Botany* 65(9), (2014): 2545–2556.

Wu, Han-Ying, Chao Liu, Min-Chun Li, et al. "Effects of monogalactoglycerolipid deficiency and diacylglycerol acyltransferase overexpression on oil accumulation in transgenic tobacco." *Plant Molecular Biology Reporter* 31, (2013): 1077–1088.

Xu, Xiao-Yu, Hong-Kun Yang, Surinder P Singh, et al. "Genetic manipulation of non-classic oilseed plants for enhancement of their potential as a biofactory for triacylglycerol production." *Engineering* 4 (4), (2018): 523–533.

Ye, X, S Al-Babili, A Klöti, et al. "Engineering the provitamin A (β-carotene) biosynthetic pathway into (carotenoids-free) rice endosperm." *Science* 287, (2000): 303–305.

Yousaf, Muhammad, Jifu Li, Jianwei Lu, et al. "Effects of fertilization on crop production and nutrient-supplying capacity under rice-oilseed rape rotation system." *Science Report* 7, (2017): 1270.

Zalatan, Jesse G, Michael E Lee, Ricardo Almeida, et al. "Engineering complex synthetic transcriptional programs with CRISPR RNA scaffolds." *Cell* 160, (2015): 339–350.

Zale J, J. H. Jung, J. Y. Kim, et al. "Metabolic engineering of sugarcane to accumulate energy-dense triacylglycerols in vegetative biomass." *Plant Biotechnology Journal* 14, (2016): 661–669.

Zhang Ling, F. Y. Jing, F. P. Li, M. Li, Y. Wang, G. Wang, et al. "Development of transgenic *Artemisia annua* (Chinese wormwood) plants with an enhanced content of artemisinin, an effective anti-malarial drug, by hairpin-RNA-mediated gene silencing." *Biotechnology and Applied Biochemistry* 52, (2009): 199–207.

Zhu C, S Naqvi, J Breitenbach, et al. "Combinatorial genetic transformation generates a library of metabolic phenotypes for the carotenoid pathway in maize." *Proceeding of National Academy of Sciences USA* 105, (2008): 18232–18237.

Zhu C, G Sanahuja, D Yuan, et al. "Biofortification of plants with altered antioxidant content and composition: genetic engineering strategies." *Plant Biotechnology Journal* 11, (2013): 129–141.

9

Genome Editing for Crop Improvement

Rakesh Kumar Prajapat and Manas Mathur
Suresh Gyan Vihar University

Tarun Kumar Upadhyay
Parul Institute of Applied Sciences, Parul University

Dalpat Lal
Jagannath University

S.R. Maloo
Pacific University

Deepak Sharma
JECRC University

CONTENTS

9.1 Introduction

Most of the biological processes associated with genome amplification and expression such as replication, transcription, translation and epigenetic changes were established only after the discovery of the

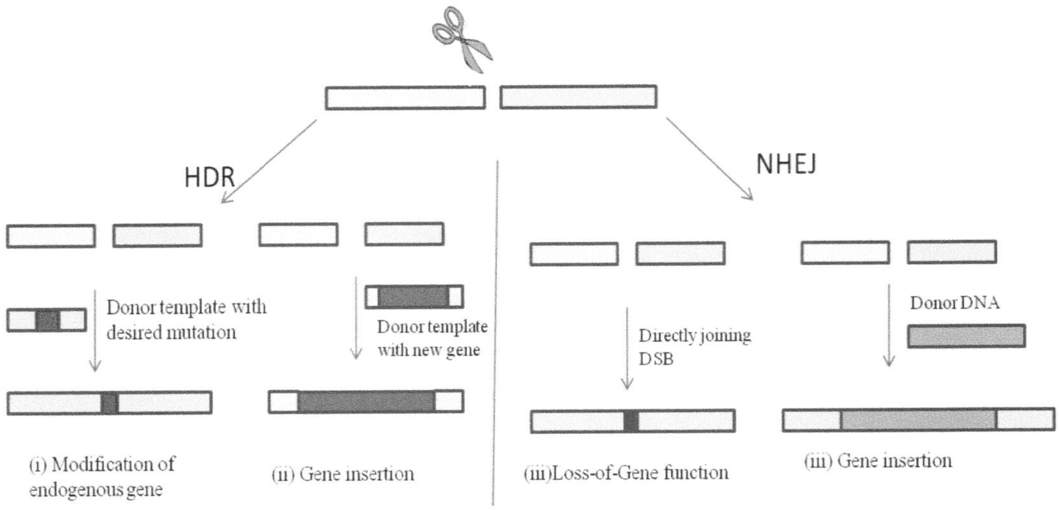

FIGURE 9.1 Repair of double-stranded DNA breaks (DSBs) through homology-directed repair (HDR) or non-homologous end joining (NHEJ).

DNA double helix in 1953. These processes were mediated through *in vitro* DNA manipulations, *in vitro* DNA synthesis, whole-genome sequencing and recombinant DNA technology. In recent times, reverse genetic approach is ideally used to analyse gene function through either gene knock-in or knockout by genome modification. Furthermore, RNA interference (RNAi) is also used to knockdown gene function (Gonczy et al. 2000; Dietzl et al. 2007; Martin and Caplen 2007) and various recombinase systems like Flp/FRT (Dymecki 1996) and Cre/loxP (Kilby et al. 1993) play an important role for gene targeting.

Genome editing refers to manipulating the nucleotide sequences of the genome with engineered nucleases. Genome editing by engineered nucleases (GEEN) acts as a valuable tool that explores artificial or edited engineered endonucleases (EENs) which induce the site-specific double-stranded DNA break (DSB) (Urnov et al. 2010). These EENs contain a DNA-binding domain or may have an RNA sequence, which identifies its template. The nucleases efficiently and exactly cleave the targeted DNA sequences. (Gaj et al. 2013; Carroll 2014). The DSBs in DNA subsequently induce the repair mechanisms of DNA leading to pave the way for gene edition at the particular sites. This DNA repair system of cells has two classes: One is non-homologous end joining (NHEJ), which is error-free, and the other is homology-directed repair (HDR) (Figure 9.1) (Wyman and Kanaar 2006).

Presently, four different genetically modified nuclease classes are accessible for editing genomes: zinc-finger nucleases (ZFNs), transcription activator-like effector nucleases (TALENs), engineered homing endonucleases or meganucleases (EMTs) and clustered regularly interspaced short palindromic repeat (CRISPR)/Cas. The CRISPR/Cas9 system is a very popular, friendly and commonly recommended tool in various living cells, including bacteria, yeast, animals and plants (Gasiunas et al. 2012; Cong et al. 2013; DiCarlo et al. 2013; Jinek et al. 2013; Sugano et al. 2014).

9.2 Zinc-Finger Nucleases (ZNFs)

The ZFNs are synthetic first-generation engineered nucleases that induce a break in the double-stranded DNA sequence at a unique site (Durai et al. 2005; Porteus and Carroll 2005; Carroll et al. 2006). These ZFNs are based on the working concept of the Cys2-His2 zinc-finger domains. The ZFN monomer is a fusion product of two domains: N-terminal made up of synthetic Cys2-His2 zinc-finger domain and C-terminal that contains a FokI DNA restriction enzyme, which induces nonspecific DNA cleavage. The individual monomer unit of Cys2-His2 domain consists of about 30 amino-acid residues that form ββα

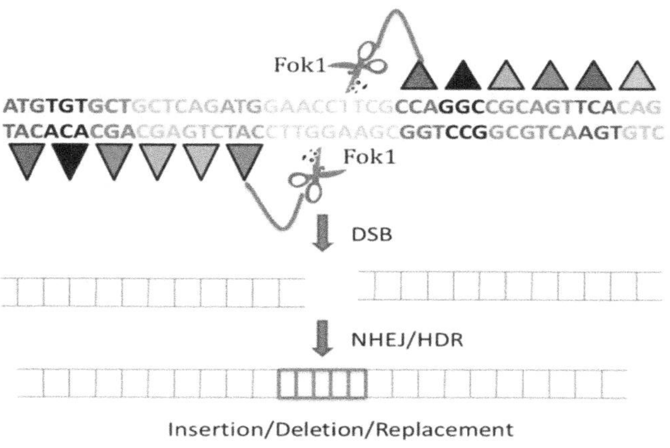

FIGURE 9.2 The structure and basic principle of ZNF-directed genomic modifications. DSB, double-stranded DNA break.

configuration. The α-helix of Cys2-His2 ZF protein facilitates binding of ZNFs to the major groove of the double helix DNA (Pavletich and Pabo 1991). The functional activation as enzyme of ZFN requires dimerisation of the FokI domain. This dimerisation of ZFN monomer interacts with target DNA in a series particular mode and induces the double-strand break. This interaction of ZFN monomer with target DNA is facilitated by flanking five to six nucleotides that allow the FokI dimer to induce the break. The characteristic domain of the zinc finger consists of three to four individual fingers like projections, which recognise three contiguous nucleotide bases. Therefore, a heterodimer of ZFN will identify either an18- or 24-nucleotide target sequence within a particular target location in the genomes (Figure 9.2). Several tools are available for the construction of novel ZFNs (Alwin et al. 2005; Scott 2005; Cathomen and Joun 2008g), for example, modular assembly utilises several web-based tools (Mandell and Barbas 2006; Sander 2007) for assembly of novel ZFNs. Today, these kinds of nucleases are developed by various companies like Sangamo BioSciences (Scott et al. 2005). Recently, an effective and robust 'open source' strategy was developed and named as the Oligomerized Pool ENgineering (OPEN) tool that includes a group of zinc-finger pools and *in vivo* assortment.

ZFNs have been successfully applied for gene manipulation in many animals and plants. However, constructing an active ZFN is a wide and time-consuming sorting process (Hsu and Zhang 2012). Further, ZFNs have a high tendency of off-target effect (Cathomen and Joung 2008) and harm the native cells. This limits the applicability of ZFNs for crop improvement.

9.3 Transcription Activator-Like Effector Nucleases (TALENs)

A new genome editing tool referred to as TALENs is a good alternative platform for ZFNs (Joung and Sander 2013). These transcription activator-like potent nucleases are based on the simple module for DNA recognition known as transcription activator-like (TAL) effectors (TALEs) which are concealed by pathogenic bacteria *Xanthomonas* spp. as a type III secretion system (Boch and Bonas 2010). TTALEs comprise a repeat of 33–35 amino acids of DNA binding domains, and each identifies a single base pair. The repeat-variable di-residues (RVDs) presented at the 12th and 13th position consist of two hypervariable amino acids that determine TALE binding specificity (Deng et al. 2012; Mak et al. 2012). There are four main RVDs, namely HD, NI, NG and NN, which are specific for the nucleotides C, A, T and G, respectively (Figure 9.3). To date many effector domains designed and fused to TALE include nucleases (Christian et al. 2010; Miller et al. 2011; Mussolino et al. 2011) and transcriptional activators (Miller et al. 2011; Zhang et al. 2011). TALE-DNA binding repeats recognise a single base that enables larger design elasticity compared with zinc-finger proteins. The TALE cloning array is a challenging job as it involves extensive identical repeat sequences. This can be overcome by Golden Gate mediated molecular cloning

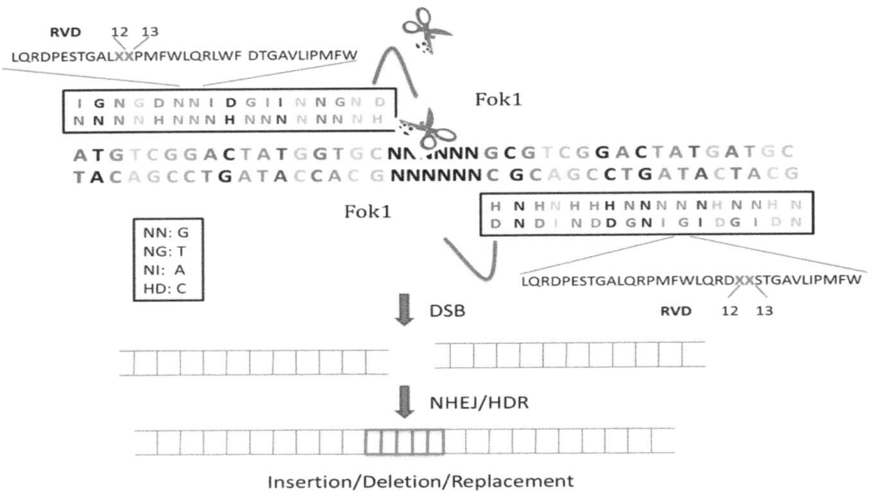

FIGURE 9.3 The structure and basic principle of TALEN-directed genomic modifications. DSB, double-stranded DNA break; HDR, homology-directed repair; NHEJ, non-homologous end joining; RVD, repeat-variable di-residue.

(Cermak et al. 2011) and by the high-throughput solid-phase congregation (Briggs et al. 2012; Reyon et al. 2012) and ligation-relied cloning (Schmid-Burgk 2013). Custom-TALE arrays can also be commercially ordered through many multinational companies like Transposagen Biopharmaceuticals (Lexington, Kentucky), Cellectis Bioresearch (Paris, France), and Life Technologies (Rockville, Maryland).

9.4 CRISPR/Cas

Earlier, the dominant genome editing tool was ZFNs (Kim et al. 1996) and TALENs (Christian et al. 2010) until 2013, before CRISPR/Cas9 came into the public interest (Table 9.1). The latest breakthrough in genome modification has relied on new engineered nucleases that are RNA guided, which has great advantages of being user friendly, simple, and flexible compared with the previous tools. The CRISPR's loci were discovered in the genome of *Escherichia coli* as an unusual sequence of 29 nucleotide

TABLE 9.1

Comparison between ZNF, TALEN and CRISPR/CAS

	ZFN	TALEN	CRISPR
Target DNA binding domain	Protein	Protein	RNA
Endonuclease activity	ZF/FokI catalytic domain	TALE/FokI catalytic domain	Guide RNA/Cas protein
Action mechanism	Target DNA sequence recognition by ZF proteins, cleavage by Fok I dimer, NHEJ or HDR-mediated repair	Target DNA sequence recognition by TALE proteins, cleavage by FokI dimer, NHEJ or HDR-mediated repair	Target DNA sequence recognition by guide RNA, cleavage by FokI dimer, NHEJ or HDR-mediated repair
Pros	Highly potent and unique	Highly potent and unique	Highly potent, simple to design and able to modify multiple sites concurrently
Cons	Large-scale selection, long duration and costly to design	Typical and long lasting to design	PAM motif next to target sequence needed

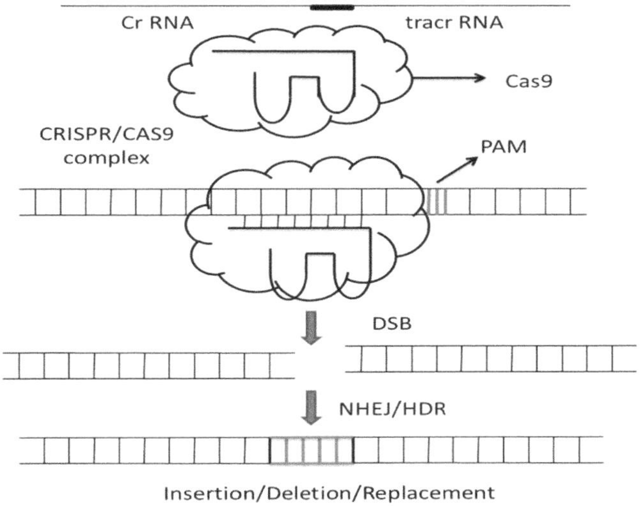

FIGURE 9.4 The structure and principle of CRISPR/CAS-mediated genomic modifications. Cr RNA, CRISPR RNA; DSB, double-stranded DNA break; HDR, homology-directed repair; NHEJ, non-homologous end joining; PAM, protospacer-adjacent motif; tracrRNA, *trans*-activating CRISPR RNA.

repeats which detached by exclusive 32 nucleotide sequences known as a 'spacer' sequence (Ishino et al. 1987; Wiedenheft et al. 2012). The CRISPR/Cas9 acts as an immune system in different bacterial cells like *Streptococcus pyogenes,* which was later developed as a genomic editing tool (Jinek et al. 2012). Prokaryotes developed this CRISPR/Cas as an adaptive immune system, which is a defensive mechanism against invading viruses, by degrading the unwanted DNA in a sequence-specific approach. This process requires the integration of viral DNA between two adjacent repeat motifs as spacers at CRISPR loci. The spacer is transcribed with CRISPR loci and subsequently amalgamated into inquisitive CRISPR RNAs (crRNAs) of nearly ˜40 nucleotides. Now, these crRNAs become associated with *trans*-activating CRISPR RNA (tracrRNA) and guide Cas9 nuclease after activating it (Barrangou et al. 2007) (Figure 9.4). Recent investigations suggested that the working principle of CRISPR's loci is similar to RNAi. Further, the conserved protospacer-adjacent motif (PAM) sequence contains sequence 5′-NGG-3′ and located downstream of target DNA induces DSBs in DNA (Gasiunas et al. 2012; Jinek et al. 2012), but infrequently (recognition sequence of PAM for CRISPR/CAS) NAG (Hsu et al. 2013). The PAM sequences also determine specificity for cleavage through the 'seed sequence', which consists of 12 nucleotides upstream of the PAM.

The present scenario of the growing world population, climate change and environmental pressure to accelerate breeding work with a high production is a major challenge. Scientists are searching the convenient and modern techniques for achieving food security by elevating nutrition for the growing human population. Plant molecular breeding and conventional tools are used to improve several crop varieties for different economic traits up to a certain extent. These methodologies are labour-intensive and time-consuming. Therefore, a new era of molecular biology through genome editing accelerated crop improvement programs by direct selection of economic traits in the field using high-throughput technologies. Genome editing tools are used to modify endogenous genes of different crops to improve a wide range of traits like yield, quality, nutritional value and disease resistance. These methods are precise, efficient and safe for the environment for modification of endogenous genes. The endogenous genes are modified through various site-specific nucleases at the desired location in the plant genome using genome editing tools. These tools have been significantly used for crop improvement either through modification or deletion of endogenous genes in several important crops like maize, wheat, rice, barley, soybean and sorghum. Summary of the genes targeted by the genome editing platform are listed in Table 9.2.

TABLE 9.2

List of Genes Targeted by Genome Editing Techniques for Crop Improvement

Crop	Genes	Genome Editing Methods	Target Traits	Reference
Rice	OsSWEET14	TALENs	Bacterial blight resistance	Li et al. (2012)
	OsQQR	ZFNs	Trait stacking	Cantos et al. (2014)
	OsBADH2	TALENs	Fragrant rice	Shan et al. (2013)
	LAZY1	CRISPR/Cas9	Tiller spreading	Miao et al. (2013)
	Gn1a, GS3, DEP1	CRISPR/Cas9	Enhanced grain number, larger grain size and dense erect panicles	Li et al. (2016)
	SBEIIb	CRISPR/Cas9	High amylose content	Sun et al. (2017)
	OsERF922	CRISPR/Cas9	Enhanced rice blast resistance	Wang et al. (2016)
	LOX3	TALEN	Enhanced storage tolerance	Ma et al. (2015)
	eIF4G	CRISPR-Cas9	Candidate rice tungro disease resistance gene	Macovei et al. (2018)
	BBM1	CRISPR-Cas9	Enables embryo formation from a fertilised egg	Khanday et al. (2019)
	Hd2, Hd4 and Hd5	CRISPR/Cas 9	Early maturity of rice varieties	Li et al. (2017)
	OsMATL	CRISPR/Cas9	Induction of haploid plants	Klap et al. (2017)
	ALS	CRISPR/Cas9	Herbicide resistance	Yao et al. (2017)
	ALS	CRISPR/Cas9	Herbicide resistance	Sun et al. (2016)
	OsNRAMP5	CRISPR/Cas 9	Low cadmium content	Tang et al. (2017)
	NRT1.1B	Base editing	Enhance nitrogen use efficiency	Lu and Zhu (2017)
	EPSPS	CRISPR/Cas9	Herbicide resistance	Li et al. (2016)
	ALS	CRISPR/Cas9	Herbicide resistance	Butt et al. (2017)
Maize	ZmIPK1	ZFNs	Herbicide tolerant and phytate reduced maize	Shukla et al. (2009)
	ZmMTL	TALENs	Induction of haploid plants	Kelliher et al. (2017)
	Wx1	CRISPR/Cas9	High amylopectin content	Pionee (2018)
Wheat	MLO	TALENs	Powdery mildew resistance	Wang et al. (2014)
	GW2	CRISPR/Cas9	Increased grain weight and protein content	Zhang et al. (2018)
	EDR1	CRISPR/Cas9	Powdery mildew resistance	Zhang et al. (2017a)
Soybean	FAD2-1A, FAD2-1B	TALENs	High oleic acid contents	Haun et al. (2014)
	FAD2-1A, FAD2-1B, FAD3A	TALENs	High oleic, low linoleic contents	Demorest et al. (2016)
Sugarcane	COMT	TALENs	Improved cell wall composition	Jung et al. 2016
	COMT	TALENs	Improved saccharification efficiency	Kannan et al. 2018
Tomato	ANT1	TALENs	Purple tomatoes with high anthocyanin	Čermák et al. 2014
	SP5G	CRISPR/Cas9	Earlier harvest time	Soyk et al. 2017
	SP, SP5G, CLV3, WUS, GGP1	CRISPR/Cas9	Tomato domestication	Li et al. 2018
	SlWUS	CRISPR/Cas9	Increase fruit size	Rodriguez-Leal et al. (2017)
Potato	Wx1	CRISPR/Cas9	High amylopectin content	Andersson et al. (2017)
	VInv	TALENs	Minimising reducing sugars	Clasen et al. (2016)
	GBSS	CRISPR/Cas9	High-amylopectin starch	Andersson et al. (2017)
Brassica oleracea	FRIGIDA	TALENs	Flowering earlier	Sun et al. (2013)
Camelina sativa	FAD2	CRISPR/Cas9	Decreased polyunsaturated fatty acids	Jiang et al. (2017)
Grapefruit	CsLOB1 promoter	CRISPR/Cas9	Alleviated citrus canker	Jia et al. (2016)
Mushroom	PPO	CRISPR/Cas9	Anti-browning phenotype	Waltz (2016)
Cucumber	eIF4E	CRISPR/Cas9	Virus resistance	Chandrasekaran et al. (2016)

9.5 Application of Genome Editing Tools for Crop Improvement

9.5.1 Bacterial Leaf Blight Resistance

This bacterial disease is a crop-devastating disease in rice triggered by *Xanthomonas oryzae* pv. *oryzae*, a key hazard for global food preservation. The microbe translocates and formulates its virulence TALEs, the type III effector proteins introduced into the host cell. TALEs bind to the promoter essentials of the host through effector binding elements resulting in the activation of the host gene for susceptibility to blight. TALEN has been recommended for interrupting the bacterial protein association arrangement in the promoter for deliberate resistance to fight bacterial blight (Li et al. 2012). Likewise, CRISPR-Cas9-based genome editing was also recommended to innovate by a mutation in rice lines with enhanced resistance to bacterial leaf blight.

9.5.2 Blast of Rice

This disease is triggered by the *Magnaporthe oryzae*. In recent scenarios, it is one of the most hazardous diseases in rice cultivation globally. CRISPR-Cas9 has been recommended to target *OsERF922*, a gene required for rice blast immunity. The beleaguered mutagenesis was effective, and a 42% mutation rate has been proved. The quantity of blast lesions was significantly decreased in the mutant lines compared with the non-treated plants (Wang et al. 2016).

9.5.3 Powdery Mildew Resistance Wheat

'S-genes' of the genome of the host are known as disease-prone genes, which can be bashed out by CRISPR/Cas9 and TALEN; the outcome of this innovation is sturdy disease-resistant crops. *MLO* genes in barley and wheat were found to be accountable for vulnerability to powdery mildew. This locus was a famous mark of mutation for researchers to discuss broad-spectrum immunity to powdery mildew in wheat and barley (Büschges et al. 1997). Since the *MLO* is incapable of stimulating disorder as a result, mildew spreading is cured from penetration into the cell wall.

9.5.4 Acrylamide-Free Potatoes

During the storage of potato, starch is changed into glucose and fructose. It caused problems in the processing of potatoes at high temperatures, reduced sugar and converted into dark-brown pigments and prominent cancer-causing agents (acrylamide types) while following the protocol. Voytas and his research team innovated TALEN constructs to mutate vascular invertase genes, which change sucrose to glucose and fructose in a potato tuber. They revealed that dispensation of genome-edited potatoes at increased temperatures followed by preservation at low temperatures showed less brown pigments and acrylamides compared with wild-type potatoes (Kim et al. 2015).

9.5.5 Non-Browning Apples

Browning of fresh-cut apples due to the enzymatic process is controlled by polyphenol oxidase (*PPO*) genes. A fresh-cut apple after some minutes loses hydrogen from polyphenol compounds and melanin formation, which is a brown colour pigment. Genetically modified apples were developed with an RNA interface (RNAi) construct to deactivate *PPO* genes in apples and initiated non-browning apples called 'arctic apples'. Similarly, *PPO* genes were mutated from the CRISPR/Cas system and generated non-browning apples.

9.5.6 Non-Browning Mushroom

The new-cut mushroom turns brown and black after some time due to an enzymatic response initiated by *PPO* genes. The knockdown in this gene achieved by genome editing can increase the shelf life.

9.5.7 Low Phytic Acid in Maize

Genome editing is recommended to secure the accretion of foreign metabolites, which possess negative effects on food and feed excellence and processing. Mustard and cabbage synthesise glucosinolates and have hazardous effects in adequate doses; cassava produces cyanide, which creates acute cyanide intoxication, goitre and partial paralysis or even ataxia. Similarly, maize possesses an adequate quantity of phosphorous in the kernels, and 75% of it is kept as phytic acid that is undigested in humans. Furthermore, phytic acid is an anti-nutritional compound that exerts a negative impact on nutrient uptake. ZFN designed to mutate the *IPK1* gene controls and coordinates the phytic acid biosynthesis route and it tremendously decreases the phytic acid concentration (Shukla et al. 2009).

9.5.8 Coffee without Caffeine

The method of eradicating caffeine from simple coffee beans is generally expensive, meanwhile synthesising toxic by-products may decrease or eradicate other flavours. RNAi constructs to deactivate the caffeine biosynthetic routes, such as xanthosine methyltransferase in *Coffea canephora*, called Robusta coffee. However, it must be described that these 'non-caffeine' coffee trees do not synthesise beans, as floral organ growth was desynchronised in the transgenic lines (Borrell 2012).

9.5.9 Waxy Corn

Maize is a maximum vital cereal crop, which is a rich source of protein for industrial feedstock, human foods and feedstuff. Maize also supplements variety of product types, like sweet corn, waxy corn and baby corn, which are very popular in consumer communities. The waxy gene encodes the enzyme granule bound starch synthase I (GBSS I) which is needed for amylose synthesis. Deletion through CRISPER Cas 9 closes down the biochemical pathway for amylose production and permits more than 96% amylopectin synthesis in maize endosperm, which has attracted a great deal of consumers. Along with that it has many broad-spectrum industrial applications (Shure et al. 1983).

9.5.10 Production of High Oleic/Low Linoleic Rice

The amount of oleic acid in rice bran oil would recover health and assist in the prevention of diseases. It is changed to linoleic acid by fatty acid desaturase 2 (*FAD2*) in plants. Rice possesses three functional *FAD2* genes, with *OsFAD2-1* displaying the maximum expression in seeds. The targeted mutation in *OsFAD2-1* via CRISPR-Cas9 leads to *OsFAD2-1* knockout rice plants with a twofold elevation in oleic acid quantity and no noticeable linoleic acid (Abe et al. 2018). Similarly, high oleic and less linoleic acid accumulated in soybean through TALEN to reduce the fatty acid desaturase genes.

9.5.11 Aroma in Rice

The aroma in rice is caused by the presence of the compound 2-acetyl-1-pyrroline (2AP) and its synthesis when betaine aldehyde dehydrogenase 2 (*Badh2*) is suppressed or non-functional. TALEN technology has been recommended to interrupt *Badh2*, resulting in an elevation in 2AP levels from 0.35 to 0.75 mg/kg. This quantity of 2AP is similar to the quantity in the aromatic rice variety recommended as a positive control (Shan et al. 2013). CRISPR-Cas9-based genome editing of *Badh2* was also recommended to develop fragrant rice where the first exon of *Badh2* was amended with an extraordinary base (T), which recommended an elevation in 2AP components (Shao et al. 2017).

9.5.12 Accumulation of β-Carotene

The enrichment of β-carotene is a vital goal for the advancement in rice variety. The *Osor* gene, an ortholog of the Orange (*Or*) gene, is a tool for β-carotene accretion in cauliflower, which was targeted in rice using CRISPR-Cas9, and outcomes possessed an increased level of β-carotene accrual in rice callus (Endo et al. 2019).

9.5.13 Red Rice

Red rice possesses large levels of healthy components like anthocyanins and proanthocyanidin 2 genes (*Rc, Rd*), engaged in deciding the red coloration of rice grains. The RcRd genotype of wild rice species, *Oryza rufipogon*, synthesises red pericarp tissue, whereas several varieties of raised rice synthesise white grains due to the mutation Rc. CRISPR-Cas9 was recommended to again revert the recessive Rc allele by changing the frameshift deletion into an in-frame mutation, thus transforming white rice varieties into red varieties (Zhu et al. 2019).

9.5.14 Herbicide Tolerance

Genome editing can induce herbicide-resistant crops. ZFN assisted particular DNA sequences into the target loci and familiarised a point mutation in the acetolactate synthase (*ALS*) gene, which synthesises imidazolinone and sulfonylurea. Furthermore, herbicide-resistant rice was innovated by using CRISPR/Cas9 to substitute the exon in the rice endogenous gene, 5-enolpyruvylshikimate-3-phosphate synthase (EPSPS), which has a 2-amino-acid replacement and deliberate immunity to the herbicide glyphosate (Li et al. 2016).

9.5.15 Improving Rice Yield

Grain yield is a typical feature for improvement, and its regulation by the number of genes is called quantitative trait loci (QTLs). In rice, productivity is mostly controlled by three major components: the number of grains per panicle, grain weight and the number of panicles per plant. Productivity has been amended by knocking out genes using CRISPR Cas9, including *Gn1a* (gene for grain number), *DEP1* (dense and erect panicle1; regulates ear head architecture) and *GS3* (regulates grain size), which are branded to be the negative controller of grain size and grain weight (Zhang et al. 2017b).

9.6 Conclusion and Future Perspectives

By genome editing, we can remove the incompatibility obstacles between species by assimilating foreign genes into target plant genomes or even acquainting *in vitro* genes to induce new varieties with required features. Currently, an innovation of new varieties by recombinant DNA technology has to use various means of risk calculation protocols before commercialisation which leads to tremendous increases in money and duration. Current investigations in genome editing technology deliver a more prominent and accurate route to either adjust endogenous genes or to insert a new gene. These methods can be subjugated to recover the economic recital of the trait by mutation breeding. Crop varieties designed through genome editing could be measured as non-transgenic, which might be more suitable than transgenic for consumer communities.

REFERENCES

Abe, K., Araki, E., Suzuki, Y., Toki, S., and Saika, H. 2018. Production of high oleic/low linoleic rice by genome editing. *Plant Physiol Biochem* 131: 58–62.

Alwin, S., Gere, M. B., Guhl, E., et al. 2005. Custom zinc-finger nucleases for use in human cells. *Mol Ther* 12: 610–617.

Andersson, M., Turesson, H., Nicolia, A., et al. 2017. Efficient targeted multiallelic mutagenesis in tetraploid potato (Solanum tuberosum) by transient CRISPR-Cas9 expression in protoplasts. *Plant Cell Rep* 36: 117–128.

Barrangou, R., Fremaux, C., Deveau, H., Richards, M., Boyaval, P., and Moineau S. 2007. CRISPR provides acquired resistance against viruses in prokaryotes. *Science* 315: 1709–1712.

Boch, J., and Bonas, U. 2010. Xanthomonas AvrBs3 family-type III effectors: discovery and function. *Annu Rev Phytopathol* 48: 419–436.

Borrell, B. 2012. Plant biotechnology: make it a decaf. *Nat News* 483: 264–266.

Briggs, A. W., Rios, X., Chari, R., et al. 2012. Iterative capped assembly: rapid and scalable synthesis of repeat-module DNA such as TAL effectors from individual monomers. *Nucleic Acids Res* 40: e117.

Büschges, R., Hollricher, K., Panstruga, R., et al. 1997. The barley *Mlo* gene: a novel control element of plant pathogen resistance. *Cell* 88: 695–705.

Butt, H., Eid, A., Ali, Z., et al. 2017. Efficient CRISPR/Cas9-mediated genome editing using a chimeric single-guide RNA molecule. *Front Plant Sci* 8: 1441.

Cantos, C., Francisco, P., Trijatmiko, K. R., Slamet-Loedin, I., and Chadha-Mohanty, P. K. 2014. Identification of "safe harbor" loci in indica rice genome by harnessing the property of zinc-finger nucleases to induce DNA damage and repair. *Fron Plant Sci* 5: 302.

Carroll, D. 2014. Genome engineering with targetable nucleases. *Annu Rev Biochem* 83: 409–439.

Carroll, D., Morton, J. J., Beumer, K. J., and Segal, D. J. 2006. Design, construction and *in vitro* testing of zinc finger nucleases. *Nat Protoc* 1: 1329–1341.

Cathomen, T., and Joung, J. K. 2008. Zinc-finger nucleases: the next generation emerges. *Mol Ther* 16: 1200–1207.

Cermak, T., Doyle, E. L., Christian, M., et al. 2011. Efficient design and assembly of custom TALEN and other TAL effector-based constructs for DNA targeting. *Nucleic Acids Res* 39: e82.

Cermak, T., Baltes, N. J., Cegan, R., Zhang, Y., Voytas, D. F. 2014. High-frequency, precise modification of the tomato genome. *Genome Biol* 16, 232.

Chandrasekaran, J., Brumin, M., Wolf, D., et al. 2016. Development of broad virus resistance in non-transgenic cucumber using CRISPR/Cas9 technology. *Mol Plant Pathol* 17: 1140–1153.

Christian, M., Cermak, T., Doyle, E. L., Schmidt, C., Zhang, F., and Hummel, A. 2010. Targeting DNA double-strand breaks with TAL effector nucleases. *Genetics* 186: 757–761.

Clasen, B. M., Stoddard, T. J., Luo, S., et al. 2015. Improving cold storage and processing traits in potato through targeted gene knockout. *Plant Biotechnol J* 14: 169–176.

Cong, L., Ran, F.A., Cox, D., Lin, S., Barretto, R., Habib, N. 2013. Multiplex genome engineering using CRISPR/Cas systems. *Science* 339: 819–823.

Demorest, Z. L., Coffman, A., Baltes, N. J., et al. 2016. Direct stacking of sequence-specific nuclease-induced mutations to produce high oleic and low linolenic soybean oil. *BMC Plant Biol.*16: 225.

Deng, D., Yan, C., Pan, X., et al. 2012. Structural basis for sequence-specific recognition of DNA by TAL effectors. *Science* 335: 720–723.

DiCarlo, J. E., Norville, J. E., Mali, P., Rios, X., Aach, J., and Church, G. M. 2013. Genome engineering in *Saccharomyces cerevisiae* using CRISPR–Cas systems. *Nucleic Acids Res* 41: 4336–4343.

Dietzl, G., Chen, D., Schnorrer, F., et al. 2007. A genome-wide transgenic RNAi library for conditional gene inactivation in Drosophila. *Nature* 448(7150): 151–156.

Durai, S., Mani, M., Kandavelou, K., Wu, J., Porteus, M. H., and Chandrasegaran, S. 2005. Zinc finger nucleases: custom-designed molecular scissors for genome engineering of plant and mammalian cells. *Nucleic Acids Res* 33: 5978–5990.

Dymecki, S. M. 1996. Flp recombinase promotes site-specific DNA recombination in embryonic stem cells and transgenic mice. *Proc Natl Acad Sci USA* 93(12): 6191–6196.

Endo, A., Saika, H., Takemura, M., Misawa, N., and Toki, S. 2019. A novel approach to carotenoid accumulation in rice callus by mimicking the cauliflower orange mutation via genome editing. *Rice* 12: 1–5.

Gaj, T., Gersbach, C. A., and Barbas, C. F. 2013. ZFN, TALEN, and CRISPR/Cas-based methods for genome engineering. *Trends Biotechnol* 31: 397–405.

Gasiunas, G., Barrangou, R., Horvath, P. and Siksnys, V. 2012. Cas9–crRNA ribonucleoprotein complex mediates specific DNA cleavage for adaptive immunity in bacteria. *Proc Natl Acad Sci USA* 109: 2579–2586.

Gonczy, P., Echeverri, C., and Oegema, K. 2000. Functional genomic analysis of cell division in *C. elegans* using RNAi of genes on chromosome III. *Nature* 408(6810): 331–336.

Greisman, H. A., and Pabo, C.O. 1997. A general strategy for selecting high-affinity zinc finger proteins for diverse DNA target sites. *Science* 275: 657–661.

Haun, W., Coffman, A., Clasen, B. M., et al., 2014. Improved soybean oil quality by targeted mutagenesis of the fatty acid desaturase 2 gene family. *Plant Biotechnol J* 12(7): 934–940.

Hsu, P. D., Scott, D. A., Weinstein, J. A., Ran, F. A., Konermann, S., and Agarwala, V. 2013. DNA targeting specificity of RNA-guided Cas9 nucleases. *Nat Biotechnol* 31: 827–832.

Hsu, P. D., and Zhang, F. 2012. Dissecting neural function using targeted genome engineering technologies. *ACS Chem Neurosci* 3: 603–610.

Isalan, M., and Choo, Y. 2001. Rapid, high-throughput engineering of sequence-specific zinc finger DNA-binding proteins. *Methods Enzymol* 340: 593–609.

Ishino, Y., Shinagawa, H., Makino, K., Amemura, M., and Nakata, A. 1987. Nucleotide sequence of the iap gene, responsible for alkaline phosphatase isozyme conversion in *Escherichia coli*, and identification of the gene product. *J Bacteriol* 169: 5429–5433.

Jia, H., Zhang, Y., Orbović, V., et al. 2016. Genome editing of the disease susceptibility gene CsLOB1 in citrus confers resistance to citrus canker. *Plant Biotechnol J* 15: 817–823.

Jiang, W. Z., Henry, I. M., Lynagh, P. G., Comai, L., Cahoon, E. B., Weeks, D. P. 2017. Significant enhancement of fatty acid composition in seeds of the allohexaploid, Camelina sativa, using CRISPR/Cas9 gene editing. *Plant Biotechnol J* 15: 648–657.

Jinek, M., Chylinski, K., Fonfara, I., Hauer, M., Doudna, J. A., and Charpentier, E. 2012. A programmable dual-RNA-guided DNA endonuclease in adaptive bacterial immunity. *Science* 337: 816–821.

Jinek, M., East, A., Cheng, A., Lin, S., Ma, E., and Doudna, J. 2013. RNA programmed genome editing in human cells. *eLife* 2: e00471.Joung, J. K., and Sander, J. D. 2013. TALENs: a widely applicable technology for targeted genome editing. *Mol Cell Biol* 14: 49–55.

Jung, J. H., and Altpeter, F. 2016. TALEN mediated targeted mutagenesis of the caffeic acid O-methyltransferase in highly polyploid sugarcane improves cell wall composition for production of bioethanol. *Plant Mol Biol* 92: 131–142.

Kannan, B., Jung, J. H., Moxley, G. W., Lee, S. M., and Altpeter, F. 2018. TALEN-mediated targeted mutagenesis of more than 100 COMT copies/alleles in highly polyploid sugarcane improves saccharification efficiency without compromising biomass yield. *Plant Biotechnol J* 16: 856–66.

Kelliher, T., Starr, D., Richbourg, L., et al. 2017. MATRILINEAL, a sperm-specific phospholipase, triggers maize haploid induction. *Nature* 542: 105–109.

Khanday, I., Skinner, D., Yang, B., Mercier, R., Sundaresan, V. A. 2019. Male-expressed rice embryogenic trigger redirected for asexual propagation through seeds. *Nature* 565: 91.

Kilby, N. J, Snaith, M. R., and Murray, J. A. 1993. Site-specific recombinases: tools for genome engineering. *Trends Genet* 9(12): 413–421.

Kim, H., Kim, S. T., Kim, S. G., and Kim, J. S. 2015. Targeted genome editing for crop improvement. *Plant Breed Biotech* 3(4): 283–290.

Kim, Y. G., Cha, J., and Chandrasegaran, S. 1996. Hybrid restriction enzymes: zinc finger fusions to Fok I cleavage domain. *Proc Natl Acad Sci USA* 93: 1156–1160.

Klap, C., Yeshayahou, E., Bolger, A. M., et al. 2017. Tomato facultative parthenocarpy results from SlAGAMOUS-LIKE 6 loss of function. *Plant Biotechnol J* 15(5): 634–647.

Li, J., Meng, X., Zong, Y., et al. 2016. Gene replacements and insertions in rice by intron targeting using CRISPR-Cas9. *Nat Plants* 12: 16139.

Li, T., Liu, B., Spalding, M. H., Weeks, D. P., and Yang, B. 2012. High-efficiency TALEN-based gene editing produces disease-resistant rice. *Nat Biotechnol* 30: 390–392.

Li, X., Zhou, W., Ren, Y., et al. 2017. High-efficiency breeding of early-maturing rice cultivars via CRISPR/Cas9-mediated genome editing. *J Genet Genom* 44(3): 175–178.

Li, C., Zong, Y., Wang, Y., et al. 2018. Expanded base editing in rice and wheat using a Cas9-adenosine deaminase fusion. *Genome Biol* 19: 59.

Lu, Y., Zhu, J.K. 2017. Precise editing of a target base in the rice genome using a modified CRISPR/Cas9 system. *Mol Plant* 10: 523–525.

Ma, X., Zhang, Q., Zhu, Q., et al. 2015. A robust CRISPR/Cas9 system for convenient, high-efficiency multiplex genome editing in monocot and dicot plants. *Mol Plant*8: 1274–1284.

Macovei, A., Sevilla, N. R., Cantos, C., Jonson, G.B., Slamet-Loedin, I., Čermák, T., Voytas, D. F., Choi, I. R., Chadha-Mohanty, P. 2018. Novel alleles of rice eIF4G generated by CRISPR/Cas9-targeted mutagenesis confer resistance to Rice tungro spherical virus. *Plant Biotechnol J* 16: 918–927.

Mak, A. N., Bradley, P., Cernadas, R. A., Bogdanove, A. J., and Stoddard, B. L. 2012. The crystal structure of TAL effector PthXo1 bound to its DNA target. *Science* 335: 716–719.

Mandell, J. G., and Barbas, C. F. 3rd. 2006. Zinc finger tools: custom DNA-binding domains for transcription factors and nucleases. *Nucleic Acids Res* 34: 516–523.

Martin, S. E., and Caplen, N. J. 2007. Applications of RNA interference in mammalian systems. *Annu Rev Genomics Hum Genet* 8: 81–108.

Miao, J., Guo, D., Zhang, J., et al. 2013 Targeted mutagenesis in rice using CRISPR-Cas system. *Cell Res* 23: 1233–1236.

Miller, J. C., Tan, S., Qiao, G., et al. 2011. A TALE nuclease architecture for efficient genome editing. *Nat Biotechnol* 29: 143–148.

Mussolino, C., Morbitzer, R., Lutge, F., Dannemann, N., Lahaye, T., and Cathomen, T. 2011. A novel TALE nuclease scaffold enables high genome editing activity in combination with low toxicity. *Nucleic Acids Res* 39: 9283–9293.

Pavletich, N. P., and Pabo, C. O. 1991. Zinc finger-DNA recognition: crystal structure of a Zif268- DNA complex at 2.1 A. *Science* 252: 809–817.

Porteus, M. H., and Carroll, D. 2005. Gene targeting using zinc finger nucleases. *Nat Biotechnol* 23: 967–973.

Reyon, D., Shengdar, Q. T., Cyd, K., Jennifer, A. F., Jeffry, D. S., and Keith, J. 2012. FLASH assembly of TALENs for high throughput genome editing. *Nat Biotechnol* 30: 460–465.

Rodriguez-Leal, D., Lemmon, Z. H., Man, J., Bartlett, M. E., Lippman, Z. B. 2017. Engineering quantitative trait variation for crop improvement by genome editing. *Cell* 171: 470–480.

Sander, J. D. 2007. Zinc finger targeter (ZiFiT): an engineered zinc finger/target site design tool. *Nucleic Acids Res* 35: 599–605.

Schmid-Burgk, J. L. 2013. A ligation-independent cloning technique for high-throughput assembly of transcription activator like effector genes. *Nat Biotechnol* 31: 76–81.

Scott, C. T. 2005. The zinc finger nuclease monopoly. *Nat Biotechnol* 23: 915–918.

Shan, Q., Wang, Y., Li, J., et al. 2013. Targeted genome modification of crop plants using a CRISPR-Cas system. *Nat Biotechnol* 686.

Shao, G., Xie, L., Jiao, G., et al. 2017. CRISPR/CAS9- mediated editing of the fragrant gene Badh2 in rice. *Chin J Rice Sci* 31: 216–222.

Shukla, V. K., Doyon, Y., Miller, J. C., et al. 2009. Precise genome modification in the crop species *Zea mays* using zinc-finger nucleases. *Nature* 459: 437–441.

Shure, M., Wessler, S., and Fedoroff, N. 1983. Molecular identification and isolation of the Waxy locus in maize. *Cell* 35: 225–233.

Soyk, S., Muller, N. A., Park, S. J., et al. 2017. Variation in the flowering gene SELF PRUNING 5G promotes day-neutrality and early yield in tomato. *Nat Genet* 49: 162–168.

Sugano, S. S., Shirakawa, M., Takagi, J., Matsuda, Y., Shimada, T., and Hara-Nishimura, I. 2014. CRISPR/Cas9-mediated targeted mutagenesis in the liverwort *Marchantia polymorpha* L. *Plant Cell Physiol* 55: 475–481.

Sun, Z., Li, N., Huang, G., 2013. Site-specific gene targeting using transcription activator-like effector (TALE)-based nuclease in *Brassica oleracea*. *J Integr Plant Biol* 55: 1092–1103.

Sun, Y., Li, J., & Xia, L. 2016. Precise genome modification via sequence-specific nucleases-mediated gene targeting for crop improvement. *Front Plant Sci* 7: 1928.

Sun, Y., Jiao, G., Liu, Z., et al. 2017. Generation of high-amylose rice through CRISPR/Cas9-mediated targeted mutagenesis of starch branching enzymes. *Front Plant Sci* 8: 298.

Tang, X., Lowder, L.G, Zhang, T., et al. 2017. A CRISPR-Cpf1 system for efficient genome editing and transcriptional repression in plants. *Nat Plants* 3: 17018.

Urnov, F. D., Rebar, E. J., Holmes, M. C., Zhang, H. S., and Gregory, P. D. 2010. Genome editing with engineered zinc finger nucleases. *Nat Rev Genet* 11: 636–646.

Voytas, D. F. 2013. Plant genome engineering with sequence-specific nucleases. *Annu Rev Plant Biol* 64: 327–350.

Waltz, E. 2016. Gene-edited CRISPR mushroom escapes US regulation. *Nature* 532: 293.

Wang, Y., Cheng, X., Shan, Q., et al. 2014. Simultaneous editing of three homoeoalleles in hexaploid bread wheat confers heritable resistance to powdery mildew. *Nat Biotechnol* 32: 947–951.

Wang, F., Wang, C., Liu, P., et al. 2016. Enhanced rice blast resistance by CRISPR/Cas9-targeted mutagenesis of the ERF transcription factor gene OsERF922. *PLoS One* 11: e0154027.

Wiedenheft, B., Sternberg, S. H., and Doudna, J. A. 2012. RNA-guided genetic silencing systems in bacteria and archaea. *Nature* 482: 331–338.

Wyman, C., and Kanaar, R. 2006. DNA double-strand break repair: all's well that ends well. *Annu Rev Genet* 40: 363–383.

Yao, L., Zhang, Y., Liu, C., et al. 2017. OsMATL mutation induces haploid seed formation in indica rice. *Nat Plants* 4: 530–3.

Zhang, F., Le, C., Simona, L., Sriram, K., George, M. C., and Paola, R. 2011. Efficient construction of sequence-specific TAL effectors for modulating mammalian transcription. *Nat Biotechnol* 29: 149–153.

Zhang, Y., Bai, Y., Wu, G., Zou, S., Chen, Y., Gao, C., Tang, D. 2017a. Simultaneous modification of three homoeologs of TaEDR1 by genome editing enhances powdery mildew resistance in wheat. *Plant J* 91: 714–724.

Zhang, H., Zhang, J., Lang, Z., Ramón, J. B., and Zhu, J. K. 2017b. Genome editing—principles and applications for functional genomics research and crop improvement. *Crit Rev Plant Sci* 36: 291–309.

Zhang, Y., Li, D., Zhang, D., et al. 2018. Analysis of the functions of TaGW2 homoeologs in wheat grain weight and protein content traits. *Plant J* 94: 857–866.

Zhu, Y., Lin, Y., Chen, S., et al. 2019. CRISPR/Cas9-mediated functional recovery of the recessive rc allele to develop red rice. *Plant Biotechnol J* 17: 2096–2105.

10

Molecular Marker-Assisted Breeding for Crop Improvement

Deepak Sharma
JECRC University

Rakesh Kumar Prajapat and Manas Mathur
Suresh Gyan Vihar University

Tarun Kumar Upadhyay
Parul Institute of Applied Sciences, Parul University

S.R. Maloo
Pacific University

Arunabh Joshi
RCA, MPUAT

Deepak Kumar Surolia
ICAR-CIAH

R.S. Dadarwal
CCSHAU

Nisha Khatik
Maharshi Dayanand Saraswati University

CONTENTS

10.1 Introduction

The recent discipline of plant breeding is a science and art which has utilised the available genetic variations by using a wide range of approaches, techniques and methods to enhance the quality and quantity of plants that are better tailored to human needs. In plant breeding, selection and variations are ancient approaches for the advancement of plants. Genetic diversity in available plant varieties is the primary source of the varietal identification, plant individuality, genetic mapping, development of upcoming breeding programs, identification of loci linked with different traits and effective exploitation of genetic resources by providing basic information of progeny performance and collection of superior parental compositions (Pandey et al. 2013).

The selection process can be direct or indirect; the direct selection methodology based on the phenotypic values or performance of target traits often takes several years for the transformation of a single genotype with desirable qualitative and quantitative traits. Plant breeding has proved an imperative position in crop improvement, but the major limitations of breeding methods are unpredictable variation in the environment besides several biotic and abiotic factors. Recent advancement in biotechnology helps to overcome the problem by introducing systematic, sustained, and reliable 'molecular marker' technology for indirect selection of target traits through linked molecular markers, which are not subjected to environmental variations, and their high-frequency number and high structural diversity allow the sequencing, detection and mapping of the targeted genes (Fu et al. 2017). Thus, this novel combination of modern molecular marker technology and ancient plant breeding practices augments novel possibilities of smart breeding and can be called 'molecular breeding' or 'marker-assisted breeding'(MAB) or 'marker-assisted selection' (MAS).

MAS, evaluated as a recent approach and a robust technology for crop improvement, is the first direct advantage of molecular markers in identifying varieties, protecting and maintaining genetic stock collections, preparing genetic and physical maps and determining genes with chromosomal location and the number of genes controlling traits (Nadeem et al. 2018; Platten et al. 2019). DNA-based molecular markers and indirect selections were applied in plant breeding programs in early 1980.Since then, several advanced molecular marker protocols have evolved and been used. MAS is useful in varietal and genomic selection, marker-based backcrossing, biotic and abiotic stress, germplasm characterisation, multiple gene pyramiding and phylogenetic analysis in addition to environmentally unaffected extremely precise varietal selection.

10.2 Molecular Markers in Crop Improvement

Molecular markers are heritable genetic characters that are used as markers or tags and identified by several forms of genes for the selected individuals among breeding populations. The highly developed molecular-assisted breeding technology utilises genomic DNA in addition to the newer class of DNA like mitochondrial and chloroplast-based microsatellites for exhibiting genetic variation (Sharma and Sharma 2018).

10.2.1 Important Properties of an Ideal Marker for MAS

- Easy and fast to recognise in all possible phenotypes
- Highly polymorphic

- Highly reproducible
- Co-dominant inheritance to allow differentiation between homozygous and heterozygous
- Evenly and frequently distributed throughout the genome
- Non-epistatic and minimum pleiotropic
- Easy in assay and availability
- Phenotypically neutral to environmental conditions and laboratory practices
- Produce exchangeable data between laboratories
- Cheap and reliable

10.3 Types of Markers

10.3.1 Classical Markers

10.3.1.1 Morphological Markers

Morphological markers can visually differentiate variations in character like size, shape, colour, surface, growth habits and other agronomic characters. These markers are straightforward to apply without any specific instrument and biochemical or molecular methodology. Breeders have successfully used morphological markers in the breeding program of several crops, viz., wheat, maize, rice, soybean and tomato for a long time and they are still relevant for genetic and breeding applications. The chief drawback of using these markers are that they are dominant in nature, are affected by plant growth phases, they exhibit pleiotropy and epistasis and they are highly influenced by environmental variations (Kadirvel et al. 2015).

10.3.1.2 Cytological Markers

Cytological markers are related to variations in banding pattern and structural features of chromosomes, viz., shape, size, numbers and position. They refer to the chromosomal banding pattern generated by contrasting stains like C, Q, G, Rand T banding. Different banding patterns unfold the variation in the distribution of euchromatin and heterochromatin (Grzywacz et al. 2019). These chromosome markers are useful in chromosomal characterisation, physical mapping and identification of linkage groups (Bharadwaj 2019).

10.3.1.3 Biochemical Markers

The enzymes that differ in amino acid sequence present in multiple forms and catalyse the same reaction in an individual and they are called 'isozymes'. These enzymes usually exhibit different charges, kinetic properties, heat stability and electrophoretic mobility. Since biochemical markers are co-dominant, cost-effective and easy to use, they have been effectively used in the estimation of phylogenetic analysis, evaluation of plant genetic assets, population structure, genetic diversity, population genetics and in developmental biology. The main weaknesses of biochemical marker are low polymorphism, low abundance and stability across tissue, organs, different plant growth stages and various extraction methods (Hazra et al. 2018; Nadeem et al. 2018; Dheer et al. 2020).

10.3.2 DNA Markers

DNA markers are DNA sequences with a well-known physical position on the chromosome and associated with a specific gene or feature that is used for the identification of individuals among and between species which follow a basic Mendelian pattern of genetics. Morphological, cytological and biochemical markers have some limitations, viz., low polymorphism, epistasis, pleiotropy, unusual occurrence, laborious procedure, instability and high environmental influence (Adhikari et al. 2017). Particularly in last three decades, the evolution of molecular markers is a landmark for the detection and utilisation of genetic assortment, and DNA polymorphism is most important. Molecular markers revolutionised expansion in the area of crop improvement, plant breeding and biotechnology since they are stable,

TABLE 10.1
List of Different DNA Markers Systems in Plants

Marker Technology	Acronym	Reference
Amplified fragment length polymorphism	AFLP	Liersch et al. (2019)
Cleaved amplified polymorphic sequence	CAPS	Jo et al. (2018)
Restriction fragment length polymorphism	RFLP	Maharajan et al. (2018)
Short tandem repeats	STR	Gymrek (2017)
Allele-specific polymerase chain reaction	AS-PCR	Bui et al. (2017)
Randomly amplified polymorphic DNA	RAPD	Balazova et al. (2016)
Expressed sequence tags	EST	Bushakra et al. (2015)
Arbitrarily primed polymerase chain reaction	AP-PCR	Singh and Singh (2015)
Simple sequence length polymorphism	SSLP	Passaro et al. (2017)
Simple sequence repeats	SSR	Jaiswal et al. (2017)
Sequence-tagged microsatellite sites	STMS	Caballo et al. (2018)
Sequence characterised amplified region	SCAR	Cho et al. (2020)
DNA amplification fingerprinting	DAF	Babu et al. (2014)
Inter-simple sequence repeats	ISSR	Gelotar et al. (2019)
Single nucleotide polymorphisms	SNP	Morgil et al. (2020)
Strand displacement amplification	SDA	Zhou et al. (2018b)
Random amplified microsatellite polymorphisms	RAMP	Cardona et al. (2018)
Selective amplification of microsatellite polymorphic loci	SAMPL	Alsamman et al. (2017)
Inter-retrotransposon amplified polymorphism	IRAP	Ghonaim et al. (2020)
Sequence-related amplified polymorphism	SRAP	Bhatt et al. (2017)
Diversity arrays technology	DArT	Mogga et al. (2018)
Start codon targeted polymorphism	SCoT	Etminan et al. (2016)
Inter-SINE amplified polymorphism	ISAP	Wenke et al. (2015)

highly polymorphic, cost-effective and evenly distributed throughout the genome of genetic assortment and DNA polymorphism. Molecular markers revolutionized expansion in the area of crop improvement, plant breeding and biotechnology since they are stable, highly polymorphic, cost-effective and evenly distributed throughout the genome (Grover and Sharma 2016). Several DNA-based marker techniques (Table 10.1) are versatile tools in the area of genome sequencing; genetic engineering; genetic mapping; MAB and taxonomical, evolutionary, phylogenetic and ecological research (Nadeem et al. 2018). The major characteristics of different available markers are listed in Table 10.2.

10.3.2.1 Hybridisation-Based DNA Marker

10.3.2.1.1 Restriction Fragment Length Polymorphism

DNA marker-based molecular biology began with the establishment of a molecular map of the human genome by using restriction fragment length polymorphism (RFLP) markers (Botstein et al. 1980). In RFLP, the restricted DNA is hybridised with a radioactively labelled, short, single copy of genomic DNA succeeded by agarose gel electrophoresis and Southern blotting and visualisation by autoradiography.

RFLP detect variation in individuals is caused by point mutation, DNA substitution and rearrangement, insertion/deletion, inversion and translocation (Yang et al. 2015; Grover and Sharma 2016). The restricted DNA is hybridised with a radioactively labelled, short, single copy of genomic DNA and visualised by using autoradiography after agarose gel electrophoresis and Southern blotting. RFLPs are co-dominant, informative, highly polymorphic, reproducible and reliable DNA markers for the detection of genetic variability among genotypes. Although RFLP was the first set of DNA markers used in the development of human genetic maps, they also have some limitations such as they are time-consuming; require high quality and quantity of DNA; involve expensive, toxic and radioactive reagents and require sequence information for probe generation (Chukwu et al. 2019).

TABLE 10.2

Molecular Markers Developed for Different Traits

Trait	Gene	Crop	Reference
Agronomic Traits			
Grain weight	*TaSus2-2B*	Wheat	Liu et al. (2012)
Semi-dwarf stature	*Rht-B1 andRht-D1*	Wheat	Zhang et al. (2006)
Grain weight	*TaGW2*	Wheat	Su et al. (2011)
Semi dwarf	*sd1*	Rice	Wang et al. (2009a)
Photoperiod response	*pms3 (p/tms12-1)*	Rice	Qi et al. (2017)
Plant height	*tb1*	Maize	Doebley et al. (1995)
Nodule formation	*Rj2 and Rfg1*	Soya bean	Yang et al. (2010)
Fruit size	*w2.2*	Tomato	Nesbitt and Tanksley (2002)
Male sterility	*ms3*	Capsicum	Naresh et al. (2018)
Flowering response	*BoFLC1.C9*	Cabbage	Abuyusuf et al. (2019)
Male sterility	*CDMs399-3*	Cabbage	Chen et al. (2013)
Quality Traits			
Yellow pigment content	*Psy1*	Wheat	He et al. (2008)
LMW-gluten	*Glu-B3*	Wheat	Wang et al. (2009b)
Yellow pigment content	*TaZds-D1*	Wheat	Zhang et al. (2011)
Lipoxygenase content	*Talox-B1*	Wheat	Geng et al. (2012)
Low glutenin content	*Lgc1*	Rice	Chen et al. (2010)
Fe and Zn content	*OsNAS3, OsNRAMP1*	Rice	Anuradha et al. (2012)
Amylose content	*Wx-in*	Rice	Zhou et al. (2018a)
Oil content	*DGAT1-2*	Maize	Chai et al. (2012)
Provitamin A	*ZmcrtRB3*	Maize	Zhou et al. (2012)
Sweetness	*sugary1*	Maize	Chhabra et al. (2019)
Provitamin A	*crtRB1 and LcyE*	Maize	Obeng-Bio et al. (2019)
Soluble acid invertase	*SAI-1*	Sorghum	Liu et al. (2014)
Fragrance	*SbBADH2*	Sorghum	Zanan et al. (2016)
Flavonoids	*AgFNSI*	Celery	Yan et al. (2020)
Anthocyanin content	*VfTTG1*	Faba bean	Gutierrez and Torres (2019)
β-Carotene and flesh	*QA/QC*	Sweet potato	Gemenet et al. (2020)
Carotenoids	*b_CHY-1*	Sweet potato	Arizio et al. (2014)
Biotic Stress			
Powdery mildew	*Pm3*	Wheat	Tommasini et al. (2006)
Stem rust resistance	*Sr45*	Wheat	Periyannan et al. (2014)
Bacterial blight resistance	*Xa3*	Rice	Hur et al. (2013)
Bacterial blight resistance	*xa5*	Rice	Hajira et al. (2016)
Bacterial blight resistance	*Xa38*	Rice	Ellur et al. (2016)
Brown plant hopper resistance	*Bph14*	Rice	Zhou et al. (2013)
Blast resistance	*Pit*	Rice	Hayashi et al. (2010)
Blast resistance	*Pi35*	Rice	Ma et al. (2015)
Tomato yellow leaf curl virus	*ACY*	Tomato	Nevame et al. (2018)
Bacterial wilt	*Bwr-6, Bwr-12*	Tomato	Kim et al. (2018)
Fusarium wilt	*Frl*	Tomato	Devran et al. (2018)
Fusarium wilt	*Fom 1*	Melon	Oumouloud et al. (2015)
Powdery mildew resistance	*Pm-2 F*	Melon	Zhang et al. (2012)
Leaf scald resistance	*Rpf*	Sugarcane	Gutierrez et al. (2018)
Powdery mildew	*Pm-s*	Cucumber	Liu et al. (2017)
Cauliflower mosaic virus	*cmv6.1*	Cucumber	Shi et al. (2018)
Powdery mildew resistance	*InDel 1*	Capsicum	Karna and Ahn (2018)

(Continued)

TABLE 10.2
(Continued)

Trait	Gene	Crop	Reference
Powdery mildew resistance	*er1-7*	Pea	Sun et al. (2016)
Mungbean yellow mosaic virus	*YR4*	Mung bean	Maiti et al. (2011)
Resistance to weevils	*VrPDF1*	Mung bean	Lan et al. (2017)
Abiotic Stress			
Dehydration tolerance	*TaMYB2*	Wheat	Garg et al. (2012)
Drought stress tolerance	*TaAQP*	Wheat	Pandey et al. (2013)
Drought stress tolerance	*DREB1*	Wheat	Huseynova (2018)
Salt tolerance	*TtASR1*	Wheat	Hamdi et al. (2020)
Tolerance to phosphorus (P) deficiency	*Pup 1*	Rice	Chin et al. (2011)
Drought stress tolerance	*OsSAPK2*	Rice	Lou et al. (2017)
Drought tolerance	*MYBE1*	Maize	Assenov et al. (2013)
Aluminium stress tolerance	*SbMATE*	Sorghum	Too et al. (2018)
Salinity tolerance	*Salt indexQTL 1*	Field pea	Leonforte et al. (2013)

Source: From Salgotra and Stewart (2020).

10.3.2.2 PCR-Based DNA Marker

The discovery of polymerase chain reaction (PCR) by Kary Mullis was important and revolutionary in the area of molecular biology; it allows amplification of a small quantity of DNA *in vitro* without any living system. PCR-based DNA markers are more rapid and require a small quantity of DNA. A PCR mixture uses a DNA template, thermostable *Taq* polymerase, dNTPs and a suitable reaction buffer followed by a thermal cycling procedure of denaturation, annealing and extension results in the commensurate escalating amplification of template DNA (Adhikari et al. 2017).

10.3.2.2.1 Random Amplified Polymorphic DNA

Random amplified polymorphic DNA (RAPD) is a modified simple PCR technique introduced by Williams et al. (1990). It detects variation in DNA between individuals by applying a single, short, 10-bp random primer with a minimum 50–60% GC content. Cost-effective RAPD analysis can be performed by using a universal non-specific primer in a few hours with a small amount of DNA, non-radioactive assay and without prior sequence information of the primer (Adhikari et al. 2017). However, the low reproducibility and incapacity to differentiate heterozygous and homozygous individuals are the major limitations for RAPD suitability in plant breeding programs.

10.3.2.2.2 Amplified Fragment Length Polymorphism

Amplified fragment length polymorphism (AFLP) techniques generate discriminatory PCR amplification of the DNA restriction sequence. The AFLP marker was introduced (Vos et al. 1995) because of the limitations of RAPD and RFLP techniques, in which PCR is performed after the digestion of DNA. AFLP is a robust technique to study genetic linkages by analysing individuals from segregating populations since it shows high polymorphism, rapid generation, high reproducibility and is easy to apply to all organisms without any preceding information of the sequence. The major limitations of ALFP are the dominant mode of inheritance, cost and requirement of technical expertise and radioactive assay, which makes it less preferable in MAS (Liersch et al. 2019).

10.3.2.2.3 Inter-Simple Sequence Repeat

Inter-simple sequence repeat (ISSR) is an attractive marker technology introduced by Zietkiewicz et al. (1994).It uses 16- to 25-bp simple sequence repeats (SSRs) as a primer to amplify mainly inter-SSR regions. The ISSR technique is simple, quick and with high genomic abundance and variability, which makes it popular for genetic diversity studies among individuals as well as other types of studies. These markers show variations in mode of inheritance, locus specificity, economic investment and technical requirement (Omondi et al. 2016; Gelotar et al. 2019).

10.3.2.2.4 Simple Sequence Repeat

SSRs are also called microsatellite and short tandem repeats (STRs); they represent variations in the repetitive sequence in the specific region of the genome (Litt and Luty 1989). SSRs are microsatellite sequence-based markers that use specific primers flanking simple repeats consisting of 1–5 nucleotides. Due to reliability, co-dominant inheritance, robustness, high genomic abundance, high reproducibility, high polymorphism and locus specificity, this technique is ideally suited for genetic mapping and population studies. However, the SSR technique also has some limitations such as high start-up cost, skilled and labour-intensive marker development and requirement of nucleotide information for primer designing. This marker is the first choice in the plant breeding program and has proved to be extensively attractive in all DNA-based markers and has been successfully used several times in quantitative trait loci (QTL) mapping, development of linkage maps, MAS and germplasm characterisation (Jiang 2015).

10.3.2.2.5 Single Nucleotide Polymorphism

Single nucleotide polymorphism (SNP) refers to the difference in a single nucleotide of individuals of a population caused by point mutation, results in transition, transversion and deletion/addition (Brooks 1999). SNPs mainly reside in non-coding genomic regions and are abundant in plants with a frequency of 1 in 100–300 bp. SNPs are the definitive form of DNA markers with the smallest unit of inheritance providing the simplest and maximum number of markers. For SNP markers, several different gel-based genotyping techniques have been introduced for analysing the product of allelic discrimination available. High genomic abundance, co-dominance and availability of high-throughput technologies of genotyping such as allele-specific PCR, next-generation sequencing and gene detection by selection makes SNPs the most efficient markers for genotyping, gene isolation, construction of a genetic map and MAB (Adhikari et al. 2017; Nadeem et al. 2018).

10.3.2.2.6 Diversity Array Technology

Diversity array technology (Wenzl et al. 2004) is based on microarray hybridisation that allows mapping of several polymorphic sequences throughout the genome and facilitates the generation of whole-genome fingerprints in a single assay. The method is cost-effective and highly reproducible and the low requirement of DNA makes it useful in gene tagging, molecular mapping studies and so forth.

10.4 Marker-Assisted Selection

MAS, also referred to as MAB, is a novel strategy in molecular biotechnology that uses a DNA banding pattern for the selection of unique plants with desired characters. MAS is based on the same principle concept as plant breeding in selecting specific genotypes more efficiently for crop improvement. This technology uses DNA-based specific sequence/markers, which are tightly linked to the targeted gene responsible for the expression of a unique phenotype (Ramalingam et al. 2017).

10.4.1 Prerequisite of Marker-Assisted Selection

DNA-based molecular MAB requires more complex machinery, facilities and skilled personnel compared with traditional breeding. The specific requirements mandatory for MAS are given below (Jiang 2015):

1. **Suitable marker system:** A suitable and reliable marker system is critically important in MAS. As discussed, the marker should be polymorphic, reproducible, cost-effective, evenly distributed throughout the genome and have co-dominant inheritance.

2. **Marker trait association:** For a successful MAS practice, the marker should be closely linked with genes responsible for the trait. This information can be obtained by mutant, linkage and recombination analysis; QTL analysis; gene mapping and genome association studies.
3. **Data processing and management:** In MAB, scientists have to access an adequate quantity of samples for detection with different markers at the same instance. So, appropriate and functional reports and analysis can be produced by rapid and proficient data management.
4. **Rapid DNA extraction and marker detection:** Working with a broad-scale population of plants for investigation by several markers, breeders need a rapid DNA extraction and marker validation system.
5. **Financial viability:** Compared with conventional breeding, MAS requires additional costs to establish a MAS lab equipped with necessary equipment and kits such as PCR thermal cycler, gel electrophoresis unit and gel documentation system. Therefore, the cost of vital chemicals and equipment is crucial in reference to net return.

10.4.2 Procedure of Marker-Assisted Selection

MAS involves the following important steps for the identification and characterisation of QTLs (Francia et al. 2005; Nadeem et al. 2018):

1. **Selection and development of the breeding population:** Selection of an appropriate parent is an imperative primary step in which homozygous parents of diverse origin and contrasting characters should be chosen and crossed up to F_2 generation for producing progeny with adequate segregation.
2. **Phenotyping and genotyping:** Young seedlings should be used for the isolation of quality DNA from each individual subject to PCR and marker analysis followed by separation of scoring of PCR products by agarose gel electrophoresis.
3. **QTL mapping and QTL validation:** Selection of plants or individuals by connecting QTL and character/phenotype using available marker data.
4. **Marker validation:** For marker validation, the abovementioned practices should be repeated and results can be applied for crop improvement.

10.5 Application of Marker-Assisted Selection in Plant Breeding and Crop Improvement

MAS overcomes the problem breeders are facing in conventional breeding, that is, the selection of phenotype indirectly based on the DNA banding pattern of individuals. For breeders, MAS has proven the most powerful tool for using DNA-based markers in the identification of unique loci in segregating progenies of different generations. In molecular breeding, MAS can be used for marker-assisted recurrent selection, germplasm evaluation and QTL mapping, gene pyramiding and genomic selection for crop improvement (Ramalingam et al. 2017; Salgotra and Stewart 2020).

10.5.1 MAS in Germplasm Characterisation

Plant genetic resources and information of genetic diversity are fundamental material for quality enhancement, increased resistance and disease and pest resistance. MAS facilitates development and improves plant breeding programs of available germplasm such as wild varieties, traditional cultivars and landraces by their screening and characterisation more efficiently and accurately (Nadeem et al. 2018). Development of several functional markers identified for a particular trait affords a broad genetic base and assists the identification and evaluation of varieties and hybrids and can be successfully used in the development of new agronomically important cultivars and collection of new plant genetic resources (Salgotra and Stewart 2020).

10.5.2 MAS in Agronomic Trait, Qualitative Traits and Stress Resistance

Recently, molecular markers are also applied for the enhancement of different agronomic and biotic-abiotic stress resistance traits in different crops by marker-assisted backcrossing since they efficiently discriminate alleles governing characteristics. MAS has been fruitfully applied in increasing the food value of various cereals and vegetables (Lau et al. 2015). A variety of disease-resistant genes also have been identified and incorporated into selected cultivars with the help of molecular breeding to achieve enhanced protection against pests and diseases (Randhawa et al. 2019). These DNA markers are successfully applied for the improvement of different crops like wheat, rice, maize, barley and soya bean (Table 10.2) (Salgotra and Stewart 2020).

10.5.3 MAS in Gene Pyramiding

Gene pyramiding is a technique of integrating several genes from diverse donor parents into a single genotype. Marker-assisted gene pyramiding is an attractive option mainly applied to enhance the resistance of crops against disease and insects by outsourcing two or more foreign genes governing different traits. Gene pyramiding by conventional method requires a great deal of time and it is very difficult to phenotypically screen a single plant having more than one gene. Pyramiding of numerous genes or QTLs by using DNA-based functional markers impart broad-spectrum resistance with high stability and has been promoted as a robust technique to improve quantitative traits and abiotic stress in rice, wheat, maize and barley (Jiang 2015; Kage et al. 2016; Chukwu et al. 2019).

10.5.4 MAS in Genomic Selection

Genomic selection is another application of MAS for the synchronised selection of up to thousands of markers from whole-genomes associated with at least one character. Genomic selection technique applies high-density marker with high accuracy to overcome the requirement of identification of individual QTL and marker associations since detection and testing of a separate set of markers are not required in genomic selection (Heffner et al. 2009; Desta and Ortiz 2014). Genomic selection involves the assumption of genomic estimated breeding values (GEBVs) by using phenotyping and genotyping of the training population generated by diverse genotypes. Genomic selection is successfully applied in wheat, maize and brassica (Gorjanc et al. 2016)

10.6 Conclusion and Future Prospects

In the last three decades, molecular MAB has witnessed an incredible evolution and steady improvement in the various crops of economic importance. Various functional molecular marker assays have been innovated along with continuous progress in plant biotechnology. MAB is established as a powerful tool for genetic manipulation of crops for crop improvement by means of agronomic traits, qualitative and quantitative traits and biotic and abiotic stress resistance. MAB can be efficiently used in germplasm characterisation, QTL mapping, gene pyramiding, genetic diversity and evolutionary and phylogenetic studies. Association of MAS with next-generation sequencing, cisgenetics, epigenetics and CRISPR technology for genome editing can lead to establishing a new platform of low-cost high-throughput crop improvement in the coming years. In the future, innovations in cost-effective and more precise molecular breeding are likely to be seen for the precise and quick development of new potent plant varieties by effective incorporation of novel traits and improvement in economically important plants.

REFERENCES

Abuyusuf, M., Nath, U. K., Kim, H., Islam, R. M., Park, J. I. and Nou, I. S.2019. Molecular markers based on sequence variation in *BoFLC1.C9* for characterizing early- and late-flowering cabbage genotypes. *BMC Genet.* 20: 42.

Adhikari, S., Saha, S., Biswas, A., Rana, T. S., Bandyopadhyay, T. K. and Ghosh, P. 2017. Application of molecular markers in plant genome analysis: a review. *Nucleus* 60: 283–97. doi: 10.1007/s13237-017-0214-7.

Alsamman, M., Alsamman, S. S., Adawy, S. D., Ibrahim, B. A. and Hussein, E. H. A. 2017. Selective Amplification of Start codon Polymorphic Loci (SASPL): a new PCR-based molecular marker in olive. *Plant Omics J.* 10(2): 64–77. doi: 10.21475/poj.10.02.17.pne385.

Anuradha, K., Agarwal, S., Rao, Y. V., Rao, K. V., Viraktamath, B. C. and Sarla, N. 2012. Mapping QTLs and candidate genes for iron and zinc concentrations in unpolished rice of Madhukar × Swarna RILs. *Gene* 508: 233–240.

Arizio, C. M., Costa-Tártara, S. M. and Manifesto, M. M. 2014. Carotenoids gene markers for sweet potato (*Ipomoea batatas* L. Lam): Applications in genetic mapping, diversity evaluation and cross-species transference. *Mol. Genet. Genom.* 289: 237–251.

Assenov, B., Andjelkovic, V., Ignjatovic-Micic, D. and Pagnotta, M. A. 2013. Identification of SNP mutations in *MYBF-1* gene involved in drought stress tolerance in maize. *Bulg. J. Agric. Sci.* 19: 181–85.

Babu, K. N., Rajesh, M. K., Samsudeen, K. et al. 2014. Randomly amplified polymorphic DNA (RAPD) and derived techniques. In: Besse P. (ed), *Molecular Plant Taxonomy. Methods in Molecular Biology (Methods and Protocols)*, vol 1115. Humana Press, Totowa, NJ.

Balazova, Z., Vivodik, M. and Galova, Z. 2016. Evaluation of molecular diversity of central European maize cultivars. *Emirates J. Food Agric.* 28(2): 93–98. doi: 10.9755/ejfa.2015.05.204.

Bharadwaj, D. N. 2019. *Advanced Molecular Plant Breeding: Meeting the Challenge of Food Security.* Apple Academic Press, Florida. doi:10.1201/b22473.

Bhatt, J., Kumar, S., Patel, S. and Solanki, R. 2017. Sequence-related amplified polymorphism (SRAP) markers based genetic diversity analysis of cumin genotypes. *Ann. Agrian Sci.* 15: 434–438. doi: 10.1016/j.aasci.2017.09.001.

Botstein, D., White, R. L., Skolnick, M. and Davis, R. W. 1980. Construction of genetic linkage map using restriction fragment length polymorphisms. *Am. J. Hum. Genet.* 32: 314–331.

Bui, T. G. T., Hoa, N. T. L., Yen, J. and Schafleitner, R. 2017. PCR-based assays for validation of single nucleotide polymorphism markers in rice and mungbean. *Hereditas* 154:3. doi: 10.1186/s41065-016-0024-y.

Bushakra, J. M., Lewers, K. S., Staton, M. E., Zhebentyayeva, T. and Saski, C. A. 2015. Developing expressed sequence tag libraries and the discovery of simple sequence repeat markers for two species of raspberry (*Rubus* L.). *BMC Plant Biol.* 15: 258. doi: 10.1186/s12870-015-0629-8.

Caballo, C., Castro, P., Gil, J., Izquierdo, I., Millan, T., and Rubio, J. 2018. STMS (sequence tagged microsatellite site) molecular markers as a valuable tool to confirm controlled crosses in chickpea (*Cicer arietinum* L.) breeding programs. *Euphytica* 214(12): 231.

Cardona, C. C. C., Coronado, Y. M., Conronado, A. C. M. and Ochoa, I. 2018. Genetic diversity in oil palm (*Elaeis guineensis* Jacq) using RAM (Random Amplified Microsatellites). *Bragantia* 77 (4): 546–556.

Chai, Y., Hao, X., Yang, X. et al. 2012. Validation of DGAT1-2 polymorphisms associated with oil content and development of functional markers for molecular breeding of high-oil maize. *Mol. Breeding* 29: 939–949.

Chen, C., Zhuang, M., Fang, Z. et al. 2013. A co-dominant marker *BoE332* applied to marker-assisted selection of homozygous male-sterile plants in cabbage (*Brassica oleracea var. capitata* L.). *J. Integr. Agric.* 12: 596–602.

Chen, T., Meng-xiang, T., Zhang, Y. et al. 2010. Development of simple functional markers for low glutelin content gene 1 (*Lgc1*) in rice (*Oryza sativa*). *Rice Sci.* 17: 173–178.

Chhabra, R., Hossain, F., Muthusamy, V., Baveja, A., Mehta, B. K. and Zunjare, R. U. 2019. Development and validation of breeder-friendly functional markers of sugary1 gene encoding starch-debranching enzyme affecting kernel sweetness in maize (*Zea mays*). *Crop Pasture Sci.* 70: 868–875.

Chin, J. H., Gamuyao, R., Dalid, C. et al. 2011. Developing rice with high yield under phosphorus deficiency: *Pup1* sequence to application. *Plant Physiol.* 156: 1202–1216.

Cho, K. H., Kwack, Y., Park, S. J. et al. 2020. Sequence-characterized amplified region markers and multiplex-polymerase chain reaction assays for kiwifruit cultivar identification. *Horticult. Environ.Biotechnol.* 61:395–406. doi: 10.1007/s13580-020-00227-9.

Chukwu, S. C., Rafii, M. Y., Ramlee, S. I. et al. 2019 Marker-assisted selection and gene pyramiding for resistance to bacterial leaf blight disease of rice (*Oryza sativa* L.). *Biotechnol. Biotechnologic. Equip.*33 (1): 440–455.doi: 10.1080/13102818.2019.1584054.

Desta, Z. A. and Ortiz, R. 2014. Genomic selection: genome-wide prediction in plant improvement. *Trends Plant Sci.* 19: 592–601.

Devran, Z., Kahveci, E., Hong, Y., Studholme, D. J. and Tor, M. 2018. Identifying molecular markers suitable for *Frl* selection in tomato breeding. *Theor. Appl. Genet.* 131: 2099–2105.

Dheer, P., Rautela, I., Sharma, V. et al. 2020. Evolution in crop improvement approaches and future prospects of molecular markers to CRISPR/Cas9 system. *Gene* 753: 144795. doi: 10.1016/j.gene.2020.144795.

Doebley, J., Stec, A. and Gustus, C. 1995. Teosinte branched1 and the origin of maize: Evidence for epistasis and the evolution of dominance. *Genetics* 141: 333–346.

Ellur, R. K., Khanna, A., Gopala-Krishnan, S. et al. 2016. Marker-aided incorporation of *Xa38*, a novel bacterial blight resistance gene, in PB1121 and comparison of its resistance spectrum with *xa13* + *Xa21*. *Sci. Rep.* 6: 29188.

Etminan, A., Aboughadareh, A.P., Mohammadi, R. et al. 2016. Applicability of start codon targeted (SCoT) and inter-simple sequence repeat (ISSR) markers for genetic diversity analysis in durum wheat genotypes. *Biotechnol. Biotechnologic. Equip.* 30(6): 1075–1081. doi: 10.1080/13102818.2016.1228478.

Francia, E., Tacconi, G., Crosatti, C. et al. 2005. Marker assisted selection in crop plants. *Plant Cell, Tissue and Organ Culture.* 82: 317–342.

Fu, Y. B., Yang, M. H., Zeng, F. and Biligetu, B. 2017. Searching for an accurate marker-based prediction of an individual quantitative trait in molecular plant breeding. *Front. Plant Sci.* 8: 1182. doi: 10.3389/fpls.2017.01182.

Garg, B., Lata, C. and Prasad, M. 2012. A study of the role of gene *TaMYB2* and an associated SNP in dehydration tolerance in common wheat. *Mol. Biol. Rep.* 39: 10865–10871.

Gelotar, M. J., Dharajiya, D. T., Solanki, S. D., Prajapati, N. N. and Tiwari, K. K. 2019. Genetic diversity analysis and molecular characterization of grain amaranth genotypes using inter simple sequence repeat (ISSR) markers. *Bull. Natl. Res. Cent.* 43: 103. doi: 10.1186/s42269-019-0146-2.

Gemenet, D. C., Kitavi, M. N. and David, M. 2020. Development of diagnostic SNP markers for quality assurance and control in sweet potato [*Ipomoea batatas* (L.) Lam.] breeding programs. *PLoS One* 15: e0232173.

Geng, H., Xia, X., Zhang, L., Qu, Y. and He, Z. 2012. Development of functional markers for Lipoxygenase gene *Talox-B1* on chromosome 4 BS in common wheat. *Crop Sci.* 52: 568–576.

Ghonaim, M., Kalendar, R., Barakat, H., Elsherif, N., Ashry, N. and Schulman, A. H. 2020. High-throughput retrotransposon-based genetic diversity of maize germplasm assessment and analysis. *Mol. Biol. Rep.* 47:1589–603. doi: 10.1007/s11033-020-05246-4.

Gorjanc, G., Jenko, J., Hearne, S. J. and Hickey, J. M. 2016. Initiating maize pre-breeding programs using genomic selection to harness polygenic variation from landrace populations. *BMC Genome* 17: 30. doi:10.1186/s12864-015-2345-z.

Grover, A. and Sharma, P. C. 2016. Development and use of molecular markers: past and present. *Crit. Rev. Biotechnol.* 36(2): 290–302. doi: 10.3109/07388551.2014.959891.

Grzywacz, B., Tatsuta, H., Bugrov, A. G. et al. 2019. Cytogenetic markers reveal a reinforcement of variation in the tension zone between chromosome races in the brachypterous grasshopper *Podisma sapporensis* Shir. on Hokkaido Island. *Sci. Rep.* 9:16860. doi: 10.1038/s41598-019-53416-7.

Gutierrez, A. F., Hoy, J. W., Kinbeng, C. A. and Baisakh, N.2018. Identification of genomic regions controlling leaf scald resistance in sugarcane using a bi-parental mapping population and selective genotyping by sequencing. *Front. Plant Sci.* 9: 877.

Gutierrez, N. and Torres, A. M. 2019. Characterization and diagnostic marker for *TTG1* regulating tannin and anthocyanin biosynthesis in faba bean. *Sci. Rep.* 9: 16174.

Gymrek, M. 2017. A genomic view of short tandem repeats. *Curr. Opin. Genet. Dev.* 44: 9–16. doi:10.1016/j.gde.2017.01.012.

Hajira, S. K., Sundaram, R. M., Laha, G. S. et al. 2016. A single-tube, functional marker-based multiplex PCR assay for simultaneous detection of major bacterial blight resistance genes *Xa21*, *xa13* and *xa5* in rice. *Rice Sci.* 23: 144–151.

Hamdi, K., Brini, F., Kharrat, N., Masmoudi, K. and Yakoubi, I. 2020. Abscisic acid, stress, and ripening (*TtASR1*) gene as a functional marker for salt tolerance in durum wheat. *BioMed. Res. Int.* doi: 10.1155/2020/7876357.

Hayashi, K., Yasuda, N., Fujita, Y., Koizumi, S. and Yoshida, H. 2010. Identification of the blast resistance gene Pit in rice cultivars using functional markers. *Theor. Appl. Genet.* 121: 1357–67.

Hazra, A., Dasgupta, N., Sengupta, C. and Das, S. 2018. Next generation crop improvement program: progress and prospect in tea (*Camellia sinensis* (L.) O. Kuntze). *Ann. Agrarian Sci.* 16 (2): 128–35. doi: 10.1016/j.aasci.2018.02.002.

He, X., Zhang, Y., He, Z. et al.2008. Characterization of phytoene synthase 1 gene (*Psy1*) located on common wheat chromosome 7A and development of a functional marker. *Theor. Appl. Genet.* 116: 213–221.

Heffner, E. L., Sorrells, M. E. and Jannink, J. L.2009. Genomic selection for crop improvement. *Crop Sci.* 49: 1–12.

Hur, Y. J., Jeung, J., Kim, S. Y. et al.2013. Functional markers for bacterial blight resistance gene *Xa3* in rice. *Mol. Breeding* 31: 981–985.

Huseynova, I. M.2018. Application of PCR-based functional markers for identification of *DREB1* genes in *Triticum aestivum* L. *SF Biotechnol. Bioeng. J.* 1: 1.

Jaiswal, S., Sheoran, S., Arora, V. et al. 2017. Putative microsatellite DNA marker-based wheat genomic resource for varietal improvement and management. *Front. Plant Sci.* 8: 2009. doi: 10.3389/fpls.2017.02009.

Jiang, G. L. 2015. Molecular marker-assisted breeding: A plant breeder's review. In: Al-Khayri J., Jain S., Johnson D. (eds), *Advances in Plant Breeding Strategies: Breeding, Biotechnology and Molecular Tools*. Springer, Cham, Switzerland. doi: 10.1007/978-3-319-22521-0_15.

Jo, I. H., Sung, J., Hong, C. E., Raveendar, S., Bang, K. H. and Chung, J. W. 2018. Development of cleaved amplified polymorphic sequence (CAPS) and high-resolution melting (HRM) markers from the chloroplast genome of *Glycyrrhiza* species. *3 Biotech* 8 (5): 220. doi: 10.1007/s13205-018-1245-8.

Kadirvel, P., Senthilvel, S., Geethanjali, S., Sujatha, M. and Varaprasad, K.S. 2015. Genetic markers, trait mapping and marker-assisted selection in plant breeding. In: Bahadur B., Venkat Rajam M., Sahijram L., Krishnamurthy K. (eds), *Plant Biology and Biotechnology*. Springer, New Delhi.

Kage, U., Kumar, A., Dhokane, D., Karre, S. and Kushalappa, A. C. 2016. Functional molecular markers for crop improvement. *Crit. Rev. Biotechnol.* 36: 917–930.

Karna, S. and Ahn, Y. K. 2018. Development of *InDel* markers to identify Capsicum disease resistance using whole genome resequencing. *J. Plant Biotechnol.* 45: 228–235.

Kim, B., Hwang, I. S., Lee, H. J. et al. 2018. Identification of a molecular marker tightly linked to bacterial wilt resistance in tomato by genome-wide SNP analysis. *Theor. Appl. Genet.* 131: 1017–1030.

Lan, N. T. N., Thao, H. T., Son, L. V. and Mau, C. H. 2017. Overexpression of *VrPDF1* gene confers resistance to weevils in transgenic mung bean plants. *Peer J Preprints*. doi:10.7287/PEERJ.PREPRINTS.3264V2.

Lau, W. C. P., Rafii, M. Y., Ismail, M. R., Puteh, A., Latif, M. A. and Ramli, A. 2015. Review of functional markers for improving cooking, eating, and the nutritional qualities of rice. *Front. Plant Sci.* 6: 1–11.

Leonforte, A., Sudheesh, S., Cogan, N. O. et al. 2013. SNP marker discovery, linkage map construction and identification of QTLs for enhanced salinity tolerance in field pea (*Pisum sativum* L.). *BMC Plant Biol.* 13: 161.

Liersch, A., Bocianowski, J., Popławska, W. et al. 2019. Creation of gene pools with amplified fragment length polymorphism markers for development of winter oilseed rape (*Brassica napus* L.) hybrid cultivars. *Euphytica* 215:22. doi:10.1007/s10681-019-2350-4.

Litt, M. and Luty, J. A.1989. A hypervariable microsatellite revealed by *in vitro* amplification of a dinucleotide repeat within the cardiac-muscle actin gene. *Am. J. Hum. Genet.* 44: 397–401.

Liu, P. N., Miao, H., Lu, H. W. et al.2017. Molecular mapping and candidate gene analysis for resistance to powdery mildew in *Cucumis sativus* stem. *Genet. Mol. Res.* 16: 16039680.

Liu, Y., He, Z. H., Appels, R., Xia, X. C.2012. Functional markers in wheat: current status and future prospects. *Theor. Appl. Genet.* 125:1–10.

Liu, Y., Nie, Y. D., Han, F. X. et al.2014. Allelic variation of a soluble acid invertase gene (*SAI-1*) and development of a functional marker in sweet sorghum (*Sorghum bicolor* (L.). *Mol. Breeding* 33: 721–730.

Lou, D., Wang, H., Liang, G. and Yu, D.2017. *OsSAPK2* Confers abscisic acid sensitivity and tolerance to drought stress in rice. *Front. Plant Sci.* 8: 993.

Ma, J., Ma, X. D., Zhao, Z. C. et al.2015. Development and application of a functional marker of the blast resistance gene *Pi35* in rice. *Acta Agron. Sin.* 41: 1779–1790.

Maharajan, T., Ceasar, S. A., Krishana, A. T. P. et al. 2018. Utilization of molecular markers for improving the phosphorus efficiency in crop plants. *Plant Breeding* 137: 10–26. doi:10.1111/pbr.12537.

Maiti, S., Basak, J., Kundagrami, S., Kundu, A. and Pal, A. 2011. Molecular marker-assisted genotyping of mungbean yellow mosaic India virus resistant germplasms of mungbean and urd bean. *Mol. Biotechnol.* 47: 95–104.

Mogga, M., Sibiya, J., Shimelis, H., Lamo, J. and Yao, N. 2018. Diversity analysis and genome-wide association studies of grain shape and eating quality traits in rice (*Oryza sativa* L.) using DArT markers. *PLoS One* 13(6): e0198012. doi: 10.1371/journal.pone.0198012.

Morgil, H., Gercek, Y. C. and Tulum, I. 2020. Single nucleotide polymorphisms (SNPs) in pant genetics and breeding, the recent topics in genetic polymorphisms. *Intech Open*. doi: 10.5772/intechopen.91886.

Nadeem, M. A., Nawaz, M. A., Shahid, M. Q. et al. 2018. DNA molecular markers in plant breeding: current status and recent advancements in genomic selection and genome editing. *Biotechnol. Biotechnologic. Equip.* 32(2): 261–85. doi: 10.1080/13102818.2017.1400401.

Naresh, P., Lin, S., Lin, *C.* et al. 2018. Molecular markers associated to two non-allelic genic male sterility genes in peppers (*Capsicum annuum* L.). *Front. Plant Sci.* 9: 1343.

Nesbitt, T. C. and Tanksley, S. D. 2002. Comparative sequencing in the genus *Lycopersicon*. Implications for the evolution of fruit size in the domestication of cultivated tomatoes. *Genetics* 162: 365–379.

Nevame, A. Y. M., Xia, L., Nchongboh, C. G. et al. 2018. Development of a new molecular marker for the resistance to tomato yellow leaf curl virus. *BioMed Res. Int.* 1: 1–10.

Obeng-Bio, E., Badu-Apraku, B., Elorhor Ifie, B., Danquah, A., Blay, E. T., Dadzie, M. A. 2019. Phenotypic characterization and validation of provitamin A functional genes in early maturing provitamin A-quality protein maize (*Zea mays*) inbred lines. *Plant Breeding* 139: 575–588.

Omondi, E. O., Debener, T., Linde, M., Abukutsa-Onyango, M., Dinssa, F. F. and Winkelmann, T. 2016. Molecular markers for genetic diversity studies in African leafy vegetables. *Adv. Biosci. Biotechnol.* 7(3): 188–197.

Oumouloud, A., Otmani, M. E. and Alvarez, J. 2015. Molecular characterization of Fom-1 gene and development of functional markers for molecular breeding of resistance to *Fusarium* race 2 in melon. *Euphytica* 205: 491–501.

Pandey, B., Sharma, P., Pandey, D., Sharma, I. and Chatrath, R. 2013. Identification of new aquaporin genes and single nucleotide polymorphism in bread wheat. *Evol. Bioinform.* 9: 437–452.

Passaro, M., Geuna, F., Bassi, D. et al. 2017. Development of a high-resolution melting approach for reliable and cost-effective genotyping of PPVres locus in apricot (*P. armeniaca*). *Mol. Breeding* 37:74. doi: 10.1007/s11032-017-0666-0.

Periyannan, S., Bansal, U., Bariana, H. et al. 2014. Identification of a robust molecular marker for the detection of the stem rust resistance gene *Sr45* in common wheat. *Theor. Appl. Genet.* 127: 947–955.

Platten, J. D., Cobb, J. N. and Zantua, R. E. 2019. Criteria for evaluating molecular markers: Comprehensive quality metrics to improve marker-assisted selection. *PLoS One* 14(1): e0210529. doi: 10.1371/journal.pone.0210529.

Qi, Y., Wang, L., Gui, J., Zhang, L., Liu, Q. and Wang, J. 2017. Development and validation of a functional co-dominant SNP marker for the photoperiod thermo-sensitive genic male sterility *pms3 (p/tms12-1)* gene in rice. *Breeding Sci.* 67: 535–539.

Ramalingam, J., Savitha, P., Alagarasan, G., Sarawathi, R. and Chandrababu, R. 2017. Functional marker assisted improvement of stable cytoplasmic male sterile lines of rice for bacterial blight resistance. *Front. Plant Sci.* 8: 1–9. doi:10.3389/fpls.2017.01131.

Randhawa, M. S., Bains, N. S., Sohu, V. S. et al. 2019. Marker assisted transfer of stripe rust and stem rust resistance genes into four wheat cultivars. *Agronomy* 9: 497.

Salgotra, R. K. and Stewart, C. N. 2020. Functional Markers for Precision Plant Breeding. *Int. J. Mol. Sci.* 21. doi:10.3390/ijms21134792.

Sharma, S. and Sharma, A. 2018. Molecular markers based plant breeding. *Adv. Res.*16(1): 1–15. doi: 10.9734/AIR/2018/42922.

Shi, L., Yang, Y., Xie, Q. et al. 2018. Inheritance and QTL mapping of cucumber mosaic virus resistance in cucumber. *PLoS One* 13: e0200571.

Singh, B. D. and Singh, A. K. 2015. Polymerase chain reaction-based markers. In: *Marker-Assisted Plant Breeding: Principles and Practices*. Springer, New Delhi. doi:10.1007/978-81-322-2316-0_3.

Su, Z., Hao, C., Wang, L., Dong, Y. and Zhang, X. 2011. Identification and development of a functional marker of *TaGW2* associated with grain weight in bread wheat (*T. aestivum* L.). *Theor. Appl. Genet.* 122: 211–223.

Sun, S., Deng, D., Wang, Z. et al. 2016. A novel *er1* allele and the development and validation of its functional marker for breeding pea (*Pisum sativum* L.) resistance to powdery mildew. *Theor. Appl. Genet.* 129: 909–919.

Tommasini, L., Yahiaoui, N., Srichumpa, P. and Keller, B. 2006. Development of functional markers specific for seven *Pm3* resistance alleles and their validation in the bread wheat gene pool. *Theor. Appl. Genet.* 114: 165–175.

Too, E. J., Onkware, A. O., Were, B. A., Gudu, S., Carlsson, A. and Geleta, M. 2018. Molecular markers associated with aluminium tolerance in *Sorghum bicolor. Hereditas* 155: 20.

Vos, P., Hogers, R., Bleeker, M. et al. 1995. AFLP: a new technique for DNA fingerprinting. *Nucleic Acids Res.* 23: 4407–4414.

Wang, J. Nakazaki, T., Chen, S. et al.2009a. Identification and characterization of the erect-pose panicle gene EP conferring high grain yield in rice (*Oryza sativa* L.). *Theor. Appl. Genet.* 119: 85–91.

Wang, L., Zhao, X., He, Z. et al. 2009b. Characterization of low molecular-weight glutenin subunit Glu-B3 genes and development of STS markers in common wheat (*T. aestivum* L.). *Theor. Appl. Genet.* 118: 525–539.

Wenke, T., Seibt, K. M., Dobel, T., Muders, K. and Schmidt, T. 2015. Inter-SINE amplified polymorphism (ISAP) for rapid and robust plant genotyping. In: Batley J. (ed), *Plant Genotyping. Methods in Molecular Biology (Methods and Protocols)*, vol 1245. Humana Press, New York.

Wenzl, P., Carling, J., Kudrna, D. et al. 2004. Diversity Arrays Technology (DArT) for whole-genome profiling of barley. *Proc. Natl. Acad. Sci. USA* 101: 9915–9920.

Williams, J. G. K., Kubelik, A. R., Livak, K. J., Rafalski, J. A. and Tingey, S. 1990. DNA polymorphism amplified by arbitrary primers are useful as genetic markers. *Nucl. Aci. Res.* 18: 6537–6535.

Yan, J., He, L., Xu, S. et al. 2020. Expression analysis, functional marker development and verification of *AgFNSI* in celery. *Sci. Rep.* 10: 531.

Yang, H. B., Kang, W.H., Nahm, S. H. and Kang, B.C. 2015. Methods for developing molecular markers. In: *Current Technologies in Plant Molecular Breeding*, Chapter 2, Springer, Dordrecht, pp. 15–50. doi:10.1007/978-94-017-9996-6_2.

Yang, S., Tang, F., Gao, M., Krishnan, H. B. and Zhu, H.2010. R gene-controlled host specificity in the legume–rhizobia symbiosis. *Proc. Natl. Acad. Sci. USA* 107: 18735–18740.

Zanan, R., Khandagale, K., Hinge, V., Elangovan, M., Henry, R. J. and Nadaf, A. 2016. Characterization of fragrance in sorghum (*Sorghum bicolor* L. Moench) grain and development of a gene-based marker for selection in breeding. *Mol. Breeding* 36:146. doi: 10.1007/s11032-016-0582-8.

Zhang, C., Dong, C., He, X., Zhang, L., Xia, X. and Zhonghu, H. 2011. Allelic variants at the *TaZds-D1* locus on wheat chromosome 2DL and their association with yellow pigment content. *Crop Sci.* 51: 1580–1590.

Zhang, C., Ren, Y., Guo, S. et al. 2012. Application of comparative genomics in developing markers tightly linked to the *Pm-2F* gene for powdery mildew resistance in melon (*Cucumis melo* L.). *Euphytica* 190: 157–168.

Zhang, X., Yang, S., Zhou, Y., He, Z. and Xia, X.2006. Distribution of the *Rht-B1b, Rht-D1b* and *Rht8* reduced height genes in autumn-sown Chinese wheats detected by molecular markers. *Euphytica* 152: 109–116.

Zhou, L., Chen, S., Yang, G. et al. 2018a. A perfect functional marker for the gene of intermediate amylose content Wx-in in rice (*Oryza sativa* L.). *Crop Breed. Appl. Biotechnol.* 18: 103–109.

Zhou, L., Chen, Z., Lang, X. et al. 2013. Development and validation of a PCR-based functional marker system for the brown planthopper resistance gene *Bph14* in rice. *Breed. Sci.* 63: 347–352.

Zhou, W., Hu, L., Ying, L., Zhao, Z., Chu, P.K. and Yu, X. F.2018b. A CRISPR–Cas9-triggered strand displacement amplification method for ultrasensitive DNA detection. *Nat. Commun.* 9:5012. doi:10.1038/s41467-018-07324-5.

Zhou, Y., Han, Y., Li, Z. et al. 2012. *ZmcrtRB3* encodes a carotenoid hydroxylase that affects the accumulation of a-carotene in maize kernel. *J. Integr. Plant Biol.* 54: 260–269.

Zietkiewicz, E., Rafalski, A. and Labuda, D. 1994. Genome fingerprinting by simple sequence repeat (SSR) – anchored polymerase chain reaction amplification. *Genomics* 20: 176–183.

11

Biotechnological Advances and Advanced Mutation Breeding Techniques for Crop Plants

Priyanka Sood, Saurabh Pandey and Manoj Prasad
National Institute of Plant Genome Research

CONTENTS

11.1 Introduction

Climatic fluctuations and population explosion have a great impact on crop yield and demand for food. There is a requirement to improve the existing plant characteristics like yield and stress tolerance to adapt to the changing environmental conditions and attain sustainable global food security

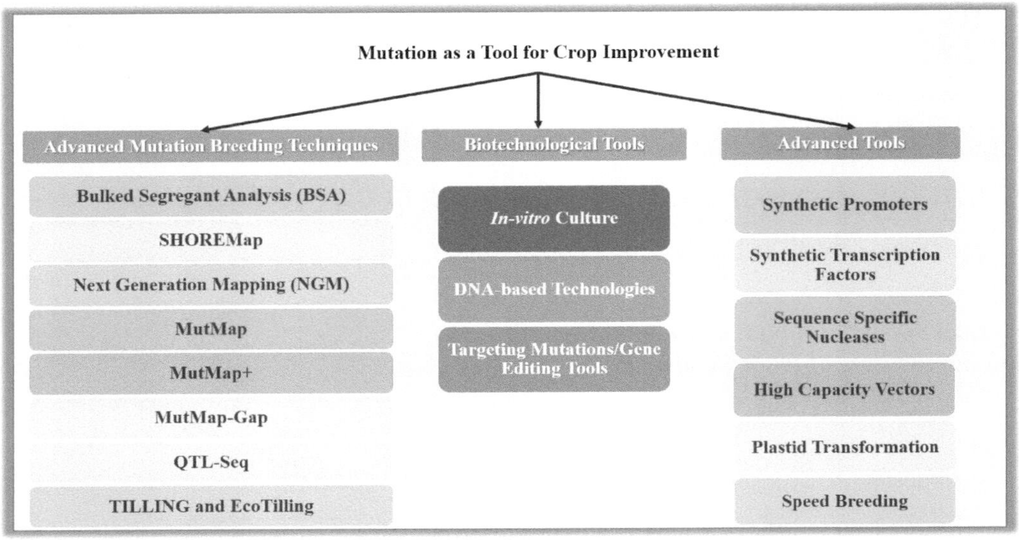

FIGURE 11.1 Schematic diagram for mutation methods used in crop improvement.

(Wollenweber et al. 2005). Genetic improvement of crops has historically led to an enormous gain in yields. However, search for the more advanced, convenient and faster biotechnology-based crop improvement technologies is an incessant journey for researchers, which could help in addressing the new world challenges (Qaim 2020). Advanced biotechnological approaches such as the use of synthetically made promoters, enhancers and repressors for native or transgene expression and regulation, multigene transfer by artificially synthesized chromosomes, site-specific integration and specifically regulated gene expression, precise genome editing, genomics-assisted mutational breeding and next-generation mapping (NGM) are potential tools for addressing agro-environmental issues. Variations in plants are created due to mutations; thus, for the generation of diverse crop populations for crop improvement relies on the generation and selection of useful mutations. This aspect has also been covered with a special focus on mutation breeding and advanced biotechnological tools (Figure 11.1).

In this chapter, we discuss the advances and next-generation tools for trait mapping, precise regulation of gene expression in plants followed by tools for multigene engineering, targeted genome modification and speed breeding (SB). Next-generation transgenic technologies like the use of engineered nucleases for precise genome editing, synthetic biology tools and components could act as essential components of the crop improvement toolbox.

11.2 Historical Perspective

The use of mutation for crop improvement dates back to 300 BC, and it used to be based on spontaneous mutations up to the start of the 19th century (Kharkwal 2012). In the year 1894, Bateson published the book *Materials for the Study of Plant Variations*, which laid the foundation of conceptualization of the use of mutations for crop improvement. Mutagens such as X-rays, β-rays and γ-rays, which led to the generation of mutations, were discovered in the early 20th century. In 1901, Hugo de Vries coined the term mutation to describe sudden heritable changes of existing traits. In the later years, he suggested the artificial induction of mutation by radiation. The first mutational experiments on living beings were performed by T. H. Morgan in the year 1909 with *Drosophila melanogaster*. Further, N. I. Vavilov's 'law of homologous series of variation' led to the foundation of the proof of induced mutations and the development of mutant varieties. In the year 1926, N. I. Vavilov proposed the concept of 'centres of origin' based on genetic diversity. The next year, C. Stuart Gager and A. F. Blakeslee reported initiation of mutations in *Datura stramonium,* which was later confirmed with H. J. Muller's mutation induction experiments by X-rays demonstrating the likelihood of gaining genetically superior plants by inducing mutations. Stadler's pioneering work on mutation induction in barley and maize by use of X-rays in the subsequent years established radiation as a resource for mutation breeding in plants (Stadler 1930). The 'Chlorina' was the first commercial mutant variety developed via X-ray irradiation by D. Tollenar in tobacco and released in Indonesia during 1934 to 1938. In the year 1937, the colchicine effect of doubling the chromosome was discovered in plants. Detection of transposons in maize by Barbara McClintock in the year 1951 was a pioneering work that later found its use in plant mutation breeding. This led to the identification of different categories of mutagens and the concept of physical and chemical mutagenesis. Large-scale application of mutation breeding started in the latter half of the 20th century with government-backed projects and internationally coordinated mutation breeding research programs. These coordinated efforts at various institutes led to the Green Revolution with the development of short stature, fertiliser-responsive varieties of rice and wheat by Norman Borlaug (Kharkwal 2012; Van Harten 1998). The advent of modern biotechnological tools and cost-effective sequencing technologies, integrated with mutation breeding and genomics, has opened a new area in the field of crop improvement.

11.3 Advances in Breeding Techniques

11.3.1 Bulked Segregant Analysis

Bulked segregant analysis (BSA) is a concept of mapping created by Michelmore in the year 1991 for simply inherited traits. It is based on the creation of two isogenic lines from the contrasting set of bulks from a cross between two genotypes distinct for a trait of interest (Michelmore et al. 1991). The polymorphic marker associated with a trait of interest will show a similar form among the parents and their respective bulks. Apart from the linked group of markers, other markers will segregate randomly in the bulk created. BSA has been adopted with new modifications for a broad range of applications. It is mostly suited for simply inherited traits such as disease resistance and susceptibility. This method is the basis for all the advanced methods that developed later.

11.3.2 SHOREmap (SHOrtREad Map)

SHOREmap is a gene mapping analysis approach that uses next-generation sequencing (NGS) platforms for the reference-based assembly of the genome through obtained short reads (Schneeberger et al. 2009). It was first used in *Arabidopsis thaliana* to screen mutations responsible for pale green leaves and slow growth phenotype. Ethyl methane sulphonate (EMS) was used to create random mutations in *A. thaliana* Col-0 seeds in M0 generation, which was further used for the generation of homozygous mutant loci which led to the identification of a mutant phenotype. These mutants were further crossed with distantly related Ler-1 ecotype to weaken the distribution of false single nucleotide polymorphisms (SNPs) all over the genome and generated the F2 mapping population. DNA from 500 mutants from F2 were bulked and sequenced to obtain causal SNPs between Ler-1 and the mutant bulk. Further, progenies presenting Col-0-like phenotypes are most likely to carry SNP distributions similar to the Col-0 parent for loci controlling the Col-0 phenotype. In contrast, non-linked loci will display the random distribution of SNPs. This method is useful for pinpointing the mutation at the chromosome level and could be narrowed down to the causal SNP and candidate gene.

11.3.3 NGM (Next-Generation Mapping)

NGM is a mapping technique advanced from SHOREmap with increased power for SNP detection and small mutant population (10 mutant plant bulk). Austin et al. (2011) used this mapping method to identify genes responsible for cell wall synthesis and maintenance in the *A. thaliana* plant. Authors have mutated Col-0 seeds with EMS and subjected it to *in vitro* screening in medium containing the cellulose biosynthesis inhibitor Flupoxam (2.5 nM) to observe club root phenotype. For mapping population generation, three mutant lines were selected through the *in vitro* screen and crossed with Ler-1. DNA from 80 mutants from the F2 population were bulked and subjected to sequencing (illumina genome analyser platform) for SNPs calling between the distantly related line and the mutant line. Further, the NGM pipeline was used to narrow it down to candidates responsible for cell wall biosynthesis and root architecture.

11.3.4 MutMap

This mapping technique was developed for the monogenic recessive traits in rice by Abe et al. 2012 based on BSA analysis. MutMap is often considered as an advanced technical version of NGM and SHOREmap with a difference in mapping population generation (Etherington et al. 2014). SHOREmap and NGM utilise far related mapping populations to obtain condensed SNP dispersal between parent and mutant bulk (Jiao et al. 2018). Whereas, in MutMap, the mapping population is generated through a cross between mutant and its wild type to pinpoint the contributing SNPs related to the mutant phenotype. Its application has been demonstrated for mapping the mutation responsible for leaf colour change from dark green to pale green in rice. For that, the Hitomebore cultivar of rice was used, which was generated through EMS mutations and compared with its genome sequence. Hitomebore M1 plants were selfed up to M3 generation to increase the homozygosity at each mutant locus. Mutants with pale green leaf were selected in M3 and crossed with wild-type parents to obtain the F2 mapping population. From this population, pale green (mutant) DNA bulk of 20 plants was used for sequencing with more than 10× coverage. The selection of mutants at F2 could be confirmed by progeny testing at F3 for their phenotype. After sequencing, the mutant bulk was compared with wild type, and SNPs were called. SNPs unlinked with the mutant phenotype will segregate in a 1:1 ratio, whereas causal SNPs linked with mutant phenotype will manifest into a mutant and show reads in mutant bulks only. Here the SNP index was calculated and a value near 1 indicates causal SNPs, whereas values near 0.5 indicate unlinked SNPs. Apart from mapping recessive mutations, the MutMap can be used for mapping dominant mutations by bulking homozygous F2 plants showing dominant phenotype at progeny testing in the F3 generation.

11.3.5 MutMap+

MutMap is dependent on the development of the F2 population with crosses from wild type. However, recessive mutations could lead to lethality or sterility of the mutant, which creates problems in F2 mapping population generation. To overcome this problem, Fekih et al. 2013 developed the MutMap+ technique when they were mapping the NAP6 gene in rice, which is lethal after 3 weeks of germination. Here 20–30 mutant plants were selected in M3 generation for both mutant and wild-type bulk and subjected for sequencing at 10× coverage. SNPs and the SNP index were called similarly as in MutMap. However, here the ΔSNP index was considered by deducting the wild-type SNP index from the mutant one. The positive value of the ΔSNP index designates the causal mutations for the mutant phenotype. In MutMap+ no genetic hybridization is required, and progeny testing is compulsory to differentiate homozygous and heterozygous plants.

11.3.6 MutMap-Gap

This mapping technique is an extension of the MutMap technique, where the contributing SNPs were positioned at the gap section compared with the reference genome. Here, *de novo* assembly for the gap region was performed with the MutMap analysis pipeline. MutMap-gap was first demonstrated by Takagi et al. (2013b) in Hitomebore cultivar of rice for rice blast resistance gene (*Pii*), which was situated at the gap section of the Nipponbare genome. Mutants were generated with EMS treatment of the Hitomebore cultivar and subjected to blast resistance screening. The susceptible cultivar was advanced to the next generation by selfing to generate M2 generation. These M2 plants were crossed with a wild-type parent to generate the F2 mapping population. Other procedures remain the same as MutMap for identifying the causative SNPs through the SNP index, but due to SNPs presence at the gap region, all unassembled regions from the reference genome were exposed for *de novo* assembly, and contributing mutation was identified based on SNP index analysis.

11.3.7 QTL-Seq

QTL-seq is a mapping technique used for mapping major quantitative trait loci (QTL) with the help of NGS. It is an advancement from BSA and MutMap for mapping quantitative traits mainly related to yield and other agronomic traits (Takagi et al. 2013a). Established on BSA, the resequencing of extreme bulks was performed, and SNPs were called between them. Here mapping population was generated through a cross between cultivars having contrasting phenotypes for the trait of interest. Apart from that, recombinant inbred lines and doubled haploid could also be used as a mapping population to detect minor QTLs. In this method, 10–20 individuals from extreme phenotypes were bulked as highest and lowest bulk and subjected to DNA sequencing. Each bulk sequence was aligned to the reference genome, and the SNP index was determined. SNP index of the lowest bulk was deducted from the highest bulk to plot the peak of the SNP index, which corresponds to the QTL position.

11.3.8 TILLING and EcoTilling

Targeting Induced Local Lesions IN Genome (TILLING) was developed by Claire McCallum and co-workers in *Arabidopsis* (Borevitz et al. 2003). Chromatography was earlier used for the distinction of heteroduplex (wild type) and homoduplex (mutant) that limited its application at a larger scale (McCallum et al. 2000). For high-throughput conversion of TILLING, individual polymerase chain reaction (PCR) products were labelled and recognized by high-quality tools, for example, Lambda Instruments Corporation (LI-COR) DNA analysers (Colbert et al. 2001). TILLING utilises mutagenized populations to detect the induced mutations causing phenotypic variations. EcoTILLING, on the other hand, is a method that utilises naturally available populations to identify polymorphisms/mutations within these populations. These polymorphisms are further utilised to assess the phylogenetic diversity occurring within the population as well as alleles causing these variations in cereals. EcoTILLING has been applied in different crops such as rice, sorghum and wheat to identify allelic variation (Jawhar et al. 2018; Wang et al. 2008). Chemical and physical mutagens were utilised in seeds for cereals and cuttings, tubers and bulbs for vegetatively propagated crops (Cooper et al. 2008). These mutagens can create different mutations ranging from point mutations to insertion and deletion of any segment. However, breeders prefer to induce point mutation because other mutations can change chromosomal structures, thus impacting the phenotype globally (Cooper et al. 2008: Suzuki et al. 2008). Also, large-scale alteration in chromosome structures removes the undesired linkages and might be useful for a combined breeding program (Parry et al. 2009). EMS has been the preferred mutagen used for creating induced mutations because of its optimum rate of mutation that saves both time and resources (Serrat et al. 2014).

TILLING comprises three popular methods used to screen for mutants. The first method is LI-COR, which has been the most preferred method in which the CEL1 enzyme is used (Henikoff et al. 2004; Sestili et al. 2010). The second method is high-resolution melting (HRM), which is very recent for mutant detection in terms of application. In this method, different temperatures were used to generate higher resolution melting profiles from PCR mixtures containing heteroduplexes and homoduplexes (Botticella et al. 2011; Comai et al. 2004). It can also be attached with fluorescence probes and screened with the LightScanner system (Idaho Technology, Inc., Salt Lake City, UT) or Rotor-Gene (Qiagen, Hilden, Germany) instruments for mismatch detection in mutant pools. This method is preferred when the target gene has multiple small exons separated by long introns (Comai et al. 2004). The third method is the most advanced one, where the NGS platform is used for mutation detection (Blomstedt et al. 2012). Several studies demonstrated the use of the various previously discussed technologies for expanding the repertoire of applications for crop improvement in different plant species (Table 11.1).

TABLE 11.1

Application of Advanced Breeding and Biotechnology-Based Approaches for Crop Improvement in Plants

Technique	Species	Gene/QTLs	Trait	Remarks	Reference
SHOREmap	*Arabidopsis thaliana*	*AT4G35090*	Slow growth and pale green leaves	First gene to be identified through this technique	Schneeberger et al. (2009)
NGM	*A. thaliana*	*FPH1, FPH2, MUR11*	Cell wall biosynthesis and maintenance gene	First gene to be identified through this technique	Austin et al. (2011)
MutMap	*Oryza sativa*	*CAO1*	Leaf colour	First gene to be identified through this technique	Abe et al. (2012)
MutMap+	*O. sativa*	*NAP6*	Pale green leaves and dwarfism followed by lethality post 3 week of germination	First gene to be identified through this technique	Fekih et al. (2013)
MutMap-Gap	*O. sativa*	*Pii*	Blast resistance	First gene to be identified through this technique	Takagi et al. (2013 a, b)
QTL-seq	*O. sativa*	QTL for partial blast resistance	Blast resistance	First gene to be identified through this technique	Takagi et al. (2013 a, b)
TILLING	*Zea mays*		Chromomethylase	CEL-1-PAGE used for detection	Till et al. (2004)
	O. sativa		Phytic acid metabolism	TILLING by sequencing used for detection	Kim and Tai (2014)
	Hordeum vulgare		Floral parts regulation	dHPLC for detection	Caldwell et al. (2004)
	H. vulgare		Fungus immunity	CEL-1-PAGE used for detection	Gottwald et al. (2009)
	H. vulgare		Starch metabolism	CEL-1 Agarose gel for detection	Sparla et al. (2014)
	Triticum aestivum		Resistance against powdery mildew	HRM for detection	Acevedo-Garcia et al. (2017)
	T. aestivum		Quality of Starch	CEL-1-PAGE and HRM for detection	Botticella et al. (2011)
	T. aestivum		Development of spike	Agarose gel, PAGE	Chen and Dubcovsky (2012)
	T. aestivum		Carotenoid biosynthesis	CEL-1, agarose gel, dHPLC for detection	Colasuonno et al. (2016)
	T. aestivum		Waxy and lignin	CEL-1 for detection	Rawat et al. (2012)
	T. aestivum		Plant height	Exome sequencing for detection	Mo et al. (2018)
	T. aestivum		Gluten content	TILLING by electrophoresis for detection	Moehs et al. (2018)
	T. aestivum		Kernel hardness and starch	CEL-1-PAGE for detection	Li et al. (2017a)

EcoTILLING	*A. thaliana*	*DMMT2, DRM1C7, PIF2, AtWR*		CEL-1-PAGE for detection	Comai et al. (2004)
	T. aestivum	*VRN-A1, Pin a, Pin b*	Vernalization and kernel hardness	CEL-1-PAGE for detection	Liang et al. (2011); Ma et al. (2017)
	O. sativa	*OSCP17*	Salt tolerance	CJE/Agarose gel for detection	Negrao et al. (2011)
	O. sativa	*osCPK17, osRMC, osNHX1, osHKT1, SalT*	Salt resistance	CEl-1-PAGE for detection	Negrao et al. (2013)
	O. sativa	*GBSS1, SS1, SSIIa, SSIIIa, SBE1a, SBE1b*	Starch synthesis	EcoTILLING by sequencing used for detection	Raja et al. (2017)
	H. vulgare	*HSP17.8*	Heat shock protein	CEL-1-PAGE used for detection	Xia et al. (2013)
	Brassica sp.	*FAE1-A8, FAE1-C3*	Erucic acid content	CEL-1-PAGE used for detection	Wang et al. (2010)
	Solanum lycopersicum	Folate biosynthesis genes	Folate biosynthesis	CEL-1-PAGE used for detection	Upadhyaya et al. (2017)
	Gossypium hirsutum	Sucrose synthesis genes	Sucrose synthesis	CEL-1-PAGE used for detection	Zeng et al. (2016)
	Capsicum annum	*eIF4E* and their homologues	Virus resistance	CEL-1-PAGE used for detection	Ibiza et al. (2010)
	Glycine max	*Gy1, Gy2, Gy3, Gy4, Gy5*	Seed proteins	Agarose gel for detection	Kadaru et al. (2006)
	Beta vulgaris	*BTC1, BVFL1, BvFT1*	Winter hardiness	CEL-1-PAGE for detection	Frerichmann et al. (2013)
	Cicer arietinum	Seed weight–related genes	Seed weight	Agarose gel for detection	Bajaj et al. (2016)
CRISPR	*O. sativa*	*OsMPK2, OsPDS, OsBADH2*	Multiple stress tolerance	Knockout	Shan et al. (2013)
	O. sativa	*OsSWEET13*	Resistance against bacterial blight	*OsSWEET13:* sucrose transporter gene; knockout	Zhou et al. (2015)
	Cucumis sativus	*eIF4E*	Broad virus resistance	*eIF4E*: translation initiation factor; knockout	Chandrasekaran et al. (2016)
	O. sativa	*IPA, GS3, DEP1, Gn1a*	Improved yield	Knockout	Li et al. (2016b)
	Z. mays	*PPR, RPL*	Reduced zein protein	Knockout	Qi et al. (2016)
	O. sativa	*OsERF922*	Resistance against blast fungus	*OsERF922: ERF* transcription factor; knockout	Wang et al. (2016)
	O. sativa	*GS3, GW2, GW5, TGW6*	Grain weight	Knockout	Xu et al. (2016)
	Citrus sinensis	*CsLOB1*	Citrus canker resistant	*CsLOB1:* increase susceptibility against citrus canker; knockout	Jia et al. (2017)
	T. aestivum	*TaDREB2, TaDREB3*	Drought tolerance	*TaDREB2*: dehydration responsive element binding protein 2; *TaDREB3*: ethylene responsive factor 3; knockout	Kim et al. (2017a)

(Continued)

TABLE 11.1
(Continued)

Technique	Species	Gene/QTLs	Trait	Remarks	Reference
	Glycine max	*FAD2-1A, FAD2-1B*	Improved oil quality	Knockout	Kim et al. (2017b)
	O. sativa	*Hd2, Hd4, Hd5*	Early heading	Knockout	Li et al. (2017b)
	O. sativa	*SAPK2*	Drought and salinity tolerance	Knockout	Lou et al. (2017)
	Camelina sativa	*CsFAD2*	Reduced polyunsaturated fatty acids and increased oleic acid in oil	Knockout	Morineau et al. (2017)
	O. sativa	*OsAnn3*	Cold tolerance	Knockout	Shen et al. (2017)
	T. aestivum	*EDR1*	Improved resistant against powdery mildew	*EDR1:* encode powdery mildew resistance; knockout	Zhang et al. (2017)
	O. sativa	*OsCCD7*	High-tillering	Knockout	Butt et al. (2018)
	G. max	*GmFT2a*	Delayed flowering	Knockout	Cai et al. (2018)
	S. lycopersicum	*lncRNA1459*	Long shelf life	Knockout	Li et al. (2018a)
	S. lycopersicum	*SlCBF1*	Cold tolerance	C-repeat-binding factor-1; knockout	Li et al. (2018b)
	S. lycopersicum	*SGR1, LCY-E, Blc, LCY-B1*	Increased lycopene	Knockout	Li et al. (2018c)
	O. sativa	*eIF4G*	Resistance against *rice tungro spherical virus*	*eIF4G:*translation initiation factor; knockout	Macovei et al. (2018)
	S. lycopersicum	*SlJAZ2*	Bacterial speck resistant	*SlJAZ2:* co-receptor of coronatine; knockout	Ortigosa et al. (2018)
	T. aestivum	*α–gGliadin*	Low gluten	Knockout	Sanchez-Leon et al. (2018)
	Vitis vinifera	*VvWRKY52*	Increased resistance against *Botrytis cinerea*	*VvWRKY52:* transcription factor; knockout	Wang et al. (2018)
	Z. mays	*ZmHKT1*	Salinity tolerance	Sodium transporter HKT1; knockout	Zhang et al. (2018a)
	T. aestivum	*TaGW2*	Grain weight	Knockout	Zhang et al. (2018b)
	S. lycopersicum	*SlNPR1*	Drought tolerance	Regulatory protein NPR1; knockout	Li et al. (2019)
	O. sativa	*OsRR22*	Salinity tolerance	*OsRR22:* two-component response regulator; knockout	Zhang et al. (2019)

TALENs	*O. sativa*	*Os11N3*	Bacterial blight resistance	Knockout	Li et al. (2012)
	G. max	*FAD2-1A/B*	Improved oil quality	Knockout	Haun et al. (2014)
	T. aestivum	*TaMLO-A1, TaMLO-B1, TaMLO-D1*	Powdery mildew resistance	Knockout	Wang et al. (2014)
	Z. mays	*ZmGL2*	Reduced epicuticular wax in leaves	Knockout	Char et al. (2015)
	S. tuberosum	*VInv*	Minimizing reducing sugars	Knockout	Clasen et al. (2016)
	Saccharum spp. hybrids	*COMT*	Improvement of cell wall composition for bioethanol production	Lignin biosynthetic gene; knockout	Jung and Altpeter (2016)
	Nicotiana benthamiana	*Fuct, XylT*	Production of glycoproteins	Multiplexed gene editing; knockout	Li et al. (2016a)
Zinc-finger nucleases	*Z. mays*	*IPK1*	Herbicide tolerance	Gene replacement	Shukla et al. (2009)
	Nicotiana	*ALS, SuRA, SuRB*	Resistance to imidazolinone and sulfonylurea herbicides	Tobacco acetolactate synthase genes; knockout	Townsend et al. (2009)
	G. max	*DCL*	Heritable transmission	Knockout	Curtin et al. (2011)
	Brassica napus	*KASII*	Decrease in palmitic acid, increased total C18, and reduced total saturated fatty acid contents	Gene expression	Gupta et al. (2012)
	Z. mays	*ZmTLP*	Herbicide tolerance	Trait stacking	Ainley et al. (2013)
	O. sativa	*OsQQR*	Detection of safe harbour loci	Trait stacking	Cantos et al. (2014)
	S. lycopersicum	*L1L4*	Heterochronic phenotype, plant architecture	Development related; knockout	Hilioti et al. (2016)

11.4 Biotechnological Tools for Crop Improvement

11.4.1 *In Vitro* Culture

Biotechnological tools, over time, reinforced the mutation breeding studies. One such tool is *in vitro* culture techniques that include somatic embryogenesis, double haploidy (DH) and micropropagation. Since somatic embryos (SEs) are the progeny of single cells, these are excellent explants for clonal propagation. Scaling up of SE production and limitation of chimera formation due to embryogenesis make it ideal for mutation induction (Joseph et al. 2004; Yang and Zhang 2010; Zimmerman 1993). DH has been used as a means of production of homozygous lines in the shortest possible time since the discovery of haploid culture (anther culture) from *Datura innoxia* (Guha and Maheshwari 1964). Three approaches have been utilised for DH production, namely, androgenesis (microspore and anther culture), gynogenesis and wild crossing trailed by chromosome elimination (Germana 2012; Szarejko 2012). A combination of DH with mutation induction improves the selection efficacy and recessive mutant's recovery (Bhojwani and Dantu 2010; Szarejko 2012). The third *in vitro* technique, micropropagation by organogenesis, has been utilised for quick multiplication of plants through buds and shoot tips mainly for vegetatively propagated crops. This makes micropropagation an ideal method to generate a large population for the creation of mutation as well as mutant line multiplication in subsequent steps. Separation of chimeras has been achieved through the advancement of mutant population for four to five generations (Jankowicz-Cieslak et al. 2012; Kodym and Zapata-Arias 2001; Novak and Brunner 1992).

11.4.2 DNA-Based Technologies

Mutation breeding has been complemented by the development of NGS technologies with newer variants of each platform available every year (Hert et al. 2008; Liu et al. 2012a). These platforms enable breeders for large-scale screening in less time and cheaper cost per base as compared with Sanger methods. TILLING combined with NGS, known as TILLING by sequencing (TbS), enabled mutation discovery at individual plant levels (Tsai et al. 2011). Approaches such as exome enrichment with reverse genetic analysis for more than 10,000 genes (Kettleborough et al. 2013) could be utilised for the generation of *in silico* resources for deleterious and useful mutations available for researchers. Reduction of cost and time made NGS applicable for whole-genome resequencing targeted for a particular genomic region containing mutations responsible for phenotypes observed (Abe et al. 2012). Other advancements in this direction, such as RNA-seq, chromatin immunoprecipitation (ChIP)-seq and reduced representation libraries, can be utilised to detect the mutations (Chapman et al. 2008; Kim and Tai 2013). These advancements are particularly useful for crops with a large and complex genome.

11.4.3 Targeting Mutations/Genome Editing Tools

11.4.3.1 Targeting Mutations

Genome targeting is achieved mainly through homologous recombination (HR) with other repair pathways modification. HR is a damage protection mechanism for chromosome protection against double-strand breaks (DSBs) and inter-strand cross-links. HR, along with single strand annealing (SSA) and non-homologous end joining (NHEJ) are the repair pathways responsible for the insertion of foreign DNA into the native plant genome. These processes were adopted for inducing genome targeting at targeted site DSBs to engineer plants, for example, in rice and *Arabidopsis* (Mengiste et al. 1999).

11.4.3.2 Genome Editing Technologies

Genome editing technologies are mainly dependent on engineered nucleases for insertion, deletion or replacement of a particular DNA segment at a specific site in the genome. These enzymes create precise DSBs to a particular site that are repaired through HR/NHEJ which leads to targeted mutations. In this section, we will brief about commonly used nucleases for genome editing processes, i.e., zinc-finger nucleases (ZFNs), transcription activator-like effector nucleases (TALENs) and the clustered regularly interspersed short palindromic repeat (CRISPR)-Cas system.

11.4.3.2.1 ZFNs

These are zinc-finger (ZF) domain-containing nucleases that recognize specific DNA sequences comprised of a fusion of DNA binding domain and cleavage domain from restriction enzyme FokI (Kim et al. 1996). ZFNs are important nucleases for creating breaks where knowledge of pre-existing sites of the target are absent (Urnov et al. 2010). ZFNs bind to specific nucleotide triplets, which could be broadened by a combination of various ZFs with higher specificity and affinity (Segal et al. 2003). Plant genome modification by ZFNs are managed by homology-directed repair in site-directed HR and targeted genetic manipulation. Application of ZFNs for targeted modification to endogenous genes has been reported in *Arabidopsis*, maize, soybean and tobacco (Curtin et al. 2011; Shukla et al. 2009; Townsend et al. 2009; Zhang et al. 2010). These nucleases are limited with target sites, lower efficiency of target specificity and frequent off-targets (DeFrancesco 2012).

11.4.3.2.2 TALENs

TALENs are advanced nucleases developed for precise manipulation of the targeted site in the genome (Christian et al. 2010). It is composed of DNA binding domain TALEs (transcription activator-like effectors) and cleavage domain of FokI like ZFNs. The TALE domain is composed of repeat-variable di-residue (RVD) sequences from *Xanthomonas*, a plant pathogenic bacterial genus customizable for specific nucleotide recognition (Bogdanove and Voytas 2011). Its DNA binding domain comprises a series of 33–35 amino acid repeats with recognition of a single base by each domain, which gives flexibility for designing as only four modules will be required for recognition, i.e., A, T, G and C (Liu et al. 2012b). Transient protoplast-based assays have shown the applicability of TALENs in *Arabidopsis* and tobacco as an accurate, reliable and quick method (Zhang et al. 2013). These assays were further used in *Brachypodium* and rice.

11.4.3.2.3 CRISPR/Cas

CRISPR/CRISPR-associated (Cas) type II prokaryotic system is modified to engineer genomes (Sorek et al. 2013). This system is composed of a protein Cas9 nuclease and single-guide RNA (sgRNA) for targeted cleavage (Carroll 2013). Plant scientists have adopted this system with the development of a transient expression system consisting of protoplast and leaf tissue transformation by agro-infiltration (Jaganathan et al. 2018). These enable rapid optimization of the system in different cultivars. The CRISPR/Cas system has high target specificity because of its 3′ end of the guide sequence which is dependent on protospacer adjacent motif (PAM) which makes it a highly suitable system for plant genome targeting.

11.4.4 High-Throughput Phenotyping

With the generation of the huge amount of data related to metabolomics, proteomics, genomics and transcriptomics, advancement in analysis and detection techniques have taken place with new approaches such as epigenome, hormonome and interactome (Mochida and Shinozaki 2011). These advancements generated the need for high-throughput automated systems for bioinformatics and phenotyping to associate the omics data with plant systems. High-throughput systems for phenotyping of mutants consist of infrared, near-infrared imaging and three-dimensional (3D) imaging with growth chambers for controlled environment simulation (Schunk and Eberius 2012). Application of these systems has been reported in the rice mutant's selection for seed phenotyping, which consists of X-ray and near-infrared reflectance spectroscopy (Jankowicz-Cieslak et al. 2013). Further, in future modern phenomics techniques accompanied by advanced biotechnological tools will enable the breeders efficient selection of mutant phenotype.

11.5 Advanced Tools for Crop Improvement

Genetic engineering in plants utilising the major transformation tools as biolistics and *Agrobacterium*-mediated approaches has been in use for more than 40 years. These technologies directly manipulate the target genome by introducing single or multiple genes and regulatory elements. The random nature of T-DNA integration into the host genome can have undesirable effects and is a major concern. It can disrupt genes or genomic regions vital for the target host's growth and development as well as agronomic traits. Moreover, the inability to predict transgene expression can have undesirable effects, thereby the conventional transformation tools for metabolic pathway engineering that require more intensive manipulation of the genome are not preferred.

These traditional first-generation genetic engineering technologies which manipulate genomes randomly are, however, being overtaken by new technologies collectively termed as 'targeted genetic modification' (TagMo) techniques. TagMo is also referred to as 'targeted genome editing' and 'new biotechnology-based plant breeding techniques'. It constitutes tools that enable more specific regulation of gene expression in plants. These techniques have high potential that ease and speed up the genetic engineering of plants that fall outside the genetically modified regulations. Furthermore, TagMo techniques extend the tool for engineering more traits to a large number of plant species as well as improving the efficiency of plant transformation (Bogdanove and Voytas 2011; Porteus 2009).

11.5.1 Specifically-Regulated Gene Expression

11.5.1.1 Synthetic Promoters

Understanding the regulation of endogenous plant genes at the transcriptional level, together with precise control switches of transgene expression required for genetic and metabolic pathway engineering, are the foremost challenges in plant biotechnology. Substantial advancement is in identifying the synthetic promoters, synthetic transcriptional regulators (activators and repressors), as significant tools for the regulation of transgene spatial and temporal expression. A synthetic promoter comprises a DNA sequence that constitutes a core-promoter region (also defined as the minimal region) and random or ordered arrangement of multiple repeats or heterologous upstream *cis*-regulatory elements. The major interest behind the use of synthetic promoters is to improve the transgene expression strength by reducing undesirable leaky expressions. These synthetic promoters are developed by combinatorial engineering using *cis*-regulatory elements (like activators, enhancers, insulators and repressors) from different sources (Cai et al. 2020; Liu et al. 2013; Petolino and Davies 2013). Several studies developed a series of synthetic promoters either by the combination of specific *cis*-regulatory elements with a core-promoter region or as hybrids of multiple promoters. Similarly, synthetic promoters have also been developed by manipulating native promoters with additional functional promoter elements. The use of synthetic promoters successfully regulates transgenes expression in trait- and environment-specific manners (Bilas et al. 2016; Mohan et al. 2017). Further, research in plant synthetic promoter engineering facilitated better transgene transcriptional regulation by developing robust constitutive, inducible or bidirectional synthetic promoters. Several studies have expounded different aspects of the underlying cellular regulation mechanisms of prolific transgene expression by the use of synthetic promoters.

11.5.1.2 Synthetic Transcription Factors

The transcription factors (TFs) that bind to synthetic promoter's *cis*-regulatory elements placed upstream of the core promoter region regulate the gene transcription as well as other regulatory effects on the target gene (Spitz and Furlong 2012). Belcher et al. (2020) explored synthetic transcriptional regulators for plant engineering and developed a diverse library using various yeast TF components in combination with plant regulatory elements. These synthetic TFs or *trans*-elements are developed by directly altering TFs from different families using DNA binding and effector domains (activation or repression). Moreover, the use of additional TFs can enhance or repress the basal transcriptional activity, thereby providing multilayered regulation (Shih et al. 2016). Studies specified the influence of copy number, specificity and spacing of *cis*-elements and their corresponding TFs on spatial and temporal expression and strength of synthetic promoters making them a better choice compared with native promoters (Dey et al. 2015; Mehrotra et al. 2011). Fusion proteins with engineered DNA-binding and catalytic effector domains hold inordinate potential for targeted transgene expression as well as for precise genome editing (Bogdanove and Voytas 2011).

The use of synthetic TFs like ZF-TFS or TALE-TFs for targeted gene activation or enzymes like ZFNs or TALE nucleases, (TALENs) or CRISPR/Cas for genome editing will facilitate targeted gene engineering. The synthetic TF tools activate regulatory proteins influencing transgene expression, while genome editing tools can promote targeted modifications at genomic loci with high precision and efficiency. ZFs have simultaneously been used for developing synthetic DNA-binding modules as well as for modulating differential gene expression (Jia et al. 2013). For regulated gene expression, synthetic ZF domains with extremely specific DNA binding capabilities are first designed (Stege et al. 2002). The effector ZF protein fused to the activation domain regulates and enhances the transcriptional activity on recognition and binding of the target region (Guan et al. 2002; Lindhout et al. 2006). The applications of ZF-artificial TFs have been further extended through genome interrogation to induce novel phenotypes of interest by altering genome-wide TFs (van Tol et al. 2017). Moreover, the specificity of ZFs towards their target DNA sequences has been enhanced by the combined effect of ZF-artificial TFs engineering together with computational approaches like oligomerized pool engineering and rapid 'open source' engineering (Foley et al. 2009; Maeder et al. 2008).

On the other hand, engineered TALE-DNA binding domain comprises of identical tandem repeat arrays that recognise target sequence by simple RVD-nucleotide recognition code (Bogdanove et al. 2010). TALEs regulate gene expression by fusing either with activators or repressors (Bogdanove and Voytas 2011). These fusion proteins are customised for target specificity by varying composition and number of repeats (Liu et al. 2014; Schwartz et al. 2017; Zhang et al. 2014). In another study, Lowder et al. (2017) developed the 'mTALE-Act' system that enabled simultaneous activation of multiple genes using an engineered TALE–VP16 (transcriptional activator).

11.5.1.3 Sequence-Specific Nucleases

In contrast, the genome editing tools use sequence-specific nucleases (SSNs) which, when induced, recognise targeted DNA sequences and produce DSBs. Two endogenous repair pathways can repair nuclease-induced DNA DSBs: (1) NHEJ, which introduces short insertions or deletions creating loss-of-function mutations by generating frameshifts or (2) homology-directed recombination (HDR), which can lead to gene replacement or gene knock-in by using donor sequence homologous to DSB flanking sites (Huang and Puchta 2019). Initial stages of genome editing focussed on the use of ZFNs, TALENs or meganucleases for inducing the desired DSBs, which have their strengths but suffer low specificity due to their off-targets side effects. On the other hand, the more recent CRISPR/Cas9 technology is characterised by its efficiency, simplicity, low cost and its ability to target multiple genes (Cong et al. 2013). These characteristics made CRISPR/Cas9 an effective tool for plant breeding and, therefore, rapidly exploited in plants (Gao 2018; Li et al. 2013; Shan et al. 2013). Though the technology is mainly used for targeted gene knockout, its simple design platform with high target specificity promoted its use in transcriptional regulation (Lowder et al. 2015). Advances in CRISPR-mediated gene expression regulation focussed on the use of the mutant form of Cas9 (known as Cas9 Endonuclease Dead or dead Cas9 or dCas9) whose endonuclease activity is removed. Similar to cas9, dcas9 can be directed to target genomic locations using sgRNAs that interfere with the target site transcription leading to reversible silencing of the gene (Cheng et al. 2013; Wyvekens et al. 2015).

Different studies reported strategies to improve CRISPR-based gene activation and re-engineered cell behaviour, either by fusing dCas9 to repeats of transcriptional activation domain in the SunTag array (Tanenbaum et al. 2014) or through a concerted dCas9-activation mediator system to recruit multiple transcriptional activators for robust up-regulation of multiple endogenous genes and non-coding RNAs using sgRNA (Konermann et al. 2015). In another study, dCas9 is fused to a tripartite effector domain with multiple activators to target genes simultaneously as well as to enhance activation of endogenous coding and non-coding genes (Chavez et al. 2015). More recently, multiplex transcriptional activation in plants is attained by CRISPRAct2.0 and mTALE-Act systems (Lowder et al. 2017). However, these approaches are limited due to the increased risk of off-target effects (Braun et al. 2016). A study by Li et al. (2017c) designed a potential transcriptional activation tool-dCas9-TV (dCas9-VP128 incorporated with sequence-unrelated transcription activation domains guided by a single sgRNA) to achieve 55-fold gene activation as well as to overcome the constraints of off-target effects.

Moreover, the CRISPR technology is also used for repression of transcription at specific genomic loci using the CRISPR interference (CRISPRi) tool. CRISPRi represses gene expression by directing dCas9-sgRNA complex binding to the promoter region, thereby halting the transcript elongation by RNA polymerase II (Cheng et al. 2013; Jinek et al. 2012; Qi et al. 2013). The tool is also used to target the repression of multiple endogenous genes either by fusion of dCas9 to the transcriptional repressor domain or by using multiple sgRNAs concurrently targeting the same gene (Gilbert et al. 2013; Lowder et al. 2015). Therefore, studies suggested the use of a multiplex CRISPR-Cas9 toolbox for transcriptional activation as well as repression of endogenous genes in several plant species (Lowder et al. 2015). The tools thus emerged as extremely advantageous for plant synthetic biology mediated applications.

11.5.2 Multigene-Transfer

Engineering plants for several agronomical traits that are usually polygenic requires the coordinated introduction of multiple genes or entire pathways into the genome of interest. However, engineering multiple traits represents a unique challenge for the plant genetic engineers, often limited by traditional technologies that depend on successive transformations, traditional breeding tools and transgenes present on multiple loci. Some of the tools used for the multigene transfer include (1) sexual-crosses among transgenic lines carrying different transgenes (Lucker et al. 2004; Ma et al. 1995; Zhao et al. 2003) (2) co-transformation that involves combined delivery of several transgenes in a single transformation experiment (Li et al. 2003) and (3) re-transformation that stacks multiple transgenes by successive transfer of single genes into transgenic plants (Seitz et al. 2007; Singla-Pareek et al. 2003). Among them, sexual-crosses, as well as re-transformation, are long and labour-intensive approaches that involve several breeding generations with chances of several unlinked transgenes getting segregated in subsequent generations. In contrast, co-transformation develops transgenic plants in a single generation by simultaneous delivery of multiple transgenes that often integrate at the same locus, thereby preventing segregation (De Buck et al. 1999; De Neve et al. 1997). Therefore, co-transformation emerged as a potential tool for the delivery of multiple linked (multiple genes on the same plasmid) or unlinked (different genes on different plasmids) transgenes into target genomes using single or dual gene-transformation vectors.

11.5.2.1 High-Capacity Vectors

The ability of a single vector to inherit the transgenes together makes it a preferential tool over multiple vectors for delivering multiple transgenes. In a single-vector system, a single DNA molecule needs to be transferred into the cells that will integrate into the plant genome as a single unit enabling multi-transgenes to be inherited together. To facilitate the transfer of multi-transgenes, several vectors have been developed. Some vector systems assemble an expression cassette containing multiple protein-coding sequences (CDSs: CoDing Sequence is the region of DNA or RNA that determine amino acid sequence in protein) fused to produce a single transcriptional unit followed by processing of the polyprotein giving rise to individual proteins. The system is particularly used to express multiple enzymes in transgenic plants at various subcellular locations (Dasgupta et al. 1998; Ralley et al. 2004). However, the associated complications may hinder the use of the tool for the routine expression of multi-transgenes. Furthermore, the high-capacity vectors based on binary bacterial artificial chromosomes (BIBACs; Hamilton 1997) or transformation-competent artificial chromosome (TAC; Liu et al. 1999) are also being used. Moreover, the gateway site-specific recombination technology simplified the cloning in these high-capacity vectors (Schmidt et al. 2008). Still, another super vector platform technology that enables the management of a large number of transgenes for the next-generation genetic engineering involves the use of engineered mini-chromosomes or plant artificial chromosomes (PACs; Gaeta et al. 2012, 2013; Halpin 2005). PACs are developed with only essential components like centromeres, telomeres and origins of replications to keep them compact in size. PACs enabled integration and expression of the unlimited number of transgenes or compound gene complexes in plants (Borland et al. 2014).

Moreover, PACs are designed to contain site-specific recombination (SSR) systems like Cre-*lox* (Abremski et al. 1983; Yu et al. 2015), FLP-*frt* (Golic and Lindquist 1989; Stark et al. 1992) or phiC31-*att* integrase systems (Thorpe and Smith 1998), which enables manipulation of transgenes. Genome editing facilitated precise *in vivo* gene stacking on PACs (Ainley et al. 2013). Additionally, TALEN and CRISPR/Cas9-stimulated HR is also used for replacing genes in PAC technology (Yu et al. 2015). Furthermore, PACs avoid linkage drag when multi-transgenes are stacked or transferred to other germplasm through breeding, which is a major challenge for breeders (Young and Tanksley 1989). Linkage drag is a condition when the gene of interest is associated with deleterious genes that have adverse effects on the plant. As PACs are independent chromosomes, the genes packed together occur as one linkage group that can be transferred in a single cross without linkage to other chromosomes. Moreover, PAC-mediated genetic manipulation is not influenced by endogenous genes due to position effects as well as avoid the disruption of endogenous gene functions, which is common during traditional genetic engineering due to random integration in the genome (Gaeta et al. 2012; Yu et al. 2007).

11.5.2.2 Plastid Transformation

Plastid transformation is another tool that has been explored for multigene engineering. The approach links multiple genes into operons expressed coordinately as a single polycistronic mRNA (Lu et al. 2013; Ruiz et al. 2003). The expression of these multigene operons is, however, independent of cellular control and polycistronic transcripts do not require processing into monocistrons for efficient translation (Quesada-Vargas et al. 2005). The approach facilitates multigene engineering in both green and non-green plastids without the significant intervention of chloroplast regulation (Hasunuma et al. 2008; Lossl et al. 2005). Though the tool is expanded to a large group of plant species, host specificity and inefficiency in transferring large constructs still remains a major technical challenge.

11.5.3 Speed Breeding

Another innovative, promising tool that is being adopted in agriculture is 'speed breeding', that is, advanced crop improvement by reducing the length of breeding cycles (Watson et al. 2018). Traditional breeding for new traits in most of the crops usually takes several years. In contrast, SB reduces crop generation time that accelerates the plant development as well as harvesting and germination of immature seeds (Ghosh et al. 2018; Sysoeva et al. 2010). SB experiments are done under completely controlled environment conditions and/or applying plant hormones.

Among the different factors used for speeding up the breeding cycle of crops, prolonged photoperiod emerged as the most common tool (Ghosh et al. 2018; Sysoeva et al. 2010). The rapidly generated plants show normal development with a high seed germination rate as well as amenability for crossing. Also, the generated plants are phenotyped for associated traits, thereby proving its application in accelerating plant phenotyping as well as gene transformation pipelines. Recently, Jahne et al. (2020) extended the tool to short-day plants like rice, amaranth and soybean by using a light-emitting diode (LED)-based SB system. The approach is, however, limited due to variations in the crop species, varieties and cultivars in response to the photoperiod. Therefore, crop-specific SB protocols, as per the research objectives, need to be specified. Yet another tool in SB utilised stress treatments that facilitate early seed set in plants. Different stress factors like high or low temperature, water excess or drought, nutrient deficiency, low or high light intensity, crowding, ultraviolet treatment and pathogen infection have been screened (Samineni et al. 2019; Takeno 2016). Regulators of plant flowering, seed set, embryo development like photoperiod and vernalization as well as physiological stresses are also explored as components of the SB toolbox (Ghosh et al. 2018; Takeno 2016; Wada and Takeno 2010).

Additionally, the *in vitro* tissue culture techniques like embryo rescue and DH are also being utilised for SB (Bermejo et al. 2016; Chaikam et al. 2019; Mobini et al. 2015; Palanisamy et al. 2019; Watts et al. 2020). Embryo rescue facilitates plant embryo development by harvesting immature seeds and germinating them in the presence or absence of plant growth regulators in culture medium, thereby obtaining more plant generations per year (Bermejo et al. 2016; Mobini et al. 2015). In DH technology, haploid embryos are rescued, and homozygous lines are developed by chromosome doubling within two generations compared with six or more generations required through conventional breeding (De La Fuente et al. 2013). Another vital factor in speeding the breeding cycle is the breaking of seed dormancy. Treatments like hot water, dry heat, cold stratification, treating seeds with water or germination promoting hormones such as gibberellins could be used for overcoming the dormancy of the seeds (Penfield 2017). Merging embryo rescue, together with the overcoming of seed dormancy, will shorten the generation cycle period (Zheng et al. 2013). The SB tool has been successfully used in several crops like *brassica* species, bread wheat, durum wheat, barley, chickpea, pea, grass pea, quinoa, oat, *Brachypodium distachyon* and peanut (Ghosh et al. 2018; Gorjanc et al. 2018; Hickey et al. 2017; O'Connor et al. 2013; Watson et al. 2018). However, the application of SB is limited to artificial stimulated conditions.

An advanced crop improvement program involves the integration of SB with genome editing as well as with selection tools like MAS, genomic selection (GS) and pollen-based selection (PBS) for more specific and faster results. The coupling of SB with GS called 'SpeedGS' enables intense and frequent selection stages that allow for higher genetic gain per year. Furthermore, the integration of SB with transgenic technologies could reduce the generation time for plant phenotyping and other traits. Additionally, the merger of the tools like high-throughput and multi-trait phenotyping, CRISPR-Cas9 editing, NGS and DNA mismatch repair at different stages of the breeding cycle could allow rapid breeding of plant species (Alahmad et al. 2018; Hickey et al. 2019; Karthika et al. 2020). These tools could also enable the rapid discovery of gene and trait loci (Al-Tamimi et al. 2016; Awlia et al. 2016). The implementation of SB for a greater number of crops will govern the industrial potential of the technology.

11.6 Conclusions

Sustainable agriculture in times of changing climate is a major challenge for the scientific community. Traditional practices for crop improvement, like cross-breeding and mutation breeding, are time-consuming and labour intensive. The rapid pace of development of biotechnology and its integration with advanced mutation breeding technologies provides us an efficient solution for shortening the time for crop improvement. Next-generation breeding tools and high-throughput sequencing technologies have revolutionized crop improvement programs. Precise genome editing is another powerful tool for crop improvement that has gained importance in recent years. The development of transgene-free genome-edited crops has been on the rise mainly due to the strong regulation associated with transgenics that exists in most countries. An integrated approach utilising modern breeding tools and genetic engineering will help in addressing the challenge of feeding the increasing population in the near future.

Acknowledgements

Author's work in this area is supported by the Core Grant of National Institute of Plant Genome Research, New Delhi, India.

REFERENCES

Abe, A., Kosugi, S., Yoshida, K., et al. 2012. Genome sequencing reveals agronomically important loci in rice using MutMap. *Nature Biotechnology* 30: 174–178. https://doi.org/10.1038/nbt.2095

Abremski, K., Hoess, R. and Sternberg, N. 1983. Studies on the properties of P1 site-specific recombination: evidence for topologically unlinked products following recombination. *Cell* 32(4): 1301–1311. https://doi.org/10.1016/0092-8674(83)90311-2

Acevedo-Garcia, J., Spencer, D., Thieron, H., et al. 2017. mlo-Based powdery mildew resistance in hexaploid bread wheat generated by a non-transgenic TILLING approach. *Plant Biotechnology Journal* 15: 367. https://doi:10.1111/pbi.12631

Ainley, W. M., Sastry-Dent, L., Welter, M. E., et al. 2013. Trait stacking via targeted genome editing. *Plant Biotechnology Journal* 11(9): 1126–1134. https://doi.org/10.1111/pbi.12107

Alahmad, S., Dinglasan, E., Leung, K. M., et al. 2018. Speed breeding for multiple quantitative traits in durum wheat. *Plant Methods* 14: 36. https://doi.org/10.1186/s13007-018-0302-y

Al-Tamimi, N., Brien, C., Oakey, H., et al. 2016. Salinity tolerance loci revealed in rice using high-throughput non-invasive phenotyping. *Nature Communications* 7: 13342. https://doi.org/10.5061/dryad.3118j

Austin, R. S., Vidaurre, D., Stamatiou, G., et al. 2011. Next generation mapping of Arabidopsis genes. *The Plant Journal* 67(4): 715–725. https://doi.org/10.1111/j.1365-313X.2011.04619.x

Awlia, M., Nigro, A., Faljkus, J., et al. 2016. High-throughput non-destructive phenotyping of traits contributing to salinity tolerance in *Arabidopsis thaliana*. *Frontiers in Plant Science* 7: 1414. https://doi.org/10.3389/fpls.2016.01414

Bajaj, D., Srivastava, R., Nath, M., et al. 2016. EcoTILLING-based association mapping efficiently delineates functionally relevant natural allelic variants of candidate genes governing agronomic traits in chickpea. *Frontiers in Plant Science* 7: 450. https://doi.org/10.3389/fpls.2016.00450

Bateson, W. 2012. *Materials for the study of variation: treated with especial regard to discontinuity in the origin of species* (Cambridge Library Collection – Darwin, Evolution and Genetics). Cambridge, Cambridge University Press. https://doi.org/10.1017/CBO9781139382069.

Belcher, M. S., Vuu, K. M., Zhou, A., et al. 2020. Design of orthogonal regulatory systems for modulating gene expression in plants. *Nature Chemical Biology* 16: 857–865. https://doi.org/10.1038/s41589-020-0547-4

Bermejo, C., Gatti, I. and Cointry, E. 2016. *In vitro* embryo culture to shorten the breeding cycle in lentil (*Lens culinaris* Medik). *Plant Cell Tissue and Organ Culture* 127: 585–590. https://doi.org/10.1007/s11240-016-1065-7

Bhojwani, S. S. and Dantu, P. K. 2010. Haploid plants. In *Plant cell culture: essential methods*, eds. Michael, R. D. and Anthony, P., 60–78. Chichester, UK, John Wiley & Sons. https://doi.org/10.1002/9780470686522.ch4

Bilas, R., Szafran, K., Hnatuszko-Konka, K. and Kononowicz, A. K. 2016. Cis-regulatory elements used to control gene expression in plants. *Plant Cell Tissue and Organ Culture* 127(2): 269–287. https://doi.org/10.1007/s11240-016-1057-7

Blomstedt, C. K., Gleadow, R. M., O'Donnell, N., et al. 2012. A combined biochemical screen and TILLING approach identifies mutations in *Sorghum bicolor* L. Moench resulting in acyanogenic forage production. *Plant Biotechnology Journal* 10: 54–66. https://doi.org/10.1111/j.1467-7652.2011.00646.x

Bogdanove, A. J., Schornack, S. and Lahaye, T. 2010. TAL effectors: finding plant genes for disease and defense. *Current Opinion in Plant Biology* 13(4): 394–401. https://doi.org/10.1016/j.pbi.2010.04.010

Bogdanove, A. J. and Voytas, D. F. 2011. TAL effectors: customizable proteins for DNA targeting. *Science* 333: 1843–1846. https://doi.org/10.1126/science.1204094

Borevitz, J. O., Liang, D., Plouffe, D., Chang, H. S., Zhu, T., Weigel, D., Berry, C. C., Winzeler, E. and Chory, J. 2003. Largescale identification of single-feature polymorphisms in complex genomes. *Genome Research* 13(3): 513–523. https://doi.org/10.1101/gr.541303

Borland, A. M., Hartwell, J., Weston, D. J., et al. 2014. Engineering crassulacean acid metabolism to improve water-use efficiency. *Trends in Plant Science* 19(5): 327–338. https://doi.org/10.1016/j.tplants.2014.01.006

Botticella, E., Sestili, F., Hernandez-Lopez, A., Phillips, A. and Lafiandra, D. 2011. High resolution melting analysis for the detection of EMS induced mutations in wheat *Sbella* genes. *BMC Plant Biology* 11: 156. https://doi.org/10.1186/1471-2229-11-156

Braun, C. J., Bruno, P. M., Horlbeck, M. A., Gilbert, L. A., Weissman, J. S. and Hemann, M. T. 2016. Versatile *in vivo* regulation of tumor phenotypes by dCas9-mediated transcriptional perturbation. *Proceedings of the National Academy of Sciences of the United States of America* 113(27): E3892–E3900. https://doi.org/10.1073/pnas.1600582113

Butt, H., Jamil, M., Wang, J. Y., Al-Babili, S. and Mahfouz, M. 2018. Engineering plant architecture via CRISPR/Cas9-mediated alteration of strigolactone biosynthesis. *BMC Plant Biology* 18: 174. https://doi.org/10.1186/s12870-018-1387-1

Cai, Y., Chen, L., Liu, X., et al. 2018. CRISPR/Cas9-mediated targeted mutagenesis of *GmFT2a* delays flowering time in soya bean. *Plant Biotechnology Journal* 16: 176–185. https://doi.org/10.1111/pbi.12758

Cai, Y., Kallam, K., Tidd, H., Gendarini, G., Salzman, A. and Patron, N. J. 2020. Rational design of minimal synthetic promoters for plants. *Nucleic Acids Research* 48: 11845–11856. https://doi.org/10.1101/2020.05.14.095406

Caldwell, D. G., McCallum, N., Shaw, P., et al. 2004. A structured mutant population for forward and reverse genetics in Barley (*Hordeum vulgare* L.). *The Plant Journal* 40: 143–150. https://doi.org/10.1111/j.1365-313X.2004.02190.x

Cantos, C., Francisco, P., Trijatmiko, K. R., Slamet-Loedin, I. and Chadha-Mohanty, P. K. 2014. Identification of "safe harbor" loci in indica rice genome by harnessing the property of zinc-finger nucleases to induce DNA damage and repair. *Frontiers in Plant Science* 5: 302. https://doi.org/10.3389/fpls.2014.00302

Carroll, D. 2013. Staying on target with CRISPR-Cas. *Nature Biotechnology* 31(9): 807–809. https://doi.org/10.1038/nbt.2684

Chaikam, V., Molenaar, W., Melchinger, A. E. and Boddupalli, P.M. 2019. Doubled haploid technology for line development in maize: technical advances and prospects. *Theoretical and Applied Genetics* 132: 3227–3243. https://doi.org/10.1007/s00122-019-03433-x

Chandrasekaran, J., Brumin, M., Wolf, D., et al. 2016. Development of broad virus resistance in non-transgenic cucumber using CRISPR/Cas9 technology. *Molecular Plant Pathology* 7: 1140–1153. https://doi.org/10.1111/mpp.12375

Chapman, M. A., Pashley, C. H., Wenzler, J., et al. 2008. A genomic scan for selection reveals candidates for genes involved in the evolution of cultivated sunflower (*Helianthus annuus*). *The Plant Cell* 20(11): 2931–2945. https://doi.org/10.1105/tpc.108.059808

Char, S. N., Unger-Wallace, E., Frame, B., et al. 2015. Heritable site-specific mutagenesis using TALENs in maize. *Plant Biotechnology Journal* 13: 1002–1010. https://doi.org/10.1111/pbi.12344

Chavez, A., Scheiman, J., Vora, S., et al. 2015. Highly efficient Cas9-mediated transcriptional programming. *Nature Methods* 12: 326–328. https://doi.org/10.1038/nmeth.3312

Chen, A. and Dubcovsky, J. 2012. Wheat TILLING mutants show that the vernalization gene VRN1 down-regulates the flowering repressor VRN2 in leaves but is not essential for flowering. *PLoS Genetics* 8: e1003134. https://doi.org/10.1371/journal.pgen.1003134

Cheng, A. W., Wang, H., Yang, H., et al. 2013. Multiplexed activation of endogenous genes by CRISPR-on, an RNA-guided transcriptional activator system. *Cell Research* 23: 1163–1171. https://doi.org/10.1038/cr.2013.122

Christian, M., Cermak, T., Doyle, E. L., et al. 2010. Targeting DNA double-strand breaks with TAL effector nucleases. *Genetics* 186(2): 757–761. https://doi.org/10.1534/genetics.110.120717

Clasen, B. M., Stoddard T. J., Luo S., et al. 2016. Improving cold storage and processing traits in potato through targeted gene knockout. *Plant Biotechnology Journal* 14: 169–176. https://doi.org/10.1111/pbi.12370

Colasuonno, P., Incerti, O., Lozito, M. L., Simeone, R., Gadaleta, A. and Blanco, A. 2016. DHPLC technology for high-throughput detection of mutations in a durum wheat TILLING population. *BMC Genetics* 17: 43. https://doi.org/10.1186/s12863-016-0350-0

Colbert, T., Till, B. J., Tompa, R. and Reynolds, S. 2001. High-throughput screening for induced point mutations. *Plant Physiology* 126: 480–484. https://doi.org/10.1104/pp.126.2.480

Comai, L., Young, K., Till, B. J., et al. 2004. Efficient discovery of DNA polymorphisms in natural populations by EcoTILLING. *The Plant Journal* 37(5): 778–786. https://doi.org/10.1111/j.0960-7412.2003.01999.x

Cong, L., Ran, F. A., Cox, D., et al. 2013. Multiplex genome engineering using CRISPR/Cas systems. *Science* 339(6): 819–823. https://doi.org/10.1126/science.1231143

Cooper, J. L., Till, B. J., Laport, R. G., et al. 2008. TILLING to detect induced mutations in soybean. *BMC Plant Biology* 8: 9. https://doi.org/10.1186/1471-2229-8-9

Curtin, S. J., Zhang, F., Sander, J. D., et al. 2011. Targeted mutagenesis of duplicated genes in soybean with zinc-finger nucleases. *Plant Physiology* 156(2): 466–473. https://doi.org/10.1104/pp.111.172981

Dasgupta, S., Collins, G. B. and Hunt, A. G. 1998. Coordinated expression of multiple enzymes in different subcellular compartments in plants. *The Plant Journal* 16(1): 107–116. https://doi.org/10.1046/j.1365-313x.1998.00255.x

De Buck, S., Jacobs, A., Van Montagu, M. and Depicker, A. 1999. The DNA sequences of T-DNA junctions suggest that complex T-DNA loci are formed by a recombination process resembling T-DNA integration. *The Plant Journal* 20: 295–304. https://doi.org/10.1046/j.1365-313X.1999.00602.x

De La Fuente, G. N., Frei, U. K. and Lubberstedt, T. 2013. Accelerating plant breeding. *Trends in Plant Science* 18(12): 667–672. https://doi.org/10.1016/j.tplants.2013.09.001

De Neve, M., De Buck, S., Jacobs, A., Van Montagu, M. and Depicker, A. 1997. T-DNA integration patterns in co-transformed plant cells suggest that T-DNA repeats originate from co-integration of separate T-DNAs. *The Plant Journal* 11(1): 15–29. https://doi.org/10.1046/j.1365-313X.1997.11010015.x

DeFrancesco, L. E. 2012. Move over ZFNs. *Nature Biotechnology* 29: 681–684. https://doi.org/10.1038/nbt.1935

Dey, N., Sarkar, S., Acharya, S. and Maiti, I.B. 2015. Synthetic promoters in planta. *Planta*242(5):1077–1094. https://doi.org/10.1007/s00425-015-2377-2

Etherington, G. J., Monaghan, J., Zipfel, C., and MacLean, D. 2014. Mapping mutations in plant genomes with the user-friendly web application CandiSNP. *Plant Methods* 10: 41. https://doi.org/10.1186/s13007-014-0041-7

Fekih, R., Takagi, H., Tamiru, M., et al. 2013. MutMap$^+$: genetic mapping and mutant identification without crossing in rice. *PLoS One* 8(7): e68529. https://doi.org/10.1371/journal.pone.0068529

Foley, J. E., Yeh, J.-R. J., Maeder, M. L., et al. 2009. Rapid mutation of endogenous zebrafish genes using zinc finger nucleases made by Oligomerized Pool ENgineering (OPEN). *PLoS One* 4: e4348. https://doi.org/10.1371/journal.pone.0004348

Frerichmann, S. L., Kirchhoff, M., Muller, A. E. et al. 2013. EcoTILLING in *Beta vulgaris* reveals polymorphisms in the *FLC*-like gene *BvFL1*that are associated with annuality and winter hardiness. *BMC Plant Biology* 13: 52. https://doi.org/10.1186/1471-2229-13-52

Gaeta, R. T., Masonbrink, R. E., Krishnaswamy, L., Zhao, C. and Birchler, J. A. 2012. Synthetic chromosome platforms in plants. *Annual Review of Plant Biology* 63: 307–330. https://doi.org/10.1146/annurev-arplant-042110-103924

Gaeta, R. T., Masonbrink, R. E., Zhao, C., Sanyal, A., Krishnaswamy, L. and Birchler, J. A. 2013. *In vivo* modification of a maize engineered minichromosome. *Chromosoma* 122(3): 221–232. https://doi.org/10.1007/s00412-013-0403-3

Gao, C. 2018. The future of CRISPR technologies in agriculture. *Nature Reviews Molecular Cell Biology* 19(5): 275–276. https://doi.org/10.1038/nrm.2018.2

Germana, M. A. 2012. Use of irradiated pollen to induce parthenogenesis and haploid production in fruit crops. In *Plant mutation breeding and biotechnology*, eds. Shu, Q. Y., Forster, B. P. and Nakagawa, H., 411–421. Oxfordshire, UK, CABI, FAO. https://doi.org/10.1079/9781780640853.0411

Ghosh, S., Watson, A., Gonzalez-Navarro, O. E., et al. 2018. Speed breeding in growth chambers and glasshouses for crop breeding and model plant research. *Nature Protocols* 13: 2944–2963. https://doi.org/10.1038/s41596-018-0072-z

Gilbert, L. A., Larson, M. H., Morsut, L., et al. 2013. CRISPR-mediated modular RNA-guided regulation of transcription in eukaryotes. *Cell* 154(2): 442–451. https://doi.org/10.1016/j.cell.2013.06.044

Golic, K. G. and Lindquist, S. 1989. The FLP recombinase of yeast catalyzes site-specific recombination in the *Drosophila* genome. *Cell* 59(3): 499–509. https://doi.org/10.1016/0092-8674(89)90033-0

Gorjanc, G., Gaynor, R. C. and Hickey, J. M. 2018. Optimal cross selection for long-term genetic gain in two-part programs with rapid recurrent genomic selection. *Theoretical and Applied Genetics* 131(9): 1953–1966. https://doi.org/10.1007/s00122-018-3125-3

Gottwald, S., Bauer, P., Komatsuda, T., Lundqvist, U. and Stein, N. 2009. TILLING in the two-rowed barley cultivar '*Barke*' reveals preferred sites of functional diversity in the gene *HvHox1*. *BMC Research Notes* 2: 258. https://doi.org/10.1186/1756-0500-2-258

Guan, X., Stege, J., Kim, M., et al. 2002. Heritable endogenous gene regulation in plants with designed polydactyl zinc finger transcription factors. *Proceedings of the National Academy of Sciences of the United States of America* 99: 13296–13301. https://doi.org/10.1073/pnas.192412899

Guha, S. and Maheshwari, S. C. 1964. *In vitro* production of embryos from anthers of *Datura*. *Nature* 204: 497–497. https://doi.org/10.1038/204497a0

Gupta, M., DeKelver, R.C., Palta, A., et al. 2012. Transcriptional activation of *Brassica napus* β-ketoacyl-ACP synthase II with an engineered zinc finger protein transcription factor. *Plant Biotechnology Journal* 10(7): 783–791. https://doi.org/10.1111/j.1467-7652.2012.00695.x.

Halpin, C. 2005. Gene stacking in transgenic plants–the challenge for 21st century plant biotechnology. *Plant Biotechnology Journal* 3(2): 141–155. https://doi.org/10.1111/j.1467-7652.2004.00113.x

Hamilton, C. M. 1997. A binary-BAC system for plant transformation with high-molecular-weight DNA. *Gene* 200: 107–116. https://doi.org/10.1016/S0378-1119(97)00388-0

Hasunuma, T., Miyazawa, S. I., Yoshimura, S., et al. 2008. Biosynthesis of astaxanthin in tobacco leaves by transplastomic engineering. *The Plant Journal* 55(5): 857–868. https://doi.org/10.1111/j.1365-313X.2008.03559.x

Haun, W., Coffman, A., Clasen, B. M., et al. 2014. Improved soybean oil quality by targeted mutagenesis of the fatty acid desaturase 2 gene family. *Plant Biotechnology Journal* 12: 934–940. https://doi.org/10.1111/pbi.12201

Henikoff, S., Till, B. J. and Comai, L. 2004. TILLING traditional mutagenesis meets functional genomics. *Plant Physiology* 135: 630–636. https://doi.org/10.1104/pp.104.041061

Hert, D. G., Fredlake, C. P. and Barron, A. E. 2008. Advantages and limitations of next-generation sequencing technologies: a comparison of electrophoresis and non-electrophoresis methods. *Electrophoresis* 29(23): 4618–4626. https://doi.org/10.1002/elps.200800456

Hickey, L. T., German, S. E., Pereyra, S. A., et al. 2017. Speed breeding for multiple disease resistance in barley. *Euphytica* 213(3): 64. https://doi.org/10.1007/s10681-016-1803-2

Hickey, L. T., Hafeez, A. N., Robinson, H., et al. 2019. Breeding crops to feed 10 billion. *Nature Biotechnology* 37: 744–754. https://doi.org/10.1038/s41587-019-0152-9

Hilioti, Z., Ganopoulos, I., Ajith, S., et al. 2016. A novel arrangement of zinc finger nuclease system for *in vivo* targeted genome engineering: the tomato *LEC1-LIKE4* gene case. *Plant Cell Reports* 35: 2241–2255. https://doi.org/10.1007/s00299-016-2031-x

Huang, T. K. and Puchta, H. 2019. CRISPR/Cas-mediated gene targeting in plants: finally a turn for the better homologous recombination. *Plant Cell Reports* 38: 443–453. https://doi.org/10.1007/s00299-019-02379-0

Ibiza, V. P., Canizares, J. and Nuez, F. 2010. EcoTILLING in Capsicum species: Searching for new virus resistances. *BMC Genomics* 11: 631. https://doi.org/10.1186/1471-2164-11-631

Jaganathan, D., Ramasamy, K., Sellamuthu, G., Jayabalan, S. and Venkataraman, G. 2018. CRISPR for crop improvement: an update review. *Frontiers in Plant Science* 9: 985. https://doi.org/10.3389/fpls.2018.00985

Jahne, F., Hahn, V., Wurschum, T. and Leiser, W. L. 2020. Speed breeding short–day crops by LED-controlled light schemes. *Theoretical and Applied Genetics* 133: 2335–2342. https://doi.org/10.1007/s00122-020-03601-4

Jankowicz-Cieslak, J., Huynh, O. A., Brozynska, M., Nakitandwe, J. and Till, B. J. 2012. Induction, rapid fixation and retention of mutations in vegetatively propagated banana. *Plant Biotechnology Journal* 10: 1056–1066. https://doi.org/10.1111/j.1467-7652.2012.00733.x

Jankowicz-Cieslak, J., Kozak-Stankiewicz, K., Seballos, G., et al. 2013. Application of soft X-ray and near-infrared reflectance spectroscopy for rapid phenotyping of mutant rice seed. In *Plant genetics and breeding technologies*, 37–40. Austria, Vienna.

Jawhar, M., Arabi, M. I. E., MirAli, N. and Till, B. J. 2018. Efficient discovery of single-nucleotide variations in *Cochliobolus sativus* vegetative compatibility groups by EcoTILLING. *Journal of Plant Biochemistry and Physiology* 6(2): 1000211. https://doi.org/10.4172/2329-9029.1000211

Jia, H., Zhang, Y., Orbovic, V., Xu, J., White, F. F., Jones, J. B. and Wang, N. 2017. Genome editing of the disease susceptibility gene *CsLOB1* in citrus confers resistance to citrus canker. *Plant Biotechnology Journal* 15: 817–823. https://doi.org/10.1111/pbi.12677

Jia, Q., Verk, M. C., Pinas, J. E., Lindhout, B. I., Hooykaas, P. J. and Zaal, B. J. 2013. Zinc finger artificial transcription factor-based nearest inactive analogue/nearest active analogue strategy used for the identification of plant genes controlling homologous recombination. *Plant Biotechnology Journal* 11: 1069–1079. https://doi.org/10.1111/pbi.12101

Jiao, Y., Burow, G., Gladman, N., et al. 2018. Efficient identification of causal mutations through sequencing of bulked F2 from two allelic bloomless mutants of *Sorghum bicolor. Frontiers in Plant Science* 8: 2267. https://doi.org/10.3389/fpls.2017.02267

Jinek, M., Chylinski, K., Fonfara, I., Hauer, M., Doudna, J. A. and Charpentier, E. 2012. A programmable dual-RNA-guided DNA endonuclease in adaptive bacterial immunity. *Science* 337(6096): 816–821. https://doi.org/10.1126/science.1225829

Joseph, R., Yeoh, H. H. and Loh, C. S. 2004. Induced mutations in cassava using somatic embryos and the identification of mutant plants with altered starch yield and composition. *Plant Cell Reports* 23: 91–98. https://doi.org/10.1007/s00299-004-0798-7

Jung, J. H. and Altpeter, F. 2016. TALEN mediated targeted mutagenesis of the caffeic acid O-methyltransferase in highly polyploid sugarcane improves cell wall composition for production of bioethanol. *Plant Molecular Biology* 92: 131–142. https://doi.org/10.1007/s11103-016-0499-y

Kadaru, S. B., Yadav, A. S., Fjellstrom, R. G. and Oard, J. H. 2006. Alternative EcoTILLING protocol for rapid, cost-effective single-nucleotide polymorphism discovery and genotyping in rice (*Oryza sativa* L.). *Plant Molecular Biology Reporter* 24: 3–22. https://doi.org/10.1007/BF02914042

Karthika, V., Babitha, K.C., Kiranmai, K., Shankar, A. G., Vemanna, R. S. and Udayakumar, M. 2020. Involvement of DNA mismatch repair systems to create genetic diversity in plants for speed breeding programs. *Plant Physiology Reports* 25: 185–199. https://doi.org/10.1007/s40502-020-00521-9

Kettleborough, R. N. W., Busch-Nentwich, E. M., Harvey, S. A., et al. 2013. A systematic genome-wide analysis of zebrafish protein-coding gene function. *Nature* 496(7446): 494–497. https://doi.org/10.1038/nature11992

Kharkwal, M.C. 2012. A brief history of plant mutagenesis. In *Plant mutation breeding and biotechnology*, eds. Shu, Q. Y., Forster, B. P. and Nakagawa, H., 21–30. CABI, Wallingford. https://doi.org/10.1079/9781780640853.0021

Kim, D., Kim, D., Alptekin, B. and Budak, H. 2017a. CRISPR/Cas9 genome editing in wheat. *Functional and Integrative Genomics* 18: 31–41. https://doi.org/10.1007/s10142-017-0572-x

Kim, H., Kim, S. T., Ryu, J., Kang, B. C., Kim, J. S. and Kim S. G. 2017b. CRISPR/Cpf1-mediated DNA-free plant genome editing. *Nature Communications* 8: 14406. https://doi.org/10.1038/ncomms14406

Kim, S. I. and Tai, T. H. 2013. Identification of SNPs in closely related temperate Japonica rice cultivars using restriction enzyme-phased sequencing. *PLoS One* 8(3): e60176. https://doi.org/10.1371/journal.pone.0060176

Kim, S. I. and Tai, T. H. 2014. Identification of novel rice low phytic acid mutations via TILLING by sequencing. *Molecular Breeding* 34: 1717–1729. https://doi.org/10.1007/s11032-014-0127-y

Kim, Y. G., Cha, J. and Chandrasegaran, S. 1996. Hybrid restriction enzymes: zinc finger fusions to Fok I cleavage domain. *Proceedings of the National Academy of Sciences of the United States of America* 93(3): 1156–1160. https://doi.org/10.1073/pnas.93.3.1156

Kodym, A. and Zapata-Arias, F. J. 2001. Low-cost alternatives for the micropropagation of banana. *Plant Cell Tissue and Organ Culture* 66: 67–71. https://doi.org/10.1023/A:1010661521438

Konermann, S., Brigham, M. D., Trevino, A. E., et al. 2015. Genome-scale transcriptional activation by an engineered CRISPR-Cas9 complex. *Nature* 517: 583–588. https://doi.org/10.1038/nature14136

Li, J., Stoddard, T. J., Demorest, Z. L., et al. 2016a. Multiplexed, targeted gene editing in *Nicotiana benthamiana* for glyco-engineering and monoclonal antibody production. *Plant Biotechnology Journal* 14: 533–542. https://doi.org/10.1111/pbi.12403

Li, J. F., Norville, J. E., Aach, J., et al. 2013. Multiplex and homologous recombination-mediated genome editing in *Arabidopsis* and *Nicotiana benthamiana* using guide RNA and Cas9. *Nature Biotechnology* 31: 688–691. https://doi.org/10.1038/nbt.2654

Li, L., Zhou, Y., Cheng, X., et al. 2003. Combinatorial modification of multiple lignin traits in trees through multigene cotransformation. *Proceedings of the National Academy of Sciences of the United States of America* 100(8): 4939–4944. https://doi.org/10.1073/pnas.0831166100

Li, M., Li, X., Zhou, Z., et al. 2016b. Reassessment of the four yield-related genes *Gn1a*, *DEP1*, *GS3*, and *IPA1* in rice using a CRISPR/Cas9 system. *Frontiers in Plant Science* 7: 377. https://doi.org/10.3389/fpls.2016.00377

Li, R., Fu, D., Zhu, B., Luo, Y. and Zhu, H. 2018a. CRISPR/Cas9-mediated mutagenesis of lncRNA1459 alters tomato fruit ripening. *The Plant Journal* 94: 513–524. https://doi.org/10.1111/tpj.13872

Li, R., Liu, C., Zhao, R. et al. 2019. CRISPR/Cas9-Mediated *SlNPR1* mutagenesis reduces tomato plant drought tolerance. *BMC Plant Biology* 19:38. https://doi.org/10.1186/s12870- -018-1627-4

Li, R., Zhang, L., Wang, L., et al. 2018b. Reduction of tomato-plant chilling tolerance by CRISPR-Cas9-mediated *SlCBF1* mutagenesis. *Journal of Agricultural and Food Chemistry* 66: 9042–9051. https://doi.org/10.1021/acs.jafc.8b02177

Li, T., Liu, B., Spalding, M. H., Weeks, D. P. and Yang, B. 2012. High-efficiency TALEN-based gene editing produces disease-resistant rice. *Nature Biotechnology* 30: 390–392. https://doi.org/10.1038/nbt.2199

Li, W., Guo, H., Wang, Y., et al. 2017a. Identification of novel alleles induced by EMS-mutagenesis in key genes of kernel hardness and starch biosynthesis in wheat by TILLING. *Genes and Genomics* 39: 387–395. https://doi.org/10.1007/s13258-016-0504-5

Li, X., Wang, Y., Chen, S., et al. 2018c. Lycopene is enriched in tomato fruit by CRISPR/Cas9-mediated multiplex genome editing. *Frontiers in Plant Science* 9: 559. https://doi.org/10.3389/fpls.2018.00559

Li, X., Zhou, W., Ren, Y., et al. 2017b. High-efficiency breeding of early-maturing rice cultivars via CRISPR/Cas9-mediated genome editing. *Journal of Genetics and Genomics* 44: 175–178. https://doi.org/10.1016/j.jgg.2017.02.001

Li, Z., Zhang, D., Xiong, X., et al. 2017c. A potent Cas9-derived gene activator for plant and mammalian cells. *Nature Plants* 3: 930–936. https://doi.org/10.1038/s41477-017-0046-0

Liang, C., Wang, S. Q. and Hu, Y. G. 2011. Detection of SNPs in the VRN-A1 gene of common wheat (*Triticum aestivum* L.) by a modified EcoTILLING method using agarose gel electrophoresis. *Australian Journal of Crop Science* 5: 318–326.

Lindhout, B. I., Pinas, J. E., Hooykaas, P. J. and Van Der Zaal, B. J. 2006. Employing libraries of zinc finger artificial transcription factors to screen for homologous recombination mutants in *Arabidopsis*. *The Plant Journal* 48: 475–483. https://doi.org/10.1111/j.1365-313X.2006.02877.x

Liu, J., Li, C., Yu, Z., et al. 2012b. Efficient and specific modifications of the *Drosophila* genome by means of an easy TALEN strategy. *Journal of Genetics and Genomics* 39(5): 209–215. https://doi.org/10.1016/j.jgg.2012.04.003

Liu, L., Li, Y. H., Li, S. L., et al. 2012a. Comparison of next-generation sequencing systems. *Journal of Biomedicine and Biotechnology* 2012: 251364. https://doi.org/10.1155/2012/251364

Liu, W., Mazarei, M., Peng, Y., et al. 2014. Computational discovery of soybean promoter cis-regulatory elements for the construction of soybean cyst nematode-inducible synthetic promoters. *Plant Biotechnology Journal* 12(8): 1015–1026. https://doi.org/10.1111/pbi.12206

Liu, W., Yuan, J. S. and Stewart, C. N. Jr. 2013. Advanced genetic tools for plant biotechnology. *Nature Reviews Genetics* 14(11): 781–793. https://doi.org/10.1038/nrg3583

Liu, Y. G., Shirano, Y., Fukaki, H., et al. 1999. Complementation of plant mutants with large genomic DNA fragments by a transformation-competent artificial chromosome vector accelerates positional cloning. *Proceedings of the National Academy of Sciences of the United States of America* 96(11): 6535–6540. https://doi.org/10.1073/pnas.96.11.6535

Lossl, A., Bohmert, K., Harloff, H., Eibl, C., Muhlbauer, S. and Koop, H. U. 2005. Inducible trans-activation of plastid transgenes: expression of the *R. eutrophaphb* operon in transplastomic tobacco. *Plant and Cell Physiology* 46(9): 1462–1471. https://doi.org/10.1093/pcp/pci157

Lou, D., Wang, H., Liang, G. and Yu, D. 2017. *OsSAPK2* confers abscisic acid sensitivity and tolerance to drought stress in rice. *Frontiers in Plant Science* 8: 993. https://doi.org/10.3389/fpls.2017.00993

Lowder, L. G., Paul, J. W., Baltes, N. J., et al. 2015. A CRISPR/Cas9 toolbox for multiplexed plant genome editing and transcriptional regulation. *Plant Physiology* 169: 971–985. https://doi.org/10.1104/pp.15.00636

Lowder, L. G., Zhou, J., Zhang, Y., et al. 2017. Robust transcriptional activation in plants using multiplexed CRISPR-Act2.0 and mTALE-Act systems. *Molecular Plant* 11: 245–256. https://doi.org/10.1016/j.molp.2017.11.010

Lu, Y., Rijzaani, H., Karcher, D., Ruf, S. and Bock, R. 2013. Efficient metabolic pathway engineering in transgenic tobacco and tomato plastids with synthetic multigene operons. *Proceedings of the National Academy of Sciences of the United States of America* 110: E623–E632. https://doi.org/10.1073/pnas.1216898110

Lucker, J., Schwab, W., Van Hautum, B., Van Der Plas, L. H. W., Bouwmeester, H. J. and Verhoeven, H. A. 2004. Increased and altered fragrance of tobacco plants after metabolic engineering using three monoterpene synthases from lemon. *Plant Physiology* 134: 510–519. https://doi.org/10.1104/pp.103.030189

Ma, J. K., Hiatt, A., Hein, M., et al. 1995. Generation and assembly of secretory antibodies in plants. *Science* 268(5211): 716–719. https://doi.org/10.1126/science.7732380

Ma, X., Sajjad, M., Wang, J., et al. 2017. Diversity, distribution of Puroindoline genes and their effect on kernel hardness in a diverse panel of Chinese wheat germplasm. *BMC Plant Biology* 7: 158. https://doi.org/10.1186/s12870-017-1101-8

Macovei, A., Sevilla, N. R., Cantos, C., et al. 2018. Novel alleles of rice eIF4G generated by CRISPR/Cas9-targeted mutagenesis confer resistance to Rice tungro spherical virus. *Plant Biotechnology Journal* 16: 1918–1927. https://doi.org/10.1111/pbi.12927

Maeder, M. L., Thibodeau-Beganny, S., Osiak, A., et al. 2008. Rapid "open-source" engineering of customized zinc-finger nucleases for highly efficient gene modification. *Molecular Cell* 31: 294–301. https://doi.org/10.1016/j.molcel.2008.06.016

McCallum, C. M., Comai, L., Greene, E. A. and Henikoff, S. 2000. Targeting induced local lesions in genomes (TILLING) for plant functional genomics. *Plant Physiology* 123: 439–442. https://doi.org/10.1104/pp.123.2.439

McClintock, B. 1951. Mutable loci in maize. *Carnegie Institution of Washington Yearbook* 50:174–181.

Mehrotra, R., Gupta, G., Sethi, R., Bhalothia, P., Kumar, N. and Mehrotra, S. 2011. Designer promoter: an artwork of cis engineering. *Plant Molecular Biology* 75(6): 527–536. https://doi.org/10.1007/s11103-011-9755-3

Mengiste, T., Revenkova, E., Bechtold, N. and Paszkowski, J. 1999. An SMC-like protein is required for efficient homologous recombination in Arabidopsis. *EMBO Journal* 18(16): 4505–4512. https://doi.org/10.1093/emboj/18.16.4505

Michelmore, R. W., Paran, I., and Kesseli, R.V. 1991. Identification of markers linked to disease-resistance genes by bulked segregant analysis: A rapid method to detect markers in specific genomic regions by using segregating populations. *Proceedings of the National Academy of Sciences of the United States of America* 88(21): 9828–9832. https://doi.org/10.1073/pnas.88.21.9828

Mo, Y., Howell, T., Vasquez-Gross, H., de Haro, L. A., Dubcovsky, J. and Pearce, S. 2018. Mapping causal mutations by exome sequencing in a wheat TILLING population: a tall mutant case study. *Molecular Genetics and Genomics* 293: 463–477. https://doi.org/10.1007/s00438-017-1401-6

Mobini, S. H., Lulsdorf, M., Warkentin, T.D. and Vandenberg, A. 2015. Plant growth regulators improve *in vitro* flowering and rapid generation advancement in lentil and faba bean. *In Vitro Cellular and Developmental Biology Plant* 51: 71–79. https://doi.org/10.1007/s11627-014-9647-8

Mochida, K. and Shinozaki, K. 2011. Advances in omics and bioinformatics tools for systems analyses of plant functions. *Plant and Cell Physiology* 52(12): 2017–2038. https://doi.org/10.1093/pcp/pcr153

Moehs, C. P., Austill, W. J., Holm, A., et al. 2018. Development of reduced gluten wheat enabled by determination of the genetic basis of the lys3a low hordein barley mutant. *Plant Physiology* 179: 1692–1703. https://doi.org/10.1104/pp.18.00771

Mohan, C., Jayanarayanan, A. N. and Narayanan, S. 2017. Construction of a novel synthetic root-specific promoter and its characterization in transgenic tobacco plants. *3 Biotech* 7(4): 234. https://doi.org/10.1007/s13205-017-0872-9

Morgan, T.H. 1909. What are "factors" in Mendelian explanations? *American Breeder Association Reports* 5: 365–368.

Morineau, C., Bellec, Y., Tellier, F., Gissot, L., Kelemen, Z., Nogue F. and Faure, J. D. 2017. Selective gene dosage by CRISPR-Cas9 genome editing in hexaploid *Camelina sativa*. *Plant Biotechnology Journal* 15: 729–739

Negrao, S., Almadanim, C., Pires, I., McNally, K. L. and Oliveira, M. M. 2011. Use of EcoTILLING to identify natural allelic variants of rice candidate genes involved in salinity tolerance. *Plant Genetic Resources* 9: 300–304. https://doi.org/10.1017/S1479262111000566

Negrao, S., Almadanim, M. C., Pires, I. S., et al. 2013. New allelic variants found in key rice salt-tolerance genes: An association study. *Plant Biotechnology Journal* 11: 87–100. https://doi.org/10.1111/pbi.12010

Novak, F. J. and Brunner, H. 1992. Plant breeding: induced mutation technology for crop improvement. *IAEA Bulletin* 34(4): 25–33.

O'Connor, D. J., Wright, G. C., Dieters, M. J., et al. 2013. Development and application of speed breeding technologies in a commercial peanut breeding program. *Peanut Science* 40(2): 107–114. https://doi.org/10.3146/PS12-12.1

Ortigosa, A., Gimenez-Ibanez, S., Leonhardt, N. and Solano, R. 2018. Design of a bacterial speck resistant tomato by CRISPR/Cas9-mediated editing of SlJAZ2. *Plant Biotechnology Journal* 17: 665–673. https://doi.org/10.1111/pbi.13006

Palanisamy, D., Marappan, S., Ponnuswamy, R. D., Mahalingam, P. S., Bohar, R. and Vaidyanathan, S. 2019. Accelerating hybrid rice breeding through the adoption of doubled haploid technology for R-line development. *Biologia* 74: 1259–1269. https://doi.org/10.2478/s11756-019-00300-4

Parry, M. A., Madgwick, P. J., Bayon, C., et al. 2009. Mutation discovery for crop improvement. *Journal of Experimental Botany* 60(10): 2817–2825. https://doi.org/10.1093/jxb/erp189

Penfield, S. 2017. Seed dormancy and germination. *Current Biology* 27(17): R874–R878. https://doi.org/10.1016/j.cub.2017.05.050

Petolino, J. F. and Davies, J. P. 2013. Designed transcriptional regulators for trait development. *Plant Science* 201: 128–136. https://doi.org/10.1016/j.plantsci.2012.12.006

Porteus, M. H. 2009. Zinc fingers on target. *Nature* 459: 337–338. https://doi.org/10.1038/459337a

Qaim, M. 2020. Role of new plant breeding technologies for food security and sustainable agricultural development. *Applied Economic Perspectives and Policy* 42(2): 129–150. https://doi.org/10.1002/aepp.13044

Qi, L. S., Larson, M. H., Gilbert, L. A., et al. 2013. Repurposing CRISPR as an RNA-guided platform for sequence-specific control of gene expression. *Cell* 152(5): 1173–1183. https://doi.org/10.1016/j.cell.2013.02.022

Qi, W., Zhu, T., Tian, Z., Li, C., Zhang, W. and Song, R. 2016. High-efficiency CRISPR/Cas9 multiplex gene editing using the glycine tRNA-processing system-based strategy in maize. *BMC Biotechnology* 16: 58. https://doi.org/10.1186/s12896-016-0289-2

Quesada-Vargas, T., Ruiz, O. N. and Daniell, H. 2005. Characterization of heterologous multigene operons in transgenic chloroplasts: transcription, processing, and translation. *Plant Physiology* 138: 1746–1762. https://doi.org/10.1104/pp.105.063040

Raja, R. B., Agasimani, S., Jaiswal, S., et al. 2017. EcoTILLING by sequencing reveals polymorphisms in genes encoding starch synthases that are associated with low glycemic response in rice. *BMC Plant Biology* 17: 13. https://doi.org/10.1186/s12870-016-0968-0

Ralley, L., Enfissi, E. M., Misawa, N., Schuch, W., Bramley, P. M. and Fraser, P. D. 2004. Metabolic engineering of ketocarotenoid formation in higher plants. *The Plant Journal* 39(4): 477–486. https://doi.org/10.1111/j.1365-313X.2004.02151.x

Rawat, N., Sehgal, S. K., Joshi, A., et al. 2012. A diploid wheat TILLING resource for wheat functional genomics. *BMC Plant Biology* 12: 205. https://doi.org/10.1186/1471-2229-12-205

Ruiz, O. N., Hussein, H., Terry, N. and Daniell, H. 2003. Phytoremediation of organomercurial compounds via chloroplast genetic engineering. *Plant Physiology* 132: 1344–1352. https://doi.org/10.1104/pp.103.020958

Samineni, S., Sen, M., Sajja, S. B. and Gaur, P.M. 2019. Rapid generation advance (RGA) in chickpea to produce up to seven generations per year and enable speed breeding. *The Crop Journal* 8(1). https://doi.org/10.1016/j. cj.2019.08.003

Sanchez-Leon, S., Gil-Humanes, J., Ozuna, C. V., et al. 2018. Low-gluten, non-transgenic wheat engineered with CRISPR/Cas9. *Plant Biotechnology Journal* 16: 902–910. https://doi.org/10.1111/pbi.12837

Schmidt, M. A., LaFayette, P. R., Artelt, B. A. and Parrot, W. 2008. A comparison of strategies for transformation with multiple genes via microprojectile-mediated bombardment. *In Vitro Cellular and Developmental Biology Plant* 44(3): 162–168. https://doi.org/10.1007/s11627-007-9099-5

Schneeberger, K., Ossowski, S., Lanz, C., et al. 2009. SHOREmap: Simultaneous mapping and mutation identification by deep sequencing. *Nature Methods* 6: 550–551. https://doi.org/10.1038/nmeth0809-550

Schunk, C. R. and Eberius, M. 2012. Phenomics in plant biological research and mutation breeding. In *Plant mutation breeding and biotechnology*, eds. Shu, Q. Y., Forster, B. P. and Nakagawa, H., 535–559. Oxfordshire, UK, CABI, FAO. https://doi.org/10.1079/9781780640853.0000

Schwartz, A. R., Morbitzer, R., Lahaye, T. and Staskawicz, B. J. 2017. TALE-induced bHLH transcription factors that activate a pectate lyase contribute to water soaking in bacterial spot of tomato. *Proceedings of the National Academy of Sciences of the United States of America* 114: E897–E903. https://doi.org/10.1073/pnas.1620407114

Segal, D. J., Beerli, R. R., Blancafort, P., et al. 2003. Evaluation of a modular strategy for the construction of novel polydactyl zinc finger DNA-binding proteins. *Biochemistry* 42(7): 2137–2148. https://doi.org/10.1021/bi026806o

Seitz, C., Vitten, M., Steinbach, P., et al. 2007. Redirection of anthocyanin synthesis in *Osteospermum hybrida* by a two-enzyme manipulation strategy. *Phytochemistry* 68: 824–833. https://doi.org/10.1016/j.phytochem.2006.12.012

Serrat, X., Esteban, R., Guibourt, N., Moysset, L., Nogues, S. and Lalanne, E. 2014. EMS mutagenesis in mature seed-derived rice calli as a new method for rapidly obtaining TILLING mutant populations. *Plant Methods* 10: 5. https://doi.org/10.1186/1746-4811-10-5

Sestili, F., Botticella, E., Bedo, Z., Phillips, A. and Lafiandra, D. 2010. Production of novel allelic variation for genes involved in starch biosynthesis through mutagenesis. *Molecular Breeding* 25: 145. https://doi.org/10.1007/s11032-009-9314-7

Shan, Q., Wang, Y., Li, J., et al. 2013. Targeted genome modification of crop plants using a CRISPR-Cas system. *Nature Biotechnology* 31: 686–688. https://doi.org/10.1038/nbt.2650

Shen, C., Que, Z., Xia, Y., et al. 2017. Knock out of the annexin gene *OsAnn3* via CRISPR/Cas9-mediated genome editing decreased cold tolerance in rice. *Journal of Plant Biology* 60: 539–547. https://doi.org/10.1007/s12374-016-0400-1

Shih, P. M., Liang, Y. and Loque, D. 2016. Biotechnology and synthetic biology approaches for metabolic engineering of bioenergy crops. *The Plant Journal* 87: 103–117. https://doi.org/10.1111/tpj.13176

Shukla, V. K., Doyon, Y., Miller, J. C., et al. 2009. Precise genome modification in the crop species *Zea mays* using zinc-finger nucleases. *Nature* 459(7245): 437–441. https://doi.org/10.1038/nature07992

Singla-Pareek, S. L., Reddy, M. K. and Sopory, S. K. 2003. Genetic engineering of the glyoxalase pathway in tobacco leads to enhanced salinity tolerance. *Proceedings of the National Academy of Sciences of the United States of America* 100(25): 14672–14677. https://doi.org/10.1073/pnas.2034667100

Sorek, R., Lawrence, C. M. and Wiedenheft, B. 2013. CRISPR-mediated adaptive immune systems in bacteria and archaea. *Annual Review of Biochemistry* 82: 237–266. https://doi.org/10.1146/annurev-biochem-072911-172315

Sparla, F., Falini, G., Botticella, E., et al. 2014. New starch phenotypes produced by TILLING in barley. *PLoS ONE* 9: e107779. https://doi.org/10.1371/journal.pone.0107779

Spitz, F. and Furlong, E. E. 2012. Transcription factors: from enhancer binding to developmental control. *Nature Reviews Genetics* 13(9): 613. https://doi.org/10.1038/nrg3207

Stadler, L. J. 1930. Some genetic effects of X–rays in plants. *Journal of Heredity* 21(1): 3–20. https://doi.org/10.1093/oxfordjournals.jhered.a103249

Stark, W. M., Boocock, M. R. and Sherratt, D. J. 1992. Catalysis by site-specific recombinases. *Trends in Genetics* 8(12): 432–439. https://doi.org/10.1016/0168-9525(92)90327-Z

Stege, J. T., Guan, X., Ho, T., Beachy, R. N. and Barbas, C. F. 2002. Controlling gene expression in plants using synthetic zinc finger transcription factors. *The Plant Journal* 32: 1077–1086. https://doi.org/10.1046/j.1365-313X.2002.01492.x

Suzuki, T., Eiguchi, M., Kumamaru, T., et al. 2008. MNU-induced mutant pools and high performance TILLING enable finding of any gene mutation in rice. *Molecular Genetics and Genomics* 279(3): 213–223. https://doi.org/10.1007/s00438-007-0293-2

Sysoeva, M. I., Markovskaya, E. F. and Shibaeva, T. G. 2010. Plants under continuous light: A review. *Plant Stress* 4: 5–17.

Szarejko, I. 2012. Haploid mutagenesis. In *Plant mutation breeding and biotechnology*, eds. Shu, Q. Y., Forster, B. P. and Nakagawa, H., 346–387. Oxfordshire, UK, CABI, FAO. https://doi.org/10.1079/9781780640853.0411

Takagi, H., Abe, A., Yoshida, K., et al. 2013a. QTL-Seq: Rapid mapping of quantitative trait loci in rice by whole-genome resequencing of DNA from two bulked populations. *The Plant Journal* 74(1): 174–183. https://doi.org/10.1111/tpj.12105

Takagi, H., Uemura, A., Yaegashi, H., et al. 2013b. MutMap-Gap: Whole-genome resequencing of mutant F2 progeny bulk combined with de novo assembly of gap regions identifies the rice blast resistance gene *Pii*. *New Phytologist* 200: 276–283. https://doi.org/10.1111/nph.12369

Takeno, K. 2016. Stress-induced flowering: the third category of flowering response. *Journal of Experimental Botany* 67: 4925–4934. https://doi.org/10.1093/jxb/erw272

Tanenbaum, M. E., Gilbert, L. A., Qi, L. S., Weissman, J. S. and Vale, R. D. 2014. A protein-tagging system for signal amplification in gene expression and fluorescence imaging. *Cell* 159: 635–646. https://doi.org/10.1016/j.cell.2014.09.039

Thorpe, H. M. and Smith, M. C. 1998. *In vitro* site-specific integration of bacteriophage DNA catalyzed by a recombinase of the resolvase/invertase family. *Proceedings of the National Academy of Sciences of the United States of America* 95(10): 5505–5510. https://doi.org/10.1073/pnas.95.10.5505

Till, B. J., Reynolds, S. H., Weil, C., et al. 2004. Discovery of induced point mutations in maize genes by TILLING. *BMC Plant Biology* 4: 12. https://doi.org/10.1186/1471-2229-4-12

Townsend, J., Wright, D., Winfrey, R. et al. 2009. High-frequency modification of plant genes using engineered zinc-finger nucleases. *Nature* 459:442–445. https://doi.org/10.1038/nature07845

Townsend, J. A., Wright, D. A., Winfrey, R. J., et al. 2009. High-frequency modification of plant genes using engineered zinc-finger nucleases. *Nature* 459: 442–445. https://doi.org/10.1038/nature07845

Tsai, H., Howell, T., Nitcher, R., et al. 2011. Discovery of rare mutations in populations: TILLING by sequencing. *Plant Physiology* 156: 1257–1268. https://doi.org/10.1104/pp.110.169748

Upadhyaya, P., Tyagi, K., Sharma, S., Tamboli, V., Sreelakshmi, Y. and Sharma, R. 2017. Natural variation in folate levels among tomato (*Solanum lycopersicum*) accessions. *Food Chemistry* 217: 610–619. https://doi: 10.1016/j.foodchem.2016.09.031

Urnov, F. D., Rebar, E. J., Holmes, M. C., Zhang, H. S. and Gregory, P. D. 2010. Genome editing with engineered zinc finger nucleases. *Nature Reviews Genetics* 11(9): 636–646. https://doi.org/10.1038/nrg2842

Van Harten, A. M. 1998. *Mutation breeding: theory and practical applications*. Cambridge, UK, Cambridge University Press.

van Tol, N., Rolloos, M., Pinas, J. E., et al. 2017. Enhancement of Arabidopsis growth characteristics using genome interrogation with artificial transcription factors. *PLoS One* 12: e0174236. https://doi.org/10.1371/journal.pone.0174236

Vavilov, N. I. 1926. Tzentry proiskhozhdeniya kulturnykh rastenii. [The centers of origin of cultivated plants.] *Works of Applied Botany and Plant Breeding* 16:1–248.

Vries, H de. 1901. *The mutation theory. Experiments and observations on the origin of species in the plant kingdom*. Leipzig, Verlag Von Veit & Comp.

Wada, K. C. and Takeno, K. 2010. Stress-induced flowering. *Plant Signaling and Behavior* 5: 944–947. https://doi.org/10.4161/psb.5.8.11826

Wang, F., Wang, C., Liu, P., et al. 2016. Enhanced rice blast resistance by CRISPR/Cas9-Targeted mutagenesis of the ERF transcription factor gene *OsERF922*. *PLoS ONE* 11: e0154027. https://doi.org/10.1371/journal.pone.0154027

Wang, J., Sun, J., Liu, D., et al. 2008. Analysis of Pina and Pinb alleles in the micro-core collections of Chinese wheat germplasm by EcoTILLING and identification of a novel Pinb allele. *Journal of Cereal Science* 48(3): 836–842. https://doi.org/10.1016/j.jcs.2008.06.005

Wang, N., Shi, L., Tian, F., et al. 2010. Assessment of FAE1 polymorphisms in three Brassica species using EcoTILLING and their association with differences in seed erucic acid contents. *BMC Plant Biology* 10: 137. https://doi.org/10.1186/1471-2229-10-137

Wang, X., Tu, M., Wang, D., Liu, J., Li, Y., Li, Z., Wang, Y., Wang, X. 2018. CRISPR/Cas9-mediated efficient targeted mutagenesis in grape in the first generation. *Plant Biotechnology Journal* 16(4):844–855. https://doi.org/10.1111/pbi.12832.

Wang, Y., Cheng, X., Shan, Q., et al. 2014. Simultaneous editing of three homoeoalleles in hexaploid bread wheat confers heritable resistance to powdery mildew. *Nature Biotechnology* 32: 947–952. https://doi.org/10.1038/nbt.2969

Watson, A., Ghosh, S., Williams, M., et al. 2018. Speed breeding is a powerful tool to accelerate crop research and breeding. *Nature Plants* 4(1): 23–29. https://doi.org/10.1038/s41477-017-0083-8

Watts, A., Sankaranarayanan, S., Raipuria, R. K. and Watts, A. 2020. Production and application of doubled haploid in *Brassica* improvement. In *Brassica improvement*, eds. Wani, S., Thakur, A., Jeshima Khan, Y. Cham, Springer.

Wollenweber, B., Porter, J. R. and Lubberstedt, T. 2005. Need for multidisciplinary research towards a second green revolution. *Current Opinion in Plant Biology* 8(3): 337–341. https://doi.org/10.1016/j.pbi.2005.03.001

Wyvekens, N., Topkar, V. V., Khayter, C., Joung, J. K. and Tsai, S. Q. 2015. Dimeric CRISPR RNA-guided FokI-dCas9 nucleases directed by truncated gRNAs for highly specific genome editing. *Human Gene Therapy* 26: 425–431. https://doi.org/10.1089/hum.2015.084

Xia, Y., Li, R., Ning, Z., et al. 2013. Single nucleotide polymorphisms in HSP17.8 and their association with agronomic traits in barley. *PLoS ONE* 8: e56816. https://doi.org/10.1371/journal.pone.0056816

Xu, R., Yang, Y., Qin, R., et al. 2016. Rapid improvement of grain weight via highly efficient CRISPR/Cas9-mediated multiplex genome editing in rice. *Journal of Genetics and Genomics* 43: 529–532. https://doi.org/10.1016/j.jgg.2016.07.003

Yang, X. Y. and X. L. Zhang. 2010. Regulation of somatic embryogenesis in higher plants. *Critical Reviews in Plant Science* 29: 36–57. https://doi.org/10.1080/07352680903436291

Young, N. D. and Tanksley, S. D. 1989. RFLP analysis of the size of chromosomal segments retained around the Tm-2 locus of tomato during backcross breeding. *Theoretical and Applied Genetics* 77: 353–359. https://doi.org/10.1007/BF00305828

Yu, W., Han, F., Gao, Z., Vega, J. M. and Birchler, J. A. 2007. Construction and behavior of engineered minichromosomes in maize. *Proceedings of the National Academy of Sciences of the United States of America* 104(21): 8924–8929. https://doi.org/10.1073/pnas.0700932104

Yu, W., Yau, Y. Y. and Birchler, J. A. 2015. Plant artificial chromosome technology and its potential application in genetic engineering. *Plant Biotechnology Journal* 14(5): 1175–1182. https://doi.org/10.1111/pbi.12466

Zeng, Y. D., Sun, J. L., Bu, S. H., et al. 2016. EcoTILLING revealed SNPs in *GhSus* genes that are associated with fiber- and seed-related traits in upland cotton. *Scientific Reports* 6: 29250. https://doi.org/10.1038/srep29250

Zhang, A., Liu, Y., Wang, F., et al. 2019. Enhanced rice salinity tolerance via CRISPR/Cas9-targeted mutagenesis of the *OsRR22* gene. *Molecular Breeding* 39: 47. https://doi.org/10.1007/s11032-019-0954-y

Zhang, F., Maeder, M. L., Unger-Wallace, E., et al. 2010. High frequency targeted mutagenesis in Arabidopsis thaliana using zinc finger nucleases. *Proceedings of the National Academy of Sciences of the United States of America* 107(26): 12028–12033. https://doi.org/10.1073/pnas.0914991107

Zhang, H., Li, J., Hou, S., et al. 2014. Engineered TAL Effector modulators for the large-scale gain-of-function screening. *Nucleic Acids Research* 42(14): e114. https://doi.org/10.1093/nar/gku535

Zhang, M., Cao, Y., Wang, Z., et al. 2018a. Retrotransposon in an HKT1 family sodium transporter causes variation of leaf Na^+ exclusion and salt tolerance in maize. *New Phytologist* 217: 1161–1176. https://doi.org/10.1111/nph.14882

Zhang, Y., Bai, Y., Wu, G., et al. 2017. Simultaneous modification of three homoeologs of TaEDR1 by genome editing enhances powdery mildew resistance in wheat. *The Plant Journal* 91: 714–724. https://doi.org/10.1111/tpj.13599

Zhang, Y., Li, D., Zhang, D., et al. 2018b. Analysis of the functions of *TaGW2* homoeologs in wheat grain weight and protein content traits. *The Plant Journal* 94: 857–866. https://doi.org/10.1111/tpj.13903

Zhang, Y., Zhang, F., Li, X., et al. 2013. Transcription activator-like effector nucleases enable efficient plant genome engineering. *Plant Physiology* 61(1): 20–27. https://doi.org/10.1104/pp.112.205179

Zhao, J. Z., Cao, J., Li, Y., et al. 2003. Transgenic plants expressing two *Bacillus thuringiensis* toxins delay insect resistance evolution. *Nature Biotechnology* 21: 1493–1497. https://doi.org/10.1038/nbt907

Zheng, Z., Wang, H., Chen, G., Yan, G. and Liu, C. A. 2013. Procedure allowing up to eight generations of wheat and nine generations of barley per annum. *Euphytica* 191: 311–316. https://doi.org/10.1007/s10681-013-0909-z

Zhou, J., Peng, Z., Long, J., et al. 2015. Gene targeting by the TAL effector PthXo2 reveals cryptic resistance gene for bacterial blight of rice. *The Plant Journal* 82: 632–643. https://doi.org/10.1111/tpj.12838

Zimmerman, J. L. 1993. Somatic embryogenesis-a model for early development in higher plants. *The Plant Cell* 5(10): 1411–1423. https://doi.org/10.1105/tpc.5.10.1411

12

Recent Advancement of Nanotechnology in Agriculture

Deepak Sharma
JECRC University

Manas Mathur, Rakesh Kumar Prajapat, V.S. Varun Kumar and Manju Sharma
Suresh Gyan Vihar University

Fahad Khan
Noida Institute of Engineering and Technology

Arunabh Joshi
RCA, MPUAT

Tarun Kumar Upadhyay
Parul Institute of Applied Sciences, Parul University

CONTENTS

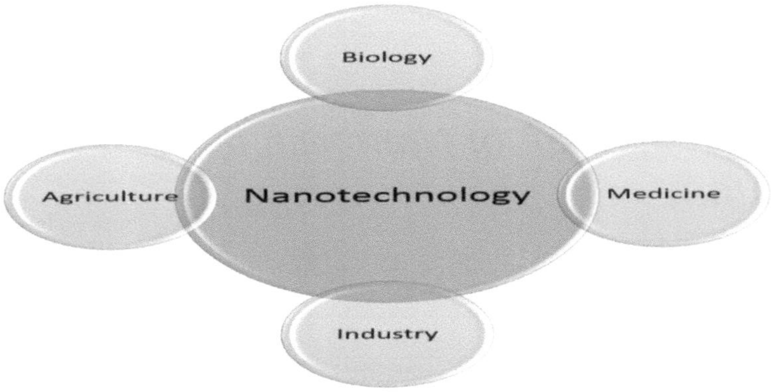

FIGURE 12.1 Graphical representation of various applications of nanotechnology.

12.1 Introduction

Nanotechnology has grown at a tremendous rate for the last few decades, and recent advances in the era of nanostructured materials and nanodevices have opened up new opportunities in a large number of applications that vary from information and communication technology (ICT) to healthcare and medicine sectors (Figure 12.1). A nanometer could be a unit of length within the system of weights and measures, adequate to1 billionth of a meter (10^{-9}). Engineering science plays a very important role within the productivity through the management of nutrients and further, because it also can monitor the quality of water and pesticides for the proper development of agriculture. The properties of nanoparticles (NPs; other than size) also have an influence on toxicity hold shape, surface structure, chemical composition, surface charge, behaviour, the extent of particle aggregation or disaggregation, etc. which could lead to building NPs. Therefore, nanomaterials with identical chemical compositions have different sizes or shapes, which will exhibit their different toxicities. The implication of engineering science analysis within the agricultural sector has become a solution issue for property development. In the case of agri-food areas, the important applications of nanotubes, fullerenes, biosensors, controlled delivery systems, nanofiltration and so forth were discovered. Nanotechnology has proven to be good for resources management of the agricultural field and targeted delivery mechanisms in plants, and it helps to require care of the soil fertility. Moreover, it was also evaluated as steady inside the employment of biomass and agricultural waste as in food process and food packaging systems and risk assessment. Recently, nanosensors are widely applied in agriculture because of their strengths and their quickness for environmental release.

12.2 Metallic Nanoparticles: Top-Down and Bottom-Up Approaches

In nanotechnology, basically there are two different ways/approaches by which metallic nanoparticles can be constructed: "bottom-up" and "top-down" (Figures 12.2 and 12.3). The bottom-up approach further comprises creating nanomaterials and objects within the same nanosphere based on atoms, molecules and aggregate grouping. This type of grouping occurs in a clear and manageable manner, which allows for an increase in the functionality of the structure of such materials. The top-down approach, which was sourced from microelectronics, breaks down systems in their current state and makes existing technologies more frequently. This approach lead to a reduction in the size of the devices on the nanoscale level. It has been reported that (Abbaci et al. 2014) the optimization and analysis of the innovated sensor were carried out using square wave voltammetry and cyclic voltammetry tools.

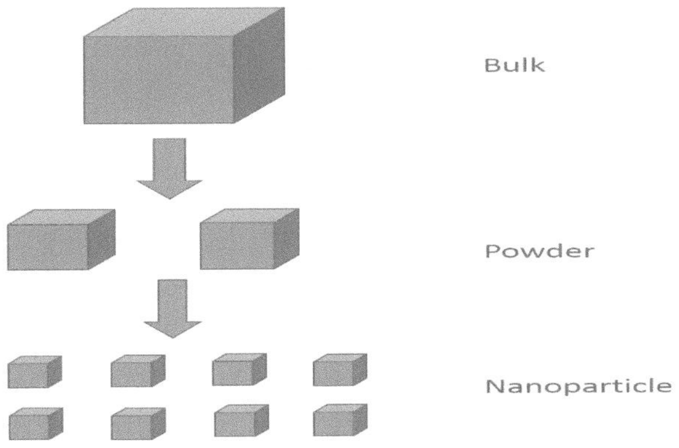

FIGURE 12.2 Top-down approach in nanotechnology to produce nanoparticles from bulk material.

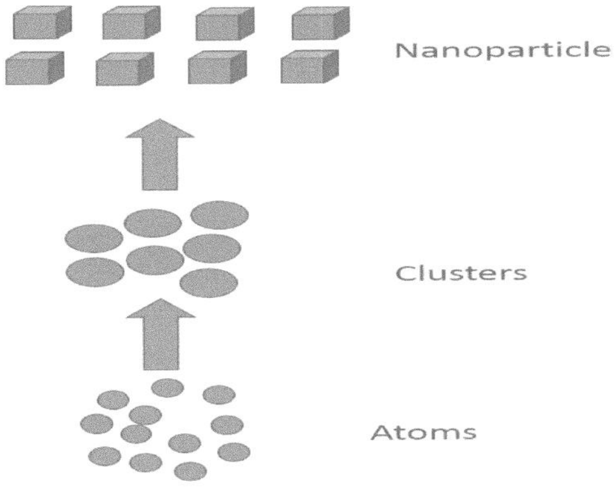

FIGURE 12.3 Bottom-up approach in nanotechnology.

With respect to size of materials, both methods are very similar as both approaches converge in terms of the size range of objects. The former approach, however, tends to be more abundant based on the type of material, design varieties and nanometric control, whereas the latter approach only acquires materials of more importance; however, control may not be as strong.

12.3 Nanoparticles as a Carrier-Based System in Agriculture

Nanoparticles can be used as eco-friendly carrier systems for targeted delivery of loaded/encapsulated substances to protect the early degradation of payload molecules. It also provides a desirable release profile such as slow and sustained release by reducing the dose by prolonged delivery, reducing toxicity into the environment and subsequently to humans. Nanotechnology, essentially through the innovation of clever transfer techniques and nanocarriers, can be used to design more well-organized and less pollutant agrochemicals. This Collection grants new linked works, casing nanodevices that recover crop defence against pests and diseases, nanoformulations for pretty plant nutrition, and nanomaterials firming the crop routine. Due to its rapid progression and development towards possible solutions, nanotechnology is considered to play its important role for the current problems in the field of agriculture (Abobatta 2018). Nanomaterials aids similarly as seasonings (mostly for skilful release) and vigorous ingredients. Product competences possibly enhanced by nanomaterials should be composed against increased ecological input fluxes. The vibrant expansion in research and its extensive public awareness are in dissimilarity with the presently still very less in quantity of NM-containing products on the market (Gogos et al. 2015).

12.4 Nanofertilisers

In the field of agriculture, nanotechnology has a more practical implication for food production, with improved nutritional value, quality, safety and security. The well-organised site-specific application of pesticides, fertilisers, herbicides and plant growth promoters is a different tool to increase crop production. Skilful application/discharge of pesticides, herbicides and plant growth promoters can be attained through nanoparticle-based delivery systems. It already has been reported by researchers that nanoparticles encapsulated with polymer (epsilon-caprolactone) have been used as a vector for atrazine (Oliveira et al. 2015). *Brassica juncea* treated with these encapsulated nanoparticles increases the herbicidal efficacy when compared with available marketable atrazine. Similarly, other nanocarriers like silica nanoparticles (Cao et al. 2018) and polymeric nanoparticles (Kumar et al. 2017) have also been designed for a release system to be used with pesticides in a well-controlled manner. These kinds of novel innovations are often termed as "precision farming", which focusses on crop production without disturbing natural resources (Duhan et al. 2017). The use of nano-encapsulation can reduce the quantity of the herbicide, without affecting potency, which results in eco-friendly conditions. Further, using recombinant DNA technology and nanoparticle-mediated gene or DNA delivery in plants was applied in producing insect-resistant varieties (Sekhon 2014; Khot et al. 2012). It has been reported that ZnO NPs are excellent sources of antifungal agents that can act against *Aspergillus flavus*, *Fusarium graminearum*, *A. fumigatus* and *A. niger*; thus remarkably they can be used as crop protective agents (Rajiv et al. 2013).

The development of nanofertilisers is an innovative remedy for drastic economic fatalities. Nanofertilisers have a vital role in overcoming nutrient loss and increasing nutrient uptake in crops and soil microflora (Dimkpa and Bindraban 2018). By using other nanomaterials such as carbon nano-onions (Tripathi et al. 2017) and chitosan, NPs (Khalifa and Hasaneen 2018) assist in enhancing crop production and quality. It is predicted that these newly designed nanofertilisers will improve recent fertiliser manufacturing industries in future decades. Preetha and Balakrishnan (2017) have reported that fertilisers played a vital role in improving agriculture production in India by 35–40%. Chemicals are inorganic fertilisers that when formulated in inappropriate concentrations and combinations for various crops supply three main nutrients: nitrogen (N), phosphorus (P) and potassium (K). However, the majority of these fertilisers are lost into the environment through the processes of physical, chemical-like leaching; volatilisation and emission. They cannot be absorbed by plants, thereby causing economic and resource losses. To augment with the nutrient use efficiency (NUE) and built up nutrients in soils, nanofertilisers are emerging as an alternative approach. For a controlled and sustained release profile of fertilisers and pesticides, nanofertilisers are expected to be far more effective by providing a large surface to volume ratio. The use of nanofertilisers helps plant roots and micro-organisms to encounter nutrient ions from solid phase of minerals in an easy and sustainable way. Thus, nanofertiliser enhances soil quality by reducing the toxic effects associated with the overdose of applied fertilisers.

Nanofertilisers are the encapsulated fertilisers in which N, P and K are entrapped within the pores of nanoparticles; a slow release of these nutrients in the environment could be achieved by coating nano-composites and sub-nano-composites with alginate or chitosan. Nano-composite-based fertilisers consisting of micronutrients (such as N, P and K), mannose and amino acids have been developed that have increased the N uptake and utilisation by grain crops. Biodegradable chitosan-based nanoparticles have also gained great attention as controlled release for NPK fertilisers due to their bio-absorbable and bactericidal nature. Nano-zeolite can act as a potential carrier for developing smart delivery fertilisers. Zeolites are a group of naturally occurring minerals and have honeycomb-like porous layered crystal structures with the ability to exchange ions and catalyse reactions. These honeycombs are made up of networks of tunnels and cages that are interconnected; they can be loaded with nutrients, such as N, P and K, combined with other slowly dissolving ingredients containing P, calcium and a complete suite of minor and trace nutrients. Thus, nano-zeolites can act as reservoirs for nutrients that release on-demand at the targeted sites. A number of studies indicate that these slowly released nanofertilisers improve grain yield and have proved to be safe for the germination of cereals. Besides the delivery of primary nutrients, sulphur-coated nanofertilisers have gained much attention due to their potential role in releasing sulphur in sulphur-deficient soils. Nanoparticles can even pass through the plant and animal cell, which is the main way nanotechnologists are able to achieve the phenomenon of delivering the required product at the cellular level.

12.5 Various Nanotechnological Approaches to Improve the Nutrient Use Efficiency

1. Encapsulation of fertiliser with nanoparticles
2. Slow delivery
3. Smart delivery system

12.5.1 Encapsulation

Studies reported by Solanki et al. (2015) proposed the packaging of the fertilisers within the tiny 'envelope' or 'shell'. Encapsulation, controlled release methods and entrapment of agro-chemicals have revolutionised the use of nano-composite material for agricultural use. Nano-encapsulated chemicals possess characteristics such as proper concentration with increased solubility, stability and effectiveness; timely release in response to environmental stimuli and enhanced targeted activity and less eco-toxicity, which increase the production of crops and cause less harmful effects to agricultural workers/farmers. Nanoparticles lie within the size range of 100–250 nm and easily dissolve in water more effectively than existing particles due to their effective surface tension quality. The use of nano-emulsions can also be a better choice in terms of the application of uniform suspensions of pesticides or herbicidal agents (either oil based or water) and contain nanoparticles from 200–400 nm. Formulations in the form of gels, creams, liquids and so forth have been performed to assess the effectiveness of the sol-gel-based nanomaterials.

12.5.2 Slow Delivery

Slow release profile (SRP) is related to nanoparticles' solubility of water and degradation of microbial and chemical hydrolysis. Control release profile (CRP), in this case, refers to soluble fertilisers that are coated with biodegradable materials that limit the exposure of the material to water and/or release the resultant nutrient to solution through the process of diffusion. Through the surface coating/sealing and binding of nano- and sub-nano-composites, the release of loaded nutrients from the fertiliser capsule can be controlled (Liu et al. 2016). Studies reported by Jinghua (2004) showed the application of a nano-composite with micronutrients (N, P and K), mannose and amino acids enhances uptake and use of nutrients by the grain crops. Fertiliser incorporation into nanotubes makes it a suitable vehicle for slow and controlled release at the required sites.

12.5.3 Smart Delivery System

Smart delivery systems include timely controlled, targeted, self-regulated, pre-programmed avoidance of biological barriers for successful targeting. In the case of a smart delivery system, a small sealed package carries the payload molecules which will open up only when the desired location or site of the plant system is reached. Similarly, implanting nanoparticles in the plants could also determine nutrient level inside the plants and take up suitable remedial measures before the difficulty causes yield reduction in crops. This system can significantly reduce the response time to sense the problem. Identification of pathogens in wastewater (Wibowo et al. 2018) makes it suitable for drinking water (Deng et al. 2016). Other nanomaterials like copper NPs (Geszke-Moritz et al. 2012), carbon nanotube gold NPs (Lin et al. 2011) and silver NPs (Jokar et al. 2016) are still under design to act as nanosensors for immediate monitoring of natural conditions including crop production and protection.

12.6 Nanoherbicides

Prasad et al. (2017) reported that removal of weeds can be achieved by destroying their seed banks in the soil and prevent them from germinating when weather and soil conditions encourage their growth. This is one of the most promising technologies. Molecular characterisation of underground plant parts for a novel target domain and an initial receptor-based herbicide molecule with a specific binding property with nanoherbicide molecules, such as carbon nanotubes, is able to kill the viable and dormant underground propagation of weed seeds. Target-specific nano-encapsulated herbicide molecules are aimed at a particular receptor inside the roots of target weeds. When these molecules enter into the weed system by forming an association with the receptor, they translocate into its parts and inhibit the glycolysis of the food reservoir in the root system, starving and ultimately killing specific weeds.

12.7 Nanopesticides

Kah et al. (2013) reported that pesticides are substances or the mixtures of substances broadly used to remove and control the harmful organisms, causing significant economic losses in the agricultural production system. There are many regulatory barriers placed on pesticides in today's agriculture. Pesticides such as dichloro-diphenyl-trichloroethane (DTT), which causes extreme environmental hazards, also increased public and regulatory awareness of the use of chemicals in farming, and shifting the industry focus on to the use of integrated pest management systems, combining smarter and more targeted use of chemicals with the granular investigation of the plant health system. The plant pest disturbing the cultured products causes serious losses by limiting the product yield. Secondary metabolites such as (phenolics, alkaloids and terpenoids) those secreted by plants as a self-preservative mechanism provide a defence and protective function against insects.

Nanotechnology-mediated formulations can also help to weaken toxic effects on non-target organisms, as well as to develop physicochemical stability and prohibit degradation of the active agent through the micro-organisms. Nanotechnology allows companies to manipulate the properties of a carrier for slow release of substances to be used as agricultural input through encapsulation or attaching to the nanomaterial (adsorption, attachment, cross-linking, or chemical bonding, etc.). In the nano-encapsulation-based techniques, the nano-sized active ingredients such as pesticide compounds are covered by a thin-walled protective shell as a protective layer. The major aims of the nano pesticide formulations are commonly similar to those of other pesticide formulations and are used (1) to increase the apparent solubility of some poorly soluble active ingredient or (2) to release the active substance in a slow/targeted manner and/or protect the active ingredient against premature degradation at the specific site. Nanotechnology grasps the capacity of measured announcement of agrochemicals and location beleaguered transfer of different macromolecules required for amended plant disease resistance, prominent nutrient application and improved plant growth (Agrawal and Rathore 2014).

Pesticide marketed under the name Karate ZEON is a quick-release microencapsulation nanopesticide formulation containing active compounds, such as α-cyhalothrin, which breaks open while coming in contact with leaves. In contrast, an encapsulated product like 'gutbuster' only breaks open to release its content when it comes into contact with alternative environments such as the stomach of a certain insect. The surplus use of chemicals and toxic pesticides has imbalanced the ecosystem causing serious health disorders. Nanoscience helps us in overcoming such concerns by providing nanomaterials with increased potency (Chhipa 2017).

TABLE 12.1
Different Metallic Elements Used in Nanotechnology for Agricultural Benefits

S. No.	Nanoparticle	Usages in Agriculture	References
1.	Carbon nanotube	Increases in root length of wheat seedlings; utilised for water purification systems	Wang et al. (2016), Rocha et al. (2017)
2.	Zeolite	Used as fertilisers, stabilisers and chelating agents; zeolites enable both inorganic- and organic-based fertilisers to slowly release their nutrients	Perez-Caballero et al. (2008)
3.	Titanium oxide	Great pathogen disinfection efficiency	Yao et al. (2007)
4.	Alumina oxide	Efficient pesticides at nanoscale	Singh et al. (2015)
5.	Nanosilver particle	Antifungal effectiveness and control of various plant diseases	Ramezani et al. (2019)
6.	Nanosilicon	Has antifungal activity and controlled powdery mildew disease of pumpkins	Park et al. (2006)

12.8 Use of Nanotechnology in Agriculture

12.8.1 Nano Silver

Studies by Singh et al. (2015) reported that nano silver is the most studied and utilised nanoparticle for the biosystem. It has long been known to possess strong inhibitory and bactericidal effects; a wide range of antimicrobial activities; high area and a high fraction of surface atoms; a high antimicrobial effect and is adhesive in the case of bacteria and fungus, serving as a better fungicide. Studies reported the antifungal effectiveness of colloidal nano silver (1.5-nm average diameter) solution against rose mildew caused by *Sphaerotheca pannosa* var. *rosae*. The role of different metallic elements used in nanotechnology for agricultural benefits are given in Table 12.1.

12.8.2 Nano Aluminosilicate

Aluminosilicate nanotubes are sprayed on the surface of the plant and are easily taken up in insect hairs. Insects actively groom and engulf pesticide-filled nanotubes. Nano aluminosilicates were biologically the most active and relatively more environmentally safe pesticides, until now.

Mesoporous silica nanoparticles can deliver DNA and chemicals into plants, creating a powerful new tool for site-specific delivery into plant cells. These are manufactured porous, silica nanoparticle systems spherical in shape with arrays of independent porous channels.

12.8.3 Perspectives on Biosynthesised and Bioinspired Nanomaterials

Eco-friendly approaches for green synthesis for the design and development of nanomaterials will attract the whole world. Various biological modes such as bacteria, fungi, virus and medicinal plants and their extracts have gained a great deal of attention for their biosynthesis. The prime advantages to adapting such protocols are that these modes act as a manufacturing host and can act as the capping, stabilising and reducing agent so that metallic toxicity will be reduced during the synthesis processes. Jain et al. (2011) reported that extracellular proteins isolated from microbes have a major role in stabilising silver NPs. Further, sparse research reveals that the role of biomolecules isolated from these living resources acts as a capping agent in synthesis. Moreover, optimum temperature and pressure and neutral pH are the major functions of biological systems. Surface functionalisation is the most common approach to design biocompatible and environmentally friendly nanomaterials and to simulate many bioprocesses in nature. Therefore, these biological modes are recommended to provide a natural environment to provide surface functionalisation on nanomaterials; however, there are some limitations in the biosynthesis of nanoparticles like low yield and low manage production pathway. It is still a point of consideration that the exact biological processes responsible for nanoparticle biosynthesis remain to be explicated. Some of the studies reported that orange extract was used to design round spherical silver NPs, whereas pineapple extract synthesised nanoparticles with sharp corners (Hyllested et al. 2015). Since the past decades, enormous quantities of products and patents in synthesis of nanomaterials into agricultural practices (e.g., nanosensors, nanofertilizers, nanopesticides) has been filed. The cumulative aim of all these tactics is to increase the efficiency and sustainability of agricultural practices by needing less input and producing less waste than traditional products and approaches (Servin et al. 2015).

It is a great point of interest to argue about the application of bioinspired nanostructures in food and agriculture-based systems. Liang et al. (2018) recently reported that bioinspired mussel avermectin NPs with a strong grip to crop dryness reduce loss and prevent contamination of soils treated with pesticides. The biologically designed nanoparticles exhibited the incredible high retaining power of avermectin, for longer delivery at a suitable site.

12.8.4 Various Designed Bioinspired Nanoparticles

Ramezani et al. (2019) has reported that silver nanoparticles (SNPs), aluminium oxide nanoparticles (ANP) and flowers of zinc and titanium oxide control black weevil and grasserie disease in silkworm (*Bombyx mori*) caused by *Sitophilus oryzae* and baculovirus *B. mori* nuclear polyhedrosis virus (BmNPV), respectively. It was already reported that hydrophilic SNPs were best on a primary day. On day 2, an infant mortality rate of 90 was obtained with SNPs and ANPs. After 7 days of exposure, an infant mortality rate of 95 and 86 was reported with hydrophilic and hydrophobic SNPs and a rate of nearly 70 was reported after the insects were killed when the rice was treated with lipophilic SNPs. However, 100% mortality was observed in the case of ANPs. Similarly, in another bioassay undertaken for grasserie disease in silkworm (*B. mori*), a big decrease in viral load was reported when leaves of *B. mori* were treated with an ethanolic suspension of hydrophobic aluminosilicate nanoparticles. Copper nanoparticles in mixture glass powder showed efficient antimicrobial activity against gram-positive, gram-negative bacteria and fungi.

12.8.5 Nanotechnology in Reducing Agriculture Wastes

Nanotechnology is also applied to reduce waste in agriculture, mainly in the cotton industry. When cotton is processed into fabric or a garment, cellulose or the fibres are discarded as waste or used for low-value products like cotton balls, yarn and cotton batting. With the utilisation of newly developed solvents and electrospinning, scientists produce 100-nm-diameter fibres which will be used as fertiliser or pesticide absorbent. These high-performance absorbents allow targeted application at the desired time and site.

Ethanol production from maize feedstocks has increased the worldwide price of maize within the last 2 years. Cellulosic feedstocks are now considered a viable option for biofuel production, and nanotechnology can enhance the performance of enzymes utilised in the conversion of cellulose into ethanol. Scientists are experimenting with nano-engineered enzymes which will allow simple and cost-effective conversion of cellulose from waste plant parts into ethanol.

Rice husk, a rice-milling by-product, is often used as a source of renewable energy. When rice husk is burned into thermal energy or biofuel, an outsized amount of high-quality nanosilica is produced which may be further utilised in making other materials like glass and concrete. Since there is an endless source of rice husk, the production of nanosilica through nanotechnology can alleviate the growing rice husk disposal concern.

12.8.6 Nanotechnology in Tillage

Sharifnasab and Abbasi (2016) reported mechanical tillage practices to improve soil structure and increase porosity leading to a better distribution of soil aggregates and eventually modify the physical properties of soil. The literature on the effect of NPs on the tilth and tillage is limited. The use of nanomaterials increases soil pH and improves soil structure. Nanomaterials also reduce the mobility, availability and toxicity of heavy metals besides reducing soil erosion. Nanoparticles in soil reduce cohesion and internal friction along with reducing the shear strength of the soil. Reduction in adhesion of soil particles allows easy crushing of lumps with less energy.

12.8.7 Role of Nanotechnology in Seed Science

Manjunatha et al. (2019) reported that seed production is a tedious process, especially in wind-pollinated crops. Detecting pollen load that will cause contamination is a sure method to ensure genetic purity. Pollen flight is determined by air temperature, humidity, wind velocity and pollen production of the crop. The use of biosensors specific to contaminating pollen can help alert for possible contamination and reduces the contamination. The same method can also prevent the genetically modified crops from contaminating field crops. Novel genes are being incorporated into seeds and sold in the market. Tracking of old seeds could be done by nanobarcodes that are encodable, machine-readable, durable and submicron-sized taggants.

12.8.8 A New Frontier in Agricultural Development: Nano-Farming

Nanomaterials engineering is one of the latest technological innovations that demonstrate unique targeted characteristics with elevated strength. Recent technological advancements in the fabrication of nanomaterials of different sizes and shapes have yielded their wide array of applications in medicine, environmental science, agriculture and food processing. Throughout history, agriculture has always benefited from these innovations (Chen et al. 2016). As agriculture faces several and unprecedented challenges, such as reduced crop yield due to biotic and abiotic stresses, including nutrient deficiency and environmental pollution, the emergence of nanotechnology has offered promising applications for precision agriculture. The term precision agriculture or farming has emerged in recent years, meaning the development of wireless networking and miniaturisation of the sensors for monitoring, assessing and controlling agricultural practices. More specifically, it is related to the site-specific crop management with a wide array of pre- and post-production aspects of agriculture, ranging from horticultural crops to field crops (Dwivedi et al. 2016). Recent advancements in tissue engineering and bioengineered nanomaterials-based target-specific delivery of clustered regularly interspaced short palindromic repeats (CRISPR)/CRISPR-associated protein (Cas) mRNA, and sgRNA for the genetic modification of crops is a noteworthy scientific achievement (Ran et al. 2017; Miller et al. 2017; Kim et al. 2017). Further, nanotechnology provides excellent solutions for an increasing number of environmental challenges. Therefore, such revolving improvements in the field of nanotechnology with special focus on the identification of problems and development of collaborative approaches for solving sustainable agricultural growth has remarkable potential to provide broad social and equitable benefits.

12.9 Conclusion

Nanotechnology revealed auspicious capacities to be broadly consumed in the agriculture and food industry. The concrete solicitation of nanotechnology and marketing nanomaterial-based products are still to be elucidated due to the unfortunate competence to govern features and interface of materials at the nanoscale, along with mysterious environmental effects and empty toxicity databases. This sometimes creates constraints for the marketing of novel nanomaterial-based products. Since executing eco-friendly practices has been crucial for success in today's biotechnology business, the bioinspired approach is becoming famous in biomedical implications. Though, in the recent era the research and development in bioinspired nanomaterials for their application in food and agriculture industries are still limited. Many nano-based products relating innovative nanotechnology have been marketed globally.

12.10 Future Perspective

Nanotechnology-based advancements in agriculture could be proven as novel tools to transform current agriculture and food-related industries through the innovation of recent techniques such as precision farming techniques, improving the capability of plants to absorb mineral nutrients, more efficiently and targeted/site-specific use of inputs, disease detection and control. Developing new science and innovation, working with the smallest molecule, nanotechnology raises new advancements in the field of science, particularly in agriculture. More advanced research is required in the area of energy production, environment management, crop production, disease diagnosis and effective use of the resources and utilisation for high yield without altering the natural environment. The Green Revolution is responsible for the usage of pesticides and chemical fertilisers which cause the infertility of soil and plants to develop resistant mechanisms against the pathogen. Cutting-edge nanotechnology-based tools and systems can possibly address the different issues of regular farming and can reform this field. This chapter focussed mainly on the potential of nanotechnology in the field of agriculture and different agricultural applications; further, more research is needed to explore their applications and their future prospects in the field of agriculture. In the near future, nanotechnology can improve nutrient absorption capacity through the use of nano-based fertilisers, yield and nutritional quality. The scrutiny and management of diseases and pests, resolving the mechanism of host-parasite interactions, preservation and packaging of food and the deduction of pollutants from soil and water bodies increase soil quality of the agricultural fields. To handle the problems of agricultural sciences, one needs a thorough understanding of nano-science along with the broad knowledge of the agricultural system. Nanotechnology in agriculture might take a few more years to progress and transit from laboratory to land, and for this to happen, sustainable government funds and plans should be provided for improving nanotechnology-based agriculture.

REFERENCES

Abbacia, A., Azzouz, N., and Bouznit, Y. 2014. A new copper doped montmorillonite modified carbon paste electrode for propine b detection. *Applied Clay Science* 90: 130e4.

Abobatta, W. F. 2018. Nanotechnology application in agriculture. *Acta Scientific Agriculture* 2(6).

Agrawal, S., and Rathore, P. 2014. Nanotechnology pros and cons to agriculture: a review. *International Journal of Current Microbiology and Applied Science* 3(3): 43–55.

Cao, L., Zhou, Z., Niu, S., Cao, C., Li, X., and Shan, Y. 2018. Positive charge functionalized mesoporous silica nanoparticles as nanocarriers for controlled 2, 4-dichlorophenoxy acetic acid sodium salt release. *Journal of Agricultural and Food Chemistry* 66: 6594e603.

Chen, Y. W., Lee, H. V., Juan, J. C., and Phang, S. M. 2016. Production of new cellulose nanomaterial from red algae marine biomass *Gelidium elegans*. *Carbohydrate Polymers* 151: 1210–1219.

Chhipa, H. 2017. Nanofertilizers and nanopesticides for agriculture. *Environmental Chemistry Letters* 15(1): 15–22.

Deng, H., Gao, Y., Dasari, T. P. S., Ray, P. C., and Yu, H. 2016. A facile 3D construct of graphene oxide embedded with silver nanoparticles and its potential application as water filter. *The Journal of the Mississippi Academy of Sciences* 61: 190e7

Dimkpa, C. O., and Bindraban, P. S. 2018. Nanofertilizers: new products for the industry? *Journal of Agricultural and Food Chemistry* 66: 6462e73.

Duhan, J. S., Kumar, R., Kumar, N., Kaur, P., Nehra, K., and Duhan, S. 2017. Nanotechnology: the new perspective in precision agriculture. *Biotechnology Reports* 15: 11e23.

Dwivedi, S., Saquib, Q., Al-Khedhairy, A. A., and Musarrat, J. 2016. Understanding the role of nanomaterials in agriculture. In *Microbial inoculants in sustainable agricultural productivity*, pp. 271–288. Springer, New Delhi.

Geszke-Moritz, M., Clavier, G., Lulek, J., and Schneider, R. 2012. Copper-or manganese-doped ZnS quantum dots as fluorescent probes for detecting folic acid in aqueous media. *Journal of Luminescence* 132: 987e91.

Gogos, A., Knauer, K., and Bucheli, T. D. 2015. Nanomaterials in plant protection and fertilization. *Journal of Agricultural and Food Chemistry* 60(39): 9781–9792.

Hyllested, J. Æ., Palanco, M. E., Hagen, N., Mogensen, K. B., and Kneipp, K. 2015. Green preparation and spectroscopic characterization of plasmonic silver nanoparticles using fruits as reducing agents. *Beilstein Journal of Nanotechnology* 6(1): 293–299.

Jain, N., Bhargava, A., Majumdar, S., Tarafdar, J., and Panwar, J. 2011. Extracellular biosynthesis and characterization of silver nanoparticles using *Aspergillus flavus* NJP08: a mechanism perspective. *Nanoscale* 3: 635e41

Jinghua, G. 2004. Synchrotron radiation, soft X-ray spectroscopy and nano-materials. *Journal of Nanotechnology* 1(1–2):193–225.

Jokar, M., Safaralizadeh, M. H., Hadizadeh, F., Rahmani, F., and Kalani, M. R. 2016. Design and evaluation of an apta-nano-sensor to detect acetamiprid *in vitro* and *in silico*. *Journal of Biomolecular Structure & Dynamics* 34: 2505e17.

Kah, M., Beulke, S., Tiede, K., and Hofmann, T. 2013. Nanopesticides: state of knowledge, environmental fate, and exposure modeling. *Critical Reviews in Environmental Science and Technology* 43(16): 1823–1867.

Khalifa, N. S., and Hasaneen, M. N. 2018. The effect of chitosane PMAAeNPK nanofertilizer on *Pisum sativum* plants. *3 Biotech* 8: 193.

Khot, L. R., Sankaran, S., Maja, J. M., Ehsani, R., and Schuster, E. W. 2012. Applications of nanomaterials in agricultural production and crop protection: a review. *Crop Protection 35*: 64–70.

Kim, D. H., Gopal, J., and Sivanesan, I. 2017. Nanomaterials in plant tissue culture: the disclosed and undisclosed. *RSC Advances* 7: 36492–36505.

Kumar, S., Kumar, D., and Dilbaghi, N. 2017. Preparation, characterization, and bio-efficacy evaluation of controlled release carbendazim-loaded polymeric nanoparticles. *Environmental Science and Pollution Research* 24: 926e37.

Liang, J., Yu, M., Guo, L., Cui, B., Zhao, X., and Sun, C. 2018. Bioinspired development of P (SteMAA) eavermectin nanoparticles with high affinity for foliage to enhance folia retention. *Journal of Agricultural and Food Chemistry* 66: 6578e84.

Lin, Y. W., Huang, C. C., and Chang, H. T. 2011. Gold nanoparticle probes for the detection of mercury, lead and copper ions. *Analyst* 136: 863e71.

Liu, R., Zhang, H., and Lal, R. 2016. Effects of stabilized nanoparticles of copper, zinc, manganese, and iron oxides in low concentrations on lettuce (*Lactuca sativa*) seed germination: nanotoxicants or nanonutrients? *Water, Air & Soil Pollution* 227(1): 42.

Manjunatha, R. L., Naik, D., and Usharani, K. V. 2019. Nanotechnology application in agriculture: a review. *Journal of Pharmacognosy and Phytochemistry* 8(3): 1073–1083.

Miller, J. B., Zhang, S., Kos, P., et al. 2017. Non-viral CRISPR/Cas gene editing *in vitro* and *in vivo* enabled by synthetic nanoparticle co-delivery of Cas9 mRNA and sgRNA. *AngewandteChemie* 56: 1059–1063.

Oliveira, H. C., Stolf-Moreira, R., Martinez, C. B. R., Grillo, R., de-Jesus, M. B., and Fraceto, L.F. 2015. Nanoencapsulation enhances the post-emergence herbicidal activity of atrazine against mustard plants. *PLoS One* 10: e0132971.

Park, H. J., Kim, S. H., Kim, H. J., and Choi, S. H. 2006. A new composition of nanosized silica-silver for control of various plant diseases. *The Plant Pathology Journal* 22(3): 295–302.

Perez-Caballero, R., Gil, J., Benitez, C., and Gonzalez, J. L. 2008. The effect of adding zeolite to soils in order to improve the NK nutrition of olive trees. Preliminary results. *American Journal of Agricultural and Biological Sciences* 2(1): 321–324.

Prasad, R., Bhattacharyya, A., and Nguyen, Q. D. 2017. Nanotechnology in sustainable agriculture: recent developments, challenges, and perspectives. *Frontiers in Microbiology* 8:1014.

Preetha, P. S., and Balakrishnan, N. 2017. A review of nano fertilizers and their use and functions in soil. *International Journal of Current Microbiology and Applied Sciences* 6(12): 3117–3133.

Rajiv, P., Rajeshwari, S., and Venckatesh, R. 2013. Bio-fabrication of zinc oxide nanoparticles using leaf extract of *Parthenium hysterophorus* L. and its size-dependent antifungal activity against plant fungal pathogens. *Spectrochima Acta A: Molecular and Biomolecular Spectroscopy* 112: 384e7.

Ramezani, M., Ramezani, F., and Gerami, M. 2019. Nanoparticles in pest incidences and plant disease control. *Nanotechnology for Agriculture: Crop Production & Protection*, pp. 233–272. Springer, Singapore.

Ran, Y., Liang, Z., and Gao, C. 2017. Current and future editing reagent delivery systems for plant genome editing. *Science China Life Sciences* 60: 490–505.

Rocha, J. D. R., Rogers, R. E., Dichiara, A. B., and Capasse, R. C. 2017. Emerging investigators series: highly effective adsorption of organic aromatic molecules from aqueous environments by electronically sorted single-walled carbon nanotubes. *Environmental Science: Water Research & Technology* 3(2): 203–212.

Rudakiya, D., Patel, Y., Chhaya, U., and Gupte, A. 2019. Carbon nanotubes in agriculture: production, potential, and prospects. In *Nanotechnology for agriculture*, pp. 121–130. Springer, Singapore.

Sekhon, B.S. 2014. Nanotechnology in agri-food production: an overview. *Nanotechnology Science and Applications* 7: 31.

Servin, A., Elmer, W., Mukherjee, A., et al. 2015. A review of the use of engineered nanomaterials to suppress plant disease and enhance crop yield. *Journal of Nanoparticle Research* 17(2): 92.

Sharifnasab, H., and Abbasi, N. 2016. Effect of nanoclay particles on some physical and mechanical properties of soils. *Journal of Agricultural Machinery* 6(1): 250–258.

Singh, S., Singh, B. K., Yadav, S. M., and Gupta, A. K. 2015. Applications of nanotechnology in agricultural and their role in disease management. *Research Journal of Nanoscience Nanotechnology* 5(1): 1–5.

Solanki, P., Bhargava, A., Chhipa, H., Jain, N., and Panwar, J. 2015. Nano-fertilizers and their smart delivery system. *In Nanotechnologies in food and agriculture*, pp. 81–101. Springer, Cham, Switzerland.

Tripathi, K. M., Bhati, A., Singh, A., Sonker, A. K., Sarkar, S., and Sonkar, S. K. 2017. Sustainable changes in the contents of metallic micronutrients in first generation gram seeds imposed by carbon nano-onions: life cycle seed to seed study. *ACS Sustainable Chemistry and Engineering* 5: 2906e16.

Wang, Y., Sun, C., Zhao, X., Cui, B., Zeng, Z., Wang, A., and Cui, H. 2016. The application of nano-TiO2 photo semiconductors in agriculture. *Nanoscale Research Letters* 11(1): 1–7.

Wibowo, K. M., Sahdan, M. Z., Ramli, N.I., Muslihati, A., Rosni, N., and Tsen, V. H. 2018. Detection of *Escherichia coli* bacteria in wastewater by using graphene as a sensing material. *Journal of Physics: Conference Series* 995: 012063.

Yao, K. S., Wang, D. Y., Ho, W. Y., Yan, J. J., and Tzeng, K. C. 2007. Photocatalytic bactericidal effect of TiO2 thin film on plant pathogens. *Surface and Coatings Technology* 201(15): 6886–6888.

13

Cryobiotechnology in Plants: Recent Advances and Prospects

Era Vaidya Malhotra and Sangita Bansal
ICAR-National Bureau of Plant Genetic Resources

CONTENTS

13.1 Introduction

Plant genetic resources form the backbone of the food and agricultural systems across the world. These represent the diversity in crops and their wild relatives, which have co-evolved over thousands of years and form the basis of crop improvement. This diversity provides sources of stress resistance, nutritional diversification and climate resilience. However, these plant species are becoming extinct at an alarming rate of nearly three species a year since 1900 due to various natural and anthropogenic factors (Humphreys et al. 2019). Conservation of biodiversity is thus of paramount importance to prevent the irreversible loss of important genes and traits for future sustainable development (Wang et al. 2012). Various conservation strategies are in place, and effective conservation of genetic resources involves a complementary approach using both traditional strategies of *in situ* conservation along with various technological interventions such as micropropagation, *in vitro* conservation and cryopreservation (Agrawal et al. 2019).

The approach of *ex situ* conservation depends on the biology of the plant to be conserved. Seed gene banks are one of the most vastly used alternatives for *ex situ* conservation of plants, where seeds can be conserved for years together at low temperature (–20°C) after drying them to reduce their inherent moisture content. However, some species are sensitive to drying and moisture loss and do not survive moisture removal. These species, referred to as recalcitrant species, cannot be conserved in conventional seed banks and require the use of some alternate strategies. The alternate approaches involve the use of biotechnological techniques such as micropropagation along with cryopreservation to conserve the species at ultra-low temperatures (–196°C) in liquid nitrogen (LN), and collectively form the sphere of cryobiotechnology (Popova et al. 2016; Pritchard et al. 2017). Cryopreservation is the storage of biological material at ultralow temperatures (–196°C) in LN (Walters et al. 2004; Day et al. 2008). Cryopreservation and cryobiology started way back in the 17th century (Boyle 1665), and this field began to pick up momentum in the 1940s (Day et al. 2008). All metabolic, biochemical and cellular processes stop at the cryogenic temperatures and living cells can be maintained at this temperature for long durations without any contamination or change in their genetic constitution (Wang et al. 2014; Reed 2017). It is now routinely used for long-term storage of recalcitrant seeds as well as vegetatively propagated plant species (Benson 2008a; Engelmann 2011; Kaviani 2011; Matsumoto 2017).

Cryopreservation studies have evolved from the efforts of understanding the basic biology of freezing (Sakai 1965) to technologically advanced systems of conserving plant tissues. Beginning initially with controlled-rate slow cooling approaches (Withers and King 1980; Towill 1983), techniques have now been developed to conserve plant tissues using various approaches of desiccation, vitrification (Sakai et al. 1990), encapsulation-dehydration (ED; Fabre and Dereuddre 1990), droplet-vitrification (DV; Panis et al. 2005) and now the more commonly used aluminium cryo-plates and stainless steel cryo-meshes (Niino et al. 2013; Funnekotter et al. 2017).

Omics based technologies are being incorporated into cryobiology studies to unravel the mechanisms underlying freeze tolerance and recalcitrance (Volk 2010; Ogawa et al. 2012; Gross et al. 2017; González-Arnao et al. 2018; Ekinci et al. 2020). Similarly, the storage stability of propagules is being studied and understood using these technological advancements under the umbrella of cryobionomics (Harding 2004).

13.2 Cryopreservation: Fundamental Principles

Cryopreservation is based on the three fundamental principles of (1) water behaviour, (2) cryoinjury and (3) cryoprotection. Depending on the ambient temperature, water exists in a liquid, vapour, ice or glassy state. All cryopreservation approaches are based on the manipulation of the liquid, ice and glassy states of water, such that intracellular ice formation is avoided. The most critical aspect of a cryopreservation protocol is control and avoidance of intracellular ice nucleation, also termed as seeding, i.e., the point at which ice crystals are initiated. Normally water can supercool to temperatures below 0°C without ice crystallisation up to a point of homogeneous ice nucleation around the temperature of –40°C. At this temperature, water molecules aggregate to form an ice nucleus which can then grow into ice crystals exponentially (Benson 2008b). Ice nucleation causes mechanical injury as well as colligative damage to cells by excessive solute concentration (Mazur 2004; Muldrew et al. 2004).

Traditionally, avoidance of intracellular ice crystallisation during cryopreservation has been carried out using the controlled rate freezing approach (Uemura and Sakai 1980; Kartha 1985; Day and McClellan 1995). As the cooling rate directly determines the cell survival, both too fast or too slow cooling results in cellular injury. On gradually cooling the cells, a water gradient across the cell membrane is created due to extracellular ice formation. This leads to the movement of intracellular water out of the cells, thus reducing the water content available for ice nucleation within cells. However, if the cooling rate is too slow, it leads to excessive concentration of cellular solutes causing damage to cellular functions. Therefore, this method requires a precise optimisation of the cooling rate, so that the right amount of water is removed from cells without causing colligative damages (Pegg 2007).

The next approach developed for cryoprotection involves the transition of water into its amorphous, non-crystalline glassy state, a process known as vitrification (Taylor et al. 2004). This is achieved by increasing the intracellular solute concentration to high viscosity, so that water vitrifies to its glassy state on exposure to freezing temperatures. The increased viscosity of cell solutes inhibits the formation of ice. As the glassy state of water is non-crystalline, it does not damage the cellular integrity. However, the glassy vitrified state is metastable, implying that it can revert to the liquid state and it can also devitrify to ice. This method of ice-free cryopreservation for conservation was first described in the animal system (Fahy et al. 1984, 1986) and has been successfully extended to plant systems.

13.2.1 Cryoprotection

The mechanical injury to cells caused by ice crystallisation can be avoided either by controlling the cooling rate or by vitrification of intracellular water; however, the excessive concentration of solutes in the cells can cause colligative damage to the cellular functions. This damage is prevented by treating the cells with cryoprotectant solutions. Cryoprotectants may either be penetrating/colligative or non-penetrating/osmotic, and they are usually used in combination (Panis and Lambardi 2006; Ciani et al. 2012). Dimethyl sulphoxide (DMSO), glycerol and low-molecular-weight glycols are membrane permeable and are commonly used as penetrating cryoprotectants. The cryoprotective role of glycerol was discovered by Polge et al. in 1949. These chemicals act as cellular solvents and protect cells from the high concentrations of intracellular solutes after water loss. Colligative cryoprotection requires the cryoprotectants to penetrate the cells without having any cytotoxic effects. These act by increasing the cellular osmolarity, thus decreasing the amount of water that needs to be removed from cells to achieve an osmotic equilibrium with the extracellular environment, thus cells can tolerate dehydration stress (Meryman and Williams 1985). Further, they depress the freezing point such that even when ice nucleation occurs, it is not very injurious (Benson 2008a).

On the other hand, completely ice-free cryogenic storage is achieved by the use of non-penetrating cryoprotectant additives. These work by increasing cellular viscosity, impairing ice nucleation by restricting water movement and reducing water to a critical level such that any remaining water directly vitrifies on exposure to ultralow temperatures (Fuller 2004; Mazur 2004). The cellular viscosity is increased either by the addition of the cryoprotectant at very high concentrations to bring about osmotic dehydration or by removal of water by evaporative desiccation. Exposure of cells to high concentrations of the cryoprotectants can result in osmotic injury, while evaporative desiccation can cause desiccation sensitivity issues. Therefore, it is advised to use a combination of both the procedures (Fahy et al. 1986).

13.2.2 Physiological Status of Plants

Juvenile plants, either *in vitro* or *in vivo* grown, are preferred for cryopreservation. This is because the meristematic cells contain fewer vacuoles, dense cytoplasm and are smaller in size, and hence can withstand freezing more efficiently than mature tissues (Engelmann 1991). The size of the meristematic cells is also a critical factor determining the success of any cryogenic procedure. Generally, 1- to 3-mm-long shoot tips are preferred (Kami 2012). Shoot tips smaller than 1 mm fail to regenerate into complete plantlets, whereas larger shoot tips tend to lose their viability post LN exposure (Kami et al. 2005, 2010; Keller et al. 2008).

13.2.3 Preconditioning and Preculture

Preconditioning is the cold acclimation of plants to enhance their endurance to LN. This is achieved by exposing plants to low temperatures (0–5°C) in a diurnal cycle for a week to months together, to activate genes for cold adaptation (Gale et al. 2013). However, this method is suitable for temperate plant species that are naturally minimally cold hardy and not for tropical species as they are highly temperature sensitive. Tropical plant species are usually precultured on high osmoticum medium for 24–48 hours to induce dehydration tolerance before LN freezing. The explants are most commonly cultured on media containing sugars such as sucrose, glucose, or fructose or sugar alcohols such as sorbitol or mannitol.

13.3 Cryopreservation Techniques

Several cryopreservation methodologies have been developed over the years. Different techniques are used for the storage of different plant material, and often, a combination of techniques is used for germplasm conservation.

13.3.1 Controlled Rate Freezing (Classical Method)

Also known as two-step or equilibrium freezing, this method requires precise control over the cooling rate and the application of cryoprotectants. Cryopreservation protocols involving controlled rate cooling consist of a series of steps. First, the cells are acclimatised to penetrating cryoprotectants such as DMSO, along with some high osmotic solutes such as sucrose and polyethylene glycol, to minimise the cellular component damaging effects of solute concentration. Samples are then gradually cooled down with a specific cooling rate, usually ranging from –0.1 to 5°C min^{-1}, until they are cooled to an intermediate transfer temperature (–35 to –40°C), and are then directly plunged into LN (Engelmann 2004). Freezing is achieved in two steps, hence the name two-step cooling. The gradual cooling can be achieved using sophisticated instruments called controlled rate freezers, or by using alcohol baths placed in a –20 or –80°C freezer. For retrieving the conserved cells, the samples are warmed in a water bath at 40°C for 1–2 minutes, cryoprotectants removed and cells re-cultured for growth. This was the first procedure adopted for plant cryopreservation and is mostly used for the preservation of dormant buds of temperate species (Table 13.1).

13.3.2 Desiccation

Some plant species can be conserved by directly immersing their tissues in LN after desiccation in either a sterile stream of air for 1–5 hours or by drying over activated silica gel for 12–18 hours (Sherlock et al. 2005). The conserved materials can be easily recovered by bringing them back to ambient conditions. This method is preferred for the conservation of pollen, seeds and embryonic axes of several crops (Table 13.1).

13.3.3 Encapsulation-Dehydration (ED)

ED involves the encapsulation of shoot tips/embryos in calcium alginate beads prior to LN exposure (Fabre and Dereuddre 1990). The encapsulated meristems are then precultured in high sucrose medium (0.5–0.75 M) to osmotically dehydrate the samples, followed by dehydration by air or silica gel to bring down the overall moisture content to 20–25% and then directly immersing in LN for conservation. For rewarming, the beads are warmed in a water bath set at 40°C for 1–2 minutes and then directly cultured onto regeneration medium (Figure 13.1). This method is used for plants that are highly sensitive to dehydration and/or to the toxic effects of cryoprotectants (González-Arnao and Engelmann 2006). This method has been applied for the conservation of several temperate and tropical plant species (Table 13.1).

13.3.4 Vitrification

Vitrification involves the removal of freezable water from plant cells by exposure to a mixture of penetrating and non-penetrating cryoprotectant solution known as vitrification solution, at 25°C or 0°C for relatively short time durations. The removal of water increases the cellular viscosity, and the consequent ultrarapid cooling by direct LN plunging leads to vitrification of the intracellular solutes. Prior to cryoprotection, dehydration tolerance is induced by treating the tissues with high levels of sucrose (normally 0.3–0.5 M) and a glycerol-sucrose containing (2 M glycerol + 0.4 M sucrose) loading solution. The most commonly used vitrification solution, known as plant vitrification solution 2 (PVS2), consists of 30% (w/v) glycerol, 15% (w/v) ethylene glycol and 15% (w/v) DMSO in liquid medium with 0.4 M sucrose (Sakai et al. 1990). PVS2-treated shoot tips are placed in cryovials and plunged in LN. For recovery,

TABLE 13.1
Cryopreservation of Different Plant Species Using Various Cryopreservation Protocols

Plant	Explant Used for Conservation	Reference
Classical/Two-Step Freezing		
Morus alba, M. indica, M. laevigata	Dormant buds	Choudhary et al. (2013)
Malus spp.	Winter-dormant buds	Yi et al. (2013)
Vaccinium sp. (blueberry)	Dormant buds from tree	Jenderek et al. (2017)
M. laevigata (Himalayan mulberry)	Dormant axillary buds	Choudhary et al. (2018)
Malus domestica	Dormant buds	Gupta and Mir (2019)
Desiccation		
Areca catechu L. (areca nut)	Pollen grains	Karun et al. (2017)
Pineapple	Pollen grains	Silva et al. (2017)
Pineapple	Pollen grains	Souza et al. (2018)
Ekebergia capensis	Seeds	Bharutha and Naidoob (2020)
Ensete glaucum	Seeds and zygotic embryos	Singh et al. (2020)
Encapsulation-Dehydration		
Mentha × *Piperita* (mint)	Shoot tips	Kremer et al. (2015)
Arundina graminifolia	Protocorms	Cordova et al. (2016)
Asparagus officinalis L. (cv. Morado de Huétor)	Rhizome buds	Carmona-Martín et al. (2018)
Vanilla siamensis	Protocorms	Chaipanich et al. (2019)
Mentha × *Piperita* (mint)	Shoot tips	Ibáñez et al. (2019)
Chrysanthemum	Shoot tips	Kulus et al. (2019)
Vitrification		
Satureja spicigera	Calli	Ghaffarzadeh-Namazi et al. (2017)
Bacopa monnieri	Shoot tips	Sharma et al. (2017).
Lotus tenuis	Adventitious bud clusters	Espasandin et al. (2019)
Artemisia sieversiana	Seeds	Yong et al. (2019)
Cynara scolymus L. (globe artichoke)	Shoot tips	Bekheet et al. (2020).
Encapsulation-Vitrification		
Chrysanthemum morifolium	Shoot tips	Jeon et al. (2015)
Ziziphora tenuior	Shoot tips	Albaba et al. (2015)
Moringa peregrina	Shoot tips	Al-Ruwaiei et al. (2017)
Solanum tuberosum (potato)	Buds of *in vitro* grown shoots	Li et al. (2017)
Hladnikia pastinacifolia	Shoot tips	Ciringer et al. (2018)
Droplet-Vitrification		
Ananas comosus (pineapple)	Shoot tips	Souza et al. (2016)
Betula lenta (cherry birch)	Seeds and shoot tips	Rathwell et al. (2016)
Rubus idaeus (red raspberry)	Shoot tips	Ukhatova et al. (2017)
Helianthus tuberosus (globe artichoke)	Shoot tips	Zhang et al. (2017)
Vitis spp. (grapevine)	Shoot tips	Bi et al. (2018)
Vaccinium corymbosum (highbush blueberry)	Adventitious buds from leaf	Chen et al. (2018)
Panax ginseng	Adventitious roots	Le et al. (2019)
Cocos nucifera (coconut)	Shoot meristems	Wilms et al. (2019)
Gentiana kurroo Royle	Shoot tips	Sharma et al. (2020)
Allium cepa var. aggregatum (Shallot)	Shoot tips	Wang et al. (2020b)
Cryo-Plates		
Clinopodium odorum	Shoot tips	Engelmann-Sylvestre and Engelmann (2015)

(Continued)

TABLE 13.1
(Continued)

Plant	Explant Used for Conservation	Reference
Diospyros kaki (Japanese persimmon)	Dormant 1-year-old Shoots	Matsumoto et al. (2015)
Saccharum officinarum (sugarcane)	Shoot tips	Rafique et al. (2015, 2016)
Ullucus tuberosus (ulluco)	Shoot tips	Arizaga et al. (2017b)
Solanum tuberosum (potato)	Shoot tips	Arizaga et al. (2017a)
Vaccinium sp. (blueberry)	Meristems	Dhungana et al. (2017)
Sechium spp. (chayote)	Shoot tips	Arizaga et al. (2019)
Vitis aestivalis and *V. jacquemontii*	Shoot tips	Bettoni et al. (2019)
Acampe rigida	Protocorms	Imsomboon et al. (2019)
Allium chinense	Shoot tips	Tanaka et al. (2019a)
M. australis and *M. alba* (mulberry)	Shoot tips	Tanaka et al. (2019b)
A. graminifolia	Seeds	Thammasiri et al. (2019)

the samples are thawed in a recovery/unloading solution containing 1.2 M sucrose and then cultured on a regeneration medium (Figure 13.2). This process is less demanding and higher recovery rates are obtained in plants cryopreserved using this technique. Several low temperature-sensitive plant species have been conserved using vitrification (Table 13.1).

13.3.5 Encapsulation-Vitrification (EV)

Encapsulation-vitrification (EV) is a combination of the two techniques explained above, namely, ED and vitrification (Sakai and Engelmann 2007). The plant tissues are first encapsulated in alginate beads,

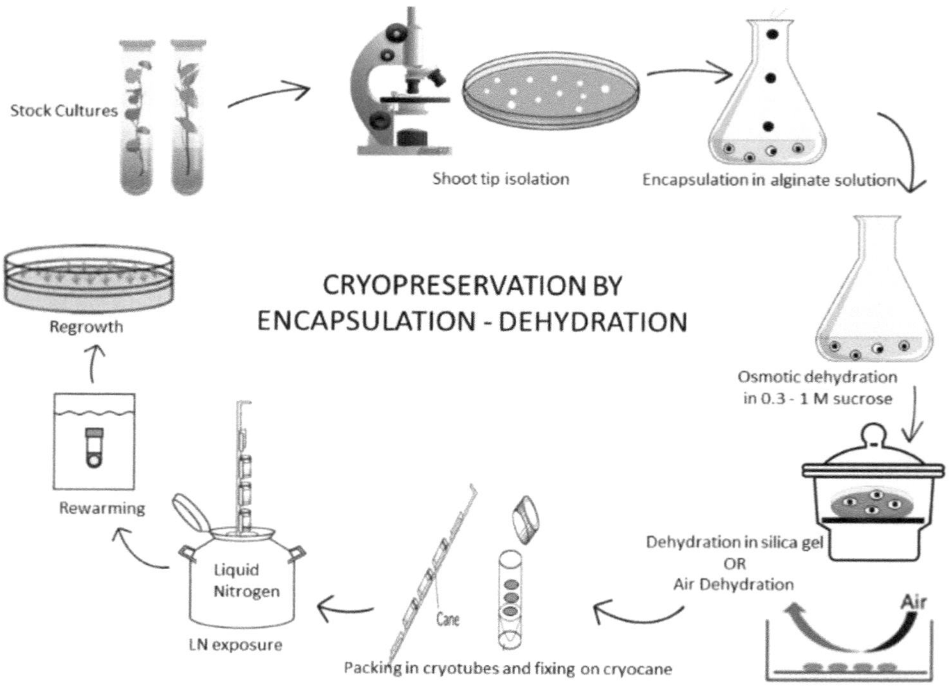

FIGURE 13.1 Process of cryopreservation by encapsulation-dehydration technique. LN, liquid nitrogen.

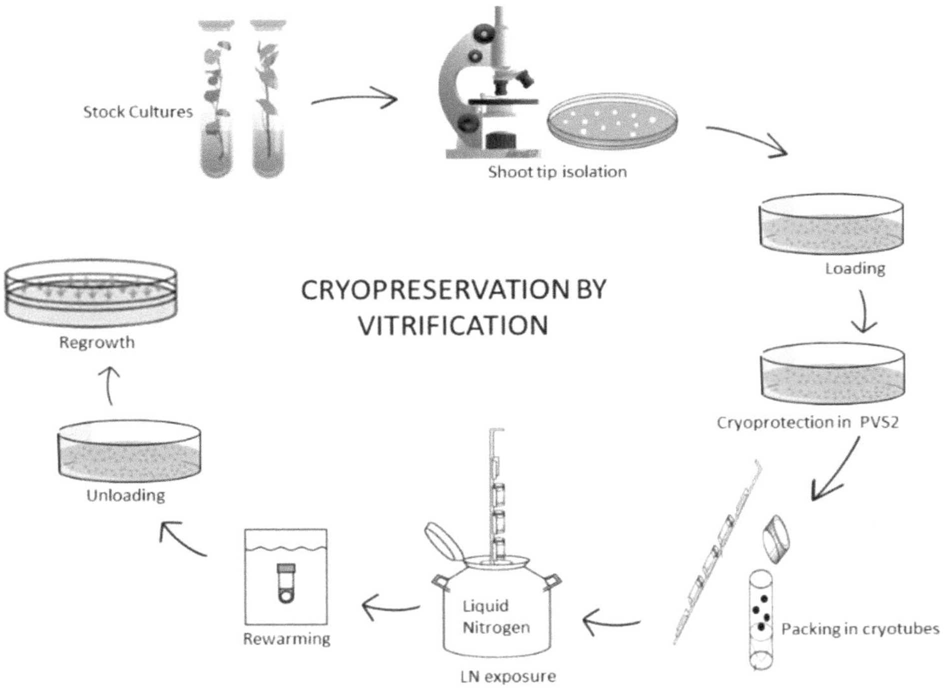

FIGURE 13.2 Process of cryopreservation by vitrification technique. LN, liquid nitrogen; PVS2, plant vitrification solution 2.

and then the beads are desiccated by treatment with PVS2, placed in cryovials and immersed in LN. Rewarming and recovery is done similar to the vitrification technique (Figure 13.3). This technique combines the advantages of both the techniques into one, i.e., chemical cryoprotection and rapidity from vitrification and ease of manipulation of delicate explants from ED. Shoot apices of many species have been conserved using this technique (Table 13.1).

13.3.6 Droplet-Vitrification (DV)

The technique of DV was applied for the first time for the conservation of potato shoot tips (Schäfer-Menuhr et al. 1994). This technique involves ultrarapid cooling of tissues by placing them in small droplets of cryoprotectant on an aluminium foil strip (Figure 13.4). The direct freezing of droplets is achieved at a much higher cooling rate (of about 130°C/min), bringing about vitrification easily. The high thermal conductivity of aluminium increases the freezing and thawing rate. Small droplets are placed on a thin aluminium foil strip, onto which cryoprotected shoot tips are placed. The strips are directly plunged into LN or into cryovials containing LN (Panis et al. 2005). For recovery, shoot tips are rapidly rewarmed, so the possibility of devitrification to form ice is avoided (Kulus 2016; Panis et al. 2016). This technique has been widely adopted for the development of cryopreservation protocols in several plant species, using a variety of explants such as shoot tips, meristems, embryogenic callus, hairy roots and embryonic axes (Table 13.1).

13.3.7 D and V Cryo-Plates

The cryo-plate technique is one of the most recent advancements in cryopreservation. These aluminium grooved plates developed by Yamamoto et al. (2011a) fit inside cryovials allowing easy manipulation of plant material. In these methods, shoot tips are adhered to the plates using alginate beads and all

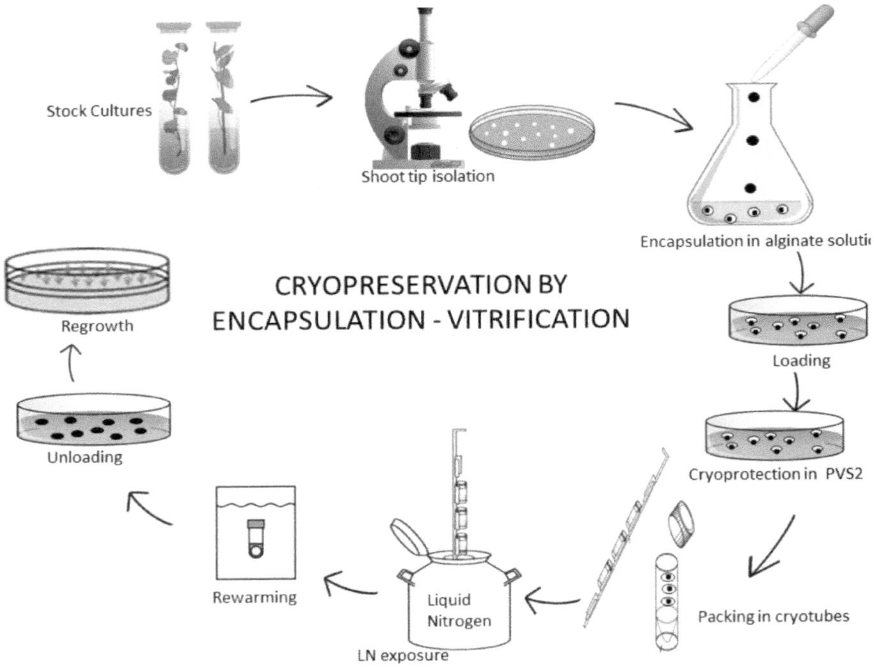

FIGURE 13.3 Process of cryopreservation by encapsulation-vitrification technique. LN, liquid nitrogen; PVS2, plant vitrification solution 2.

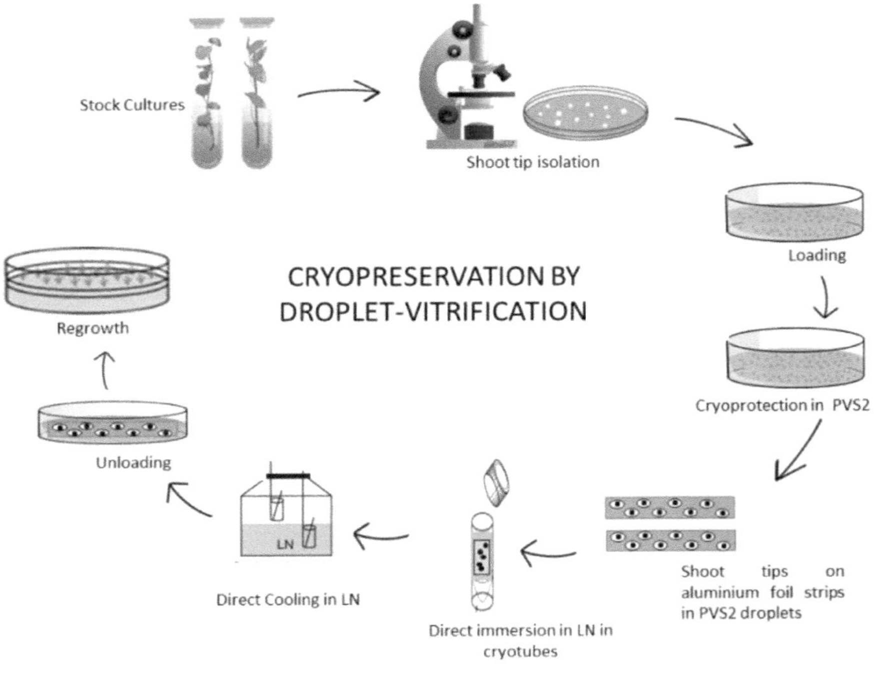

FIGURE 13.4 Process of cryopreservation by droplet-vitrification technique. LN, liquid nitrogen; PVS2, plant vitrification solution 2.

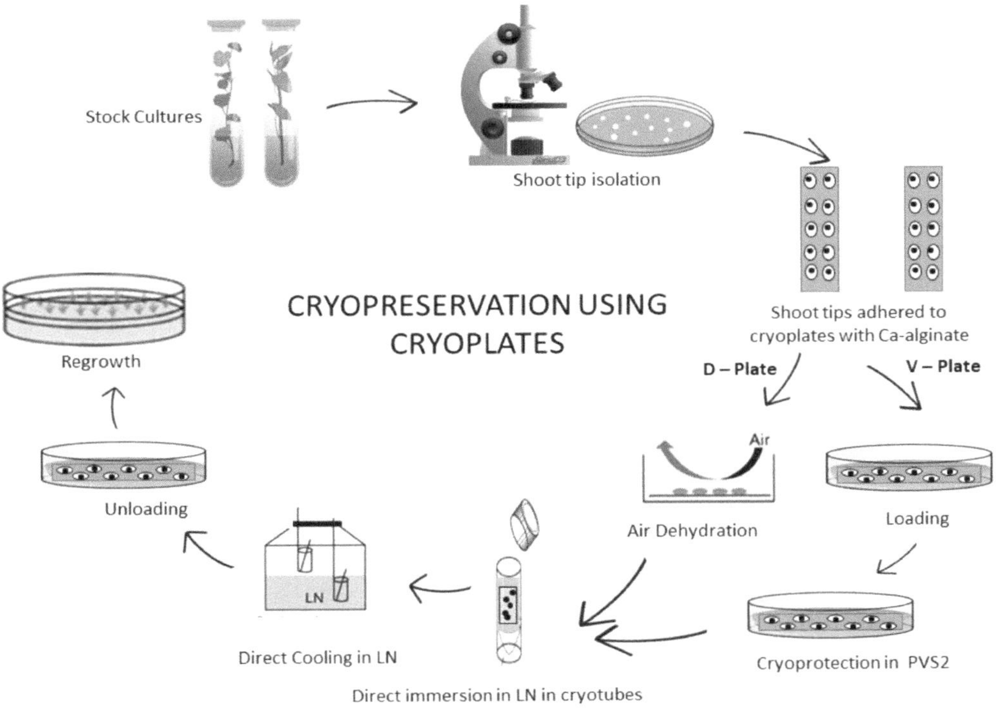

FIGURE 13.5 Process of cryopreservation using D and V cryo-plate methods. LN, liquid nitrogen; PVS2, plant vitrification solution 2.

manipulations are further carried out on the plates themselves. This method offers several advantages: (1) shoot tips are adhered to the plates, minimising any injury or loss of explants and (2) there are high regrowth rates due to rapid cooling and warming rates (Yamamoto et al. 2015).

The cryo-plates have been incorporated into two techniques: V-plate and the D-plate methods (Figure 13.5). The V-plate method employs PVS2-based vitrification-dehydration of the shoot tips (Yamamoto et al. 2011b), whereas the D-plate method involves air dehydration (Niino et al. 2013). Precultured shoot tips are adhered to the wells of the cryo-plate with calcium alginate beads and then treated with a loading solution (2 M glycerol + 0.6–1 M sucrose). They are then either dehydrated with PVS2 in the V-plate method or air dehydrated in a laminar air flow in the D-plate method. The cryo-plates containing the desiccated shoot tips are then directly immersed in LN. For recovery, the shoot tips on the cryo-plates are transferred to an unloading solution for rapid rewarming and then cultured onto regrowth media. High regrowth rates have been recorded using these methods in a large number of plant species (Table 13.1).

13.3.8 Cryo-Mesh

With the advancement of technologies, newer methods for improving post cryostorage recovery are being developed. Stainless steel meshes, known as cryo-meshes, have recently been used for preserving shoot tips of *Anigozanthos viridis* (kangaroo paw) (Funnekotter et al. 2017). Like cryo-plates, in this method too precultured shoot tips are fixed onto the cryo-mesh using alginate beads. The further step is similar to the V- plate method, and includes treatment of shoot tips with a loading solution, cryoprotection with PVS2 and then direct immersion in LN. These cryo-meshes can be used as alternatives to the cryo-plates. Here again, very high cooling and thawing rates result in high recovery rates.

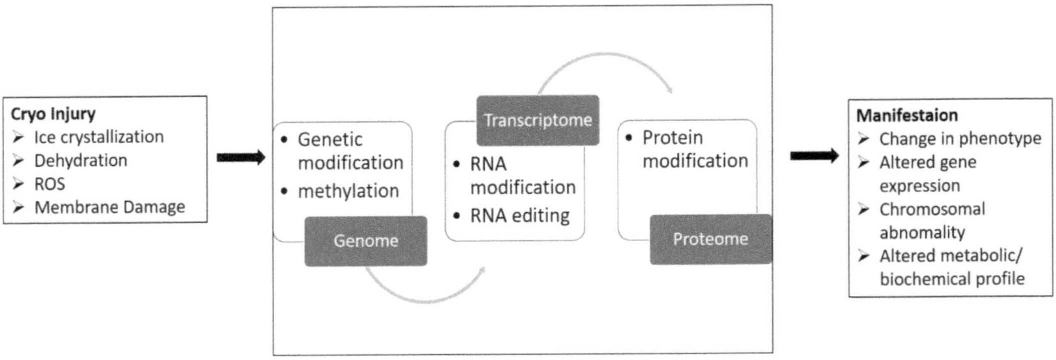

FIGURE 13.6 Linkage between the various potential damaging effects of cryoinjury on the genome, transcriptome and proteome of tissues. (Adapted from Martinez-Montero and Harding 2015.)

13.4 Omics Technologies in Cryopreservation

Cryopreservation research involves a complex interaction of a number of biological processes to achieve high levels of post-storage survival. However, plant cells are exposed to a large number of stresses during these cryogenic processes, each of which can alter the gene expression (Figure 13.6). Omics-based technologies are increasingly being employed to understand and solve various cryobiological challenges (Carpentier et al. 2007; Volk 2010).

13.4.1 Molecular Markers for Assessing Genetic Stability

A successful conservation protocol should not induce any destabilising changes in the genetic makeup of the plant cells being conserved. This is particularly important in protocols that involve an intermediate tissue culture phase, as somaclonal variations are well documented in tissue-cultured systems. Though there are few reports of induced somaclonal variations in some cryopreserved species, it is believed that such variations may be attributed to the whole culture-cryoprotection-regeneration process (Harding 2004). DNA polymorphisms between the conserved material and the same material before conservation have been studied using various molecular markers, such as randomly amplified polymorphic DNA (RAPDs), restriction fragment length polymorphisms (RFLPs), amplified fragment length polymorphism (AFLPs) and simple sequence repeats (SSRs), to identify any variations that may be induced by the cryopreservation process. At present there is no clear evidence to suggest any cytological or genetic alterations as a consequence of cryopreservation (Engelmann 2011). Several studies have been carried out to confirm the genetic integrity of conserved germplasm, such as in banana (Agrawal et al. 2014), apple (Yi et al. 2015), chrysanthemum (Bi et al. 2016), *Bacopa monnieri* (Sharma et al. 2017), *Asparagus officinalis* (Carmona-Martín et al. 2018), grapevine (Bi et al. 2018), *Hladnikia pastinacifolia* (Ciringer et al. 2018), *Thalictrum foliolosum* (Mishra et al. 2020), *Gentiana kurroo* (Sharma et al. 2020) and shallot (Wang et al. 2020a).

13.4.2 Gene Expression Analysis

During the initial stages of preparation of plant cells for cryopreservation, several signalling pathways are activated along with changes in expression patterns of related genes. Perception of cold results in activation of gene regulatory networks that help cells get acclimated to the low-temperature stress (Chinnusamy et al. 2004; Renaut et al. 2006; Volk 2010). The cold stress leads to the activation of mitogen-activated protein kinase (MAPK) cascades as well as several phytohormone-mediated pathways. Secondly, exposure to cryoprotective treatments has been found to induce molecular responses in cells. Volk and co-workers detected more than 100 genes that were either up- or down-regulated in *Arabidopsis* shoot tips post cryoprotectant exposure (Volk et al. 2011). Most of these genes were classified into groups of cold responsive, heat responsive, responsive to water deprivation and responsive to oxidative stress. Similarly, differential expression of 180 genes related to stress, defence, wounding, lipid, carbohydrate, abscisic acid, oxidation, temperature (cold/heat) or osmoregulation, on exposure to different cryoprotectants, was observed and validated in *Arabidopsis* (Gross et al. 2017).

Antioxidants are known to play an important role in the perception of cold stress and survival of plant cells in cryogenic conditions. Changes in expression levels of genes associated with the regulation of oxidative stress and programmed cell death resulted in improved post-thaw cell viability after cryopreservation of *Agapanthus praecox* embryogenic callus (Zhang et al. 2015). Several antioxidant genes, such as copper/zinc superoxide dismutase, ascorbate peroxidase, dehydroascorbate reductase, catalase and manganese superoxide dismutase, have been found to have increased expression levels in seedlings recovered after cryopreservation (Chen et al. 2015).

Several posttranscriptional regulatory mechanisms also play a crucial role in modulating protein activity and stabilising proteins. Cold-induced genes related to miRNAs (especially miR393) have a crucial role to play in the recovery of *Arabidopsis* shoot tips after cryopreservation (Ekinci et al. 2020).

13.4.3 Proteomics Insights

The high osmotic stress to which shoot tips are exposed leads to an altered gene expression pattern, resulting in significant up- or down-regulation of proteins. On comparison of the proteomes of dehydration-sensitive and dehydration-tolerant genotypes of banana, protein isoform differences associated with stress-related proteins such as phosphoglycerate kinase, phosphoglucomutase and abscisic stress ripening-like proteins were detected (Carpentier et al. 2009). A similar analysis in potato revealed 15 up- or down-regulated proteins in shoot tips grown on sorbitol medium for cryopreservation (Criel et al. 2005, 2008). Similarly, two-dimensional gel electrophoresis of vanilla shoot tips treated with PVS3 revealed 206 differentially expressed proteins (González-Arnao et al. 2011).

13.5 Conclusion

Cryopreservation is the ideal method for long-term germplasm conservation. Rapid strides are being made in the development of cryogenic protocols in several plant species; however, robust wide-spectrum protocols applicable to several genotypes of a single species need to be developed for effective and large-scale conservation in cryobanks. Moreover, emphasis should now be given to research on understanding the basic mechanisms of osmotic and dehydration stress to effectively conserve species recalcitrant to cryopreservation. Research needs to be focussed on the establishment of cryobanks for capturing wider biodiversity, which would help in avoiding loss of valuable germplasm and making it available for posterity.

REFERENCES

Agrawal, A., Sanayaima, R., Singh, R., Tandon, R., Verma, S. and Tyagi, R. K. 2014. Phenotypic and molecular studies for genetic stability assessment of cryopreserved banana meristems derived from field and *in vitro* explant sources. *In Vitro Cellular & Developmental Biology – Plant* 50: 345–356.

Agrawal, A., Singh, S., Malhotra, E. V., Meena, D. P. S. and Tyagi, R. K. 2019. *In vitro* conservation and cryopreservation of clonally propagated horticultural species, In: Rajasekharan, P. E., Rao, V. R. (Eds.), Conservation and Utilization of Horticultural Genetic Resources. Springer Singapore, Singapore, pp. 529–578.

Al-Ruwaiei, H. M., Shibli, R. A., Al Khateeb, W. M., Al Qudah, T. S., Tahtamouni, R. W. and Al-Baba, H. 2017. Rescuing endangered *Moringa peregrina* (frossk) fiori by cryopreservation using vitrification and encapsulation-vitrification protocols. *Jordan Journal of Agricultural Sciences* 13(4): 961–976.

Albaba, H. B., Shibli, R. A., Akash, M., Al-Qudah, T. S., Tahtamouni, R. W. and Al-Ruwaiei, H. 2015. Cryopreservation and genetic stability assessment of threatened medicinal plant (*Ziziphora tenuior* L.) grown wild in Jordan. *Jordan Journal of Biological Sciences* 8: 247–256.

Arizaga, M. V., Cancino, G. S., Iñiguez, J. C., Gutiérrez, E. J. C., Reyes, L. A. G., Yamamoto, S., Tanaka, D., Vazquez, S. G. O., Watanabe, K. and Niino, T. 2019. Cryopreservation of *in vitro* shoot tips of chayote (*Sechium* spp.) by D cryo-plate method. *Acta Horticulturae* 1234: 293–299.

Arizaga, M. V., Navarro, O. F., Martinez, C. R., Gutiérrez, E. J., Delgado, H. A., Yamamoto, S. I., Watanabe, K. and Niino, T. 2017a. Improvement to the D Cryo-plate protocol applied to practical cryopreservation of *in vitro* grown potato shoot tips. *Horticultural Journal* 86(2): 222–228.

Arizaga, M. V., Yamamoto, S. I., Tanaka, D. et al. 2017b. Cryopreservation of *in vitro* shoot tips of ulluco (*Ullucus tuberosus* Cal.) using D cryo-plate method. *Cryo Letters* 38(6): 419–427.

Bekheet, S. A., Sota, V., El-Shabrawi, H. M. et al. 2020. Cryopreservation of shoot apices and callus cultures of globe artichoke using vitrification method. *Journal of Genetic Engineering and Biotechnology* 18: 2.

Benson, E. E. 2008a. Cryopreservation of phytodiversity: a critical appraisal of theory and practice. *Critical Reviews in Plant Sciences* 27: 141–219.

Benson, E. E. 2008b. Cryopreservation theory. In: Reed, B. M. (Eds.), Plant Cryopreservation: A Practical Guide. Springer, Berlin, pp. 15–32.

Bettoni, J. C., Bonnart, R., Shepherd, A. N., Kretzschmar, A. A. and Volk, G. M. 2019. Modifications to a vitis shoot tip cryopreservation procedure: effect of shoot tip size and use of cryoplates. *Cryo Letters* 40(2): 103–112.

Bharutha, V. and Naidoob, C. 2020. Responses to cryopreservation of recalcitrant seeds of *Ekebergia capensis* from different provenances. *South African Journal of Botany* 132: 1–14.

Bi, W., Hao, X., Cui, Z. et al. 2018. Droplet-vitrification cryopreservation of *in vitro*-grown shoot tips of grapevine (*Vitis* spp.). *In Vitro Cellular & Developmental Biology – Plant* 54: 590–599.

Bi, W., Pan, C., Liu, J. et al. 2016. Greenhouse performance, genetic stability and biochemical compounds in *Chrysanthemum morifolium* 'Hangju' plants regenerated from cryopreserved shoot tips. *Acta Physiologiae Plantarum* 38: 268.

Boyle, R. 1665. New experiments and observations touching cold. Royal Society of London. J. Crook, London.

Carmona-Martín, E., Regalado, J. J., Perán-Quesada, R. et al. 2018. Cryopreservation of rhizome buds of *Asparagus officinalis* L. (cv. Morado de Huétor) and evaluation of their genetic stability. *Plant Cell Tissue & Organ Culture* 133: 395–403.

Carpentier, S. C., Vertommen, A., Swennen, R. and Panis, B. 2009. Will proteomics contribute to a better understanding of cryopreservation survival. In: Panis, B (Ed.), *Proceedings of the 1st International Symposium on Cryopreservation in Horticultural Species*, 5–8. Leuven.

Carpentier, S. C., Witters, E., Laukens, K., Van Onckelen, H., Swennen, R. and Panis, B. 2007. Banana (*Musa spp.*) as a model to study the meristem proteome: acclimation to osmotic stress. *Proteomics* 7: 92–105.

Chaipanich, V. V., Roberts, D. L., Yenchon, S., Te-chato, S. and Divakaran, M. 2019. Development of a cryopreservation protocol for *Vanilla siamensis*: an endangered orchid species in Thailand. *Cryo Letters* 40(5): 305–311.

Chen, G. Q., Ren, L., Zhang, J., Reed, B. M., Zhang, D. and Shen, X. H. 2015. Cryopreservation affects ROS-induced oxidative stress and antioxidant response in Arabidopsis seedlings. *Cryobiology* 70(1): 38–47.

Chen, H., Liu, J., Pan, C. et al. 2018. *In vitro* regeneration of adventitious buds from leaf explants and their subsequent cryopreservation in highbush blueberry. *Plant Cell Tissue & Organ Culture* 134: 193–204.

Chinnusamy, V., Schumaker, K. and Zhu, J. K. 2004. Molecular genetic perspectives on cross-talk and specificity in abiotic stress signalling in plants. *Journal of Experimental Botany* 55: 225–236.

Choudhary, R., Chaudhury, R., Malik, S. K., Kumar, S. and Pal, D. 2013. Genetic stability of mulberry germplasm after cryopreservation by two-step freezing technique. *African Journal of Biotechnology* 12(41): 5983–5993.

Choudhary, R., Malik, S. and Chaudhury, R. 2018. Development of an efficient cryoconservation protocol for Himalayan mulberry (*Morus laevigata* Wall. ex brandis) using dormant axillary buds as explants. *Indian Journal of Experimental Biology* 56: 342–350.

Ciani, F., Cocchia, N., Esposito, L. and Avallone, L. 2012. Fertility cryopreservation. In: *Advances in Embryo Transfer IntechOpen*. https://doi.org/10.5772/38511

Ciringer, T., Martín, C., Šajna, N., Kaligarič, M. and Ambrožič-Dolinšek, J. 2018. Cryopreservation of an endangered *Hladnikia pastinacifolia* Rchb. by shoot tip encapsulation-dehydration and encapsulation-vitrification. *In Vitro Cellular Developmental Biology Plant* 54(6): 565–575.

Cordova, I. I., Luis, B. and Thammasiri, K. 2016. Cryopreservation on a cryo-plate of *Arundina graminifolia* protocorms, dehydrated with silica gel and drying beads. *Cryo Letters* 37(2): 68–76.

Criel, B., Panis, B., Oufi, R. M., Swennen, R., Renaut, J. and Hausman, J. F. 2008. Protein and carbohydrate analyses of abiotic stress underlying cryopreservation in potato. In: Cryopreservation of Crop Species in Europe, Proceedings of CRYOPLANET COST Action, 871, 20–23. Oulu, MTT Agrifood Research Working Papers 153, J. Laamanen, M. Uosukainen, H. Häggman, A. Nukari, S. Rantala.

Criel, B., Panta, A., Carpentier, S., Renaut, J., Swennen, R., Panis, B. and Hausman, J. F. 2005. Cryopreservation and abiotic stress tolerance in potato: a proteomic approach. *Communications In Agricultural And Applied Biological Sciences* 70: 83–86.

Day, J. G. and McClellan, M. R. 1995. *Cryopreservation and Freeze-Drying Protocols*. Humana Press, Totowa, New Jersey.

Day, J. G., Harding, K., Nadarajan, J. and Benson, E. E. 2008. Cryopreservation conservation of bioresources and ultralow temperatures. In: Walker, J. M., Rapley, R. R. (Eds.), *Molecular Biomethods Handbook*, 2nd edn. Humana Press, Totowa, New Jersey, pp. 917–947.

Dhungana, S. A., Kunitake, H., Niino, T., Yamamoto, S. I., Fukui, K., Tanaka, D., Maki, S. and Matsumoto, T. 2017. Cryopreservation of blueberry shoot tips derived from *in vitro* and current shoots using D cryo-plate technique. *Plant Biotechnology* 34(1): 1–5.

Ekinci, M. H., Kayıhan, D. S., Kayıhan, C. et al. 2020. The role of microRNAs in recovery rates of *Arabidopsis thaliana* after short term cryo-storage. *Plant Cell Tissue & Organ Culture* 144: 281–293.

Engelmann, F. 1991. *In vitro* conservation of tropical plant germplasm – a review. *Euphytica* 57: 227–243.

Engelmann, F. 2004. Plant cryopreservation: progress and prospects. *In Vitro Cellular & Developmental Biology – Plant* 40: 427–433.

Engelmann, F. 2011. Use of biotechnologies for the conservation of plant biodiversity. *In Vitro Cellular & Developmental Biology – Plant* 47: 5–16.

Engelmann-Sylvestre, I. and Engelmann, F. 2015. Cryopreservation of *in vitro*-grown shoot tips of *Clinopodium odorum* using aluminium cryo-plates. *In Vitro Cellular & Developmental Biology – Plant* 51(2): 185–191.

Espasandin, F. D., Brugnoli, E. A., Ayala, P. G. et al. 2019. Long-term preservation of *Lotus tenuis* adventitious buds. *Plant Cell Tissue & Organ Culture* 136: 373–382.

Fabre, J. and Dereuddre, J. 1990. Encapsulation-dehydration: a new approach to cryopreservation of Solanum shoot-tips. *Cryo Letters* 11: 413–426.

Fahy, G. M., MacFarlane, D. R., Angell, C. A. and Meryman, H. T. 1984. Vitrification as an approach to cryopreservation. *Cryobiology* 21: 407–426.

Fahy, G. M., Takahashi, T. and Meryman, H. T. 1986. Practical aspects of ice-free cryopreservation. In: Smit-Sibinga, T. H., Das, P. C. (Eds.), Aspects of Ice-Free Cryopreservation. Martinus Nijhoff, Boston, Massachusetts, pp. 111–122.

Fuller, B. J. 2004. Cryoprotectants: the essential antifreezes to protect life in the frozen state. *Cryo Letters* 25: 375–388.

Funnekotter, B., Bunn, E. and Mancera, R. L. 2017. Cryo-Mesh: a simple alternative cryopreservation protocol. *Cryo Letters* 38(2): 155–159.

Gale, S., Benson, E. E. and Harding, K. 2013. A life cycle model to enable research of cryostorage recalcitrance in temperate woody species: the case of Sitka spruce (*Picea sitchensis*). *Cryo Letters* 34: 30–33.

Ghaffarzadeh-Namazi, L., Joachim Keller, E. R., Senula, A. and Babaeian, N. 2017. Investigations on various methods for cryopreservation of callus of the medicinal plant *Satureja spicigera*. *Journal of Applied Research on Medicinal and Aromatic Plants* 5: 10–15.

González-Arnao, M. T. and Engelmann, F. 2006. Cryopreservation of plant germplasm using the encapsulation–dehydration technique: review and case study on sugarcane. *Cryo Letters* 27: 155–168.

González-Arnao, M. T., Durán-Sánchez, B., Jiménez- Francisco, B., Lázaro-Vallejo, C. E., Valdés-Rodríguez, S. E. and Guerrero, A. 2011. Cryopreservation and proteomic analysis of vanilla (*V. planifolia* A.) apices treated with osmoprotectants. *Acta Horticulturae* 908: 67–72.

González-Arnao, M. T., Guerrero-Rangel, A., Martínez, O. and Valdés-Rodríguez, S. 2018. Protein changes in the shoot-tips of vanilla (*Vanilla planifolia*) in response to osmoprotective treatments. *Journal of Plant Biochemistry and Biotechnology* 27: 331–341.

Gross, B. L., Henk, A. D., Bonnart, R. and Volk, G. M. 2017. Changes in transcript expression patterns as a result of cryoprotectant treatment and liquid nitrogen exposure in Arabidopsis shoot tips. *Plant Cell Reports* 36: 459–470.

Gupta, S. and Mir, J. I. 2019. Cryopreservation of apple (*Malus domestica* 'Benoni') dormant buds using two-step freezing method. *Acta Horticulturae* 1234: 323–328.

Harding, K. 2004. Genetic integrity of cryopreserved plant cells: a review. *Cryo Letters* 25: 3–22.

Humphreys, A. M., Govaerts, R., Ficinski, S. Z., Nic Lughadha, E. and Vorontsova, M. S. 2019. Global dataset shows geography and life form predict modern plant extinction and rediscovery. *Nature Ecology Evolution* 3: 1043–1047.

Ibáñez, M. A., Alvarez-Mari, A., Rodríguez-Sanz, H., Kremer, C., González-Benito, M. E. and Martín, C. 2019. Genetic and epigenetic stability of recovered mint apices after several steps of a cryopreservation protocol by encapsulation-dehydration: a new approach for epigenetic analysis. *Plant Physiology and Biochemistry* 143: 299–307.

Imsomboon, T., Thammasiri, K., Kosiyajinda, P., Chuenboonngarm, N. and Panvisavas, N. 2019. Cryopreservation of non-precultured protocorms of *Acampe rigida* (Buch. Ham. ex Sm.) using V cryo-plate and D cryo-plate methods. *Acta Horticulturae* 1234: 269–278.

Jenderek, M. M., Tanner, J. D., Ambruzs, B. D., West, M., Postman, J. D. and Hummer, K. E. 2017. Twig pre-harvest temperature significantly influences effective cryopreservation of *Vaccinium* dormant buds. *Cryobiology* 74: 154–159.

Jeon, S. M., Arun, M., Lee, S. Y. and Kim, C. K. 2015. Application of encapsulation-vitrification in combination with air dehydration enhances cryotolerance of *Chrysanthemum morifolium* shoots tips. *Scientia Horticulturae* 194: 91–99.

Kami, D. 2012. Cryopreservation of plant genetic resources. In: Katkov, I. I. (Eds.), Current Frontiers in Cryobiology. InTech, Rijeka, Coratia. https://doi.org/10.5772/34414

Kami, D., Kido, S., Otokita, K., Suzuki, T., Sugiyama, K. and Suzuki, M. 2010. Cryopreservation of shoot apices of *Cardamine yezoensis in vitro* cultures by vitrification method. *Cryobiology and Cryotechnology* 56: 119–126.

Kami, D., Suzuki, T. and Oosawa, K. 2005. Cryopreservation of blue honeysuckle *in vitro* cultured tissues using encapsulation-dehydration and vitrification. *Cryobiology and Cryotechnology* 51(2): 63–68.

Kartha, K. K. 1985. Cryopreservation of Plant Cells and Organs. CRC Press Inc., Boca Raton, Florida.

Karun, A., Sajini, K. K., Muralikrishna, K. S., Rajesh, M. K. and Engelmann, F. 2017. Cryopreservation of areca nut (Areca catechu L.) pollen. *Cryo Letters* 38(6): 463–470.

Kaviani, B. 2011. Conservation of plant genetic resources by cryopreservation. *Australian Journal of Crop Science* 5(6): 778–800.

Keller, E. R. J., Senula, A. and Kaczmarczyk, A. 2008. Cryopreservation of herbaceous dicots. In: Reed, B. M. (Eds.), Plant Cryopreservation: A Practical Guide. Springer, Berlin, pp. 281–332.

Kremer, C., Martín, C., González, I. and González-Benito, M. E. 2015. Regeneration in mint (Mentha × Piperita) cryopreserved apices: can the cryopreservation technique, regeneration medium composition and genotype affect the final result. *Acta Horticulturae* 1099: 777–782.

Kulus, D. 2016. Application of cryogenic technologies and somatic embryogenesis in the storage and protection of valuable genetic resources of ornamental plants. In: Somatic Embryogenesis in Ornamentals and Its Applications. Springer, New Delhi, pp. 1–25.

Kulus, D., Rewers, M., Serocka, M. and Mikula, A. 2019. Cryopreservation by encapsulation-dehydration affects the vegetative growth of chrysanthemum but does not disturb its chimeric structure. *Plant Cell Tissue & Organ Culture* 138: 153–166.

Le, K., Kim, H. and Park, S. 2019. Modification of the droplet-vitrification method of cryopreservation to enhance survival rates of adventitious roots of *Panax ginseng*. *Horticulture, Environment and Biotechnology* 60: 501–510.

Li, J. W., Chen, H. Y., Li, X. Y., Zhang, Z., Blystad, D. R. and Wang, Q. C. 2017. Cryopreservation and evaluations of vegetative growth, microtuber production and genetic stability in regenerants of purple-fleshed potato. *Plant Cell Tissue and Organ Culture* 128(3): 641–653.

Martinez-Montero, M. E. and Harding, K. 2015. Cryobionomics: evaluating the concept in plant cryopreservation. In: Barh, D., Khan, M., Davies, E. (Eds.), PlantOmics: The Omics of Plant Science. Springer, New Delhi, pp. 655–682.

Matsumoto, T. 2017. Cryopreservation of plant genetic resources: conventional and new methods. *Reviews in Agricultural Science* 5: 13–20.

Matsumoto, T., Yamamoto, S. I., Fukui, K., Rafique, T., Engelmann, F. and Niino, T. 2015. Cryopreservation of persimmon shoot tips from dormant buds using the D cryo-plate technique. *The Horticulture Journal* 84(2): 106–110.

Mazur, P. 2004. Principles of cryobiology. In: Fuller, B., Lane, N., Benson, E. E. (Eds.), Life in the Frozen State. CRC Press, Boca Raton, Florida, pp. 3–66.

Meryman, H. T. and Williams, R. J. 1985. Basic principles of freezing injury to plant cells: natural tolerance and approaches to cryopreservation. In: Kartha, K. K. (Eds.), Cryopreservation of Plant Cells and Organs. CRC Press Inc., Boca Raton, Florida, pp. 14–47.

Mishra, M. K., Pandey, S., Misra, P. et al. 2020. *In vitro* propagation, genetic stability and alkaloids analysis of acclimatized plantlets of Thalictrum foliolosum. *Plant Cell Tissue & Organ Culture* 142: 441–446.

Muldrew, K., Acker, J. P., Elliott, A. W. and McGann, L. E. 2004. The water to ice transition: implications for living cells. In: Fuller, B., Lane, N., Benson, E. E. (Eds.), Life in the Frozen State. CRC Press, London, pp. 67–108.

Niino, T., Yamamoto, S. I., Fukui, K., Martínez, C. R., Arizaga, M. V., Matsumoto, T. and Engelmann, F. 2013. Dehydration improves cryopreservation of mat rush (*Juncus decipiens* Nakai) basal stem buds on cryo-plates. *Cryo Letters* 34(6): 549–560.

Ogawa, Y., Sakurai, N., Oikawa, A., Kai, K., Morishita, Y., Mori, K., Moriya, K., Fujii, F., Aoki, K., Suzuki, H., Ohta, D., Saito, K. and Shibata, D. 2012. High-throughput cryopreservation of plant cell cultures for functional genomics. *Plant and Cell Physiology* 53: 943–952.

Panis, B. and Lambardi, M. 2006. Status of cryopreservation technologies in plants (crops and forest trees). In: Ruane, J., Sonnino, A. (Eds.), The Role of Biotechnology in Exploring and Protecting Agricultural Genetic Resources. FAO, Rome,

Panis, B., Piette, B. and Swennen, R. 2005. Droplet vitrification of apical meristem: a cryopreservation protocol applicable to all *Musaceae*. *Plant Science* 168: 45–55.

Panis, B., Van den houwe, I., Swennen, R., Rhee, J. and Roux, N. 2016. Securing plant genetic resources for perpetuity through cryopreservation. *Indian Journal of Plant Genetic Resources* 29(3): 300–302.

Pegg, D. E. 2007. Principles of cryopreservation. In: Day, J. G., Stacey, G. N. (Eds.), Methods in Molecular Biology, vol. 368: Cryopreservation and Freeze-Drying Protocols, 2nd edn. Humana Press Inc., Totowa, NJ, pp. 39–57.

Polge, C., Smith, A. U. and Parkes, A. S. 1949. Revival of spermatozoa after vitrification and dehydration at low temperatures. *Nature* 164: 666.

Popova, E., Kim, H. H., Saxena, P. K., Engelmann, F. and Pritchard, H. W. 2016. Frozen beauty: the cryobiotechnology of orchid diversity. *Biotechnology Advances* 34: 380–403.

Pritchard, H. W., Nadarajan, J., Ballesteros, D., Thammasiri, K., Prasongsom, S., Malik, S. K., Chaudhury, R., Kim, H. H., Lin, L., Li, W. Q., Yang, X. Y. and Popova, E. 2017. Cryobiotechnology of tropical seeds – scale, scope and hope. *Acta Horticulturae* 1167: 37–48.

Rafique, T., Yamamoto, S. I., Fukui, K., Mahmood, Z. and Niino, T. 2015. Cryopreservation of sugarcane using the V cryo-plate technique. *Cryo Letters* 36(1): 51–59.

Rafique, T., Yamamoto, S. I., Fukui, K., Tanaka, D., Arizaga, M. V., Abbas, M., Matsumoto, T. and Niino, T. 2016. Cryopreservation of shoot-tips from different sugarcane varieties using D cryo-plate technique. *Pakistan Journal of Agricultural Sciences* 53(1): 151–158.

Rathwell, R., Popova, E., Shukla, M. R. and Saxena, P. K. 2016. Development of cryopreservation methods for cherry birch (*Betula lenta* L.), an endangered tree species in Canada. *Canadian Journal of Forest Research* 46(11): 1284–1292.

Reed, B. M. 2017. Plant cryopreservation: a continuing requirement for food and ecosystem security. *In Vitro Cellular Developmental Biology-Plant* 53: 285–288.

Renaut, J., Hausman, J. F. and Wisniewski, M. E. 2006. Proteomics and low-temperature studies: bridging the gap between gene expression and metabolism. *Physiologia Plantarum* 126: 97–109.

Sakai, A. 1965. Survival of plant tissue at super-low temperatures in relation between effective prefreezing temperatures and the degree of frost hardiness. *Plant Physiology* 40: 882–887.

Sakai, A. and Engelmann, F. 2007. Vitrification, encapsulation-vitrification and droplet- vitrification: a review. *Cryo Letters* 28(3): 151–172.

Sakai, A., Kobayashi, S. and Oiyama, I. 1990. Cryopreservation of nucellar cells of navel orange (*Citrus sinensis* Osb. var Brasiliensis Tanaka) by vitrification. *Plant Cell Reports* 9(1): 30–33.

Schäfer-Menuhr, A., Schumacher, H. M. and Mix-Wagner, G. 1994. Long-term storage of old potato varieties by cryopreservation of shoot- tips in liquid nitrogen. *Landbauforschung Völkenrode* 44: 301–313.

Sharma, N., Gowthami, R., Devi, S.V. et al. 2020. Cryopreservation of shoot tips of *Gentiana kurroo* Royle – a critically endangered medicinal plant of India. *Plant Cell Tissue & Organ Culture* 144: 67–72. https://doi.org/10.1007/s11240-020-01879-2

Sharma, N., Singh, R., Pandey, R. et al. 2017. Genetic and biochemical stability assessment of plants regenerated from cryopreserved shoot tips of a commercially valuable medicinal herb *Bacopa monnieri* (L.) Wettst. *In Vitro Cellular Developmental Biology-Plant* 53: 346–351.

Sherlock, G., Block, W. and Benson, E. E. 2005. Thermal analysis of the plant encapsulation/dehydration protocol using silica gel as the desiccant. *Cryo Letters* 26: 45–54.

Silva, R. L., Souza, E. H., Vieira, L. J. et al. 2017. Cryopreservation of pollen of wild pineapple accessions. *Scientia Horticulturae* 219: 326–334.

Singh, S., Thangjam, R., Harish, G.D. et al. 2020. Conservation protocols for *Ensete glaucum*, a crop wild relative of banana, using plant tissue culture and cryopreservation techniques on seeds and zygotic embryos. *Plant Cell Tissue & Organ Culture* 144: 195–209. https://doi.org/10.1007/s11240-020-01881-8

Souza, F. V., Kaya, E., de Jesus Vieira, L., de Souza, E. H., de Oliveira Amorim, V. B., Skogerboe, D., Matsumoto, T., Alves, A. A., da Silva Ledo, C. A. and Jenderek, M. M. 2016. Droplet-vitrification and morphohistological studies of cryopreserved shoot tips of cultivated and wild pineapple genotypes. *Plant Cell Tissue and Organ Culture* 124(2): 351–360.

Souza, F. V. D., de Souza, E. H. and da Silva, R. L. 2018. Cryopreservation of pollen grains of pineapple and other bromeliads. In: Loyola-Vargas, V., Ochoa-Alejo, N. (Eds.), Plant Cell Culture Protocols. Methods in Molecular Biology. Humana Press, New York, vol 1815.

Tanaka, D., Sakuma, Y., Yamamoto, S., Matsumoto, T. and Niino, T. 2019a. Development of the V cryo-plate method for cryopreservation of in vitro rakkyo (*Allium chinense* G. Don). *Acta Horticulturae* 1234: 287–292.

Tanaka, D., Yamamoto, S., Matsumoto, T., Arizaga, M. V. and Niino, T. 2019b. Development of effective cryopreservation protocols using aluminium cryo-plates for mulberry. *Acta Horticulturae*1234: 263–268.

Taylor, M. J., Song, Y. C. and Brockbank, K. G. M. 2004. Vitrification in tissue preservation: new developments. In: Fuller, B., Lane, N., Benson, E. E. Life in the Frozen State. CRC Press, Boca Raton, Florida, pp. 603–644.

Thammasiri, K., Prasongsom, S., Kongsawadworakul, P., Chuenboonngarm, N., Jenjittikul, T., Soonthornchainaksaeng, P., Viboonjun, U. and Muangkroot, A. 2019. Cryopreservation of *Arundina graminifolia* (D. Don) Hochr. seeds using D cryo-plate method. *Acta Horticulturae* 1234: 301–308.

Towill, L. E. 1983. Improved survival after cryogenic exposure of shoot tips derived from *in vitro* plantlet cultures of potato. *Cryobiology* 20: 567–573.

Uemura, M. and Sakai, A. 1980. Survival of carnation (*Dianthus caryophyllus* L.) shoot apices frozen to the temperature of liquid nitrogen. *Plant Cell Physiology* 21: 85–94.

Ukhatova, Y. V., Dunaeva, S. E., Antonova, O. Y., Apalikova, O. V., Pozdniakova, K. S., Novikova, L. Y., Shuvalova, L. E. and Gavrilenko, T. A. 2017. Cryopreservation of red raspberry cultivars from the VIR *in vitro* collection using a modified droplet vitrification method. *In Vitro Cellular Developmental Biology-Plant* 53: 394–401.

Volk, G. M. 2010. Application of functional genomics and proteomics to plant cryopreservation. *Current Genomics*11: 24–29.

Volk, G. M., Henk, A. D. and Chhandak, B. 2011. Gene expression in response to cryoprotectant and liquid nitrogen exposure in Arabidopsis shoot tips. *Acta Horticulturae* 908: 55–66.

Walters, C., Wheeler, L. and Stanwood, P.C. 2004. Longevity of cryogenically stored seeds. *Cryobiology* 48: 229–244.

Wang, B., Wang, R. R., Cui, Z. H., Bi, W. L., Li, J. W., Li, B. Q., Ozudogru, E. A., Volk, G. M. and Wang, Q. C. 2014. Potential applications of cryogenic technologies to plant genetic improvement and pathogen eradication. *Biotechnology Advances* 32: 583–595.

Wang, M., Hamborg, Z., Slimestad, R. et al. 2020a. Assessments of rooting, vegetative growth, bulb production, genetic integrity and biochemical compounds in cryopreserved plants of shallot. *Plant Cell Tissue & Organ Culture* 144: 123–131. https://doi.org/10.1007/s11240-020-01820-7

Wang, M. R., Zhang, Z., Zámečník, J., Bilavčík, A., Blystad, D. R., Haugslien, S. and Wang, Q. C. 2020b. Droplet-vitrification for shoot tip cryopreservation of shallot (*Allium cepa* var. aggregatum): effects of PVS3 and PVS2 on shoot regrowth. *Plant Cell Tissue & Organ Culture* 140: 185–195.

Wang, Q. C., Wang, R. R., Li, B. Q. and Cui, Z. H. 2012. Cryopreservation: a strategy technique for safe preservation of genetically transformed plant materials. *Advances in Genetic Engineering Biotechnology* 1: 1–2.

Wilms, H., Rhee, J. H., Rivera, R. L., Longin, K. and Panis, B. 2019. Developing coconut cryopreservation protocols and establishing cryogenebank at RDA; a collaborative project between RDA and Bioversity International. *Acta Horticulturae* 1234: 343–348.

Withers, L. A. and King, P. 1980. A simple freezing unit and routine cryopreservation method for plant-cell cultures. *Cryo Letters* 1: 213–220.

Yamamoto, S., Fukui, K. and Niino, T. 2011a. A new cryopreservation method for vegetatively propagated plant genetic resources using aluminum cryo-plates. *Developmental Technology* 10: 10–11.

Yamamoto, S., Fukui, K., Rafique, T., Khan, N. I., Castillo Martinez, C. R., Sekizawa, K., Matsumoto, T. and Niino, T. 2011b. Cryopreservation of *in vitro*-grown shoot tips of strawberry by the vitrification method using aluminium cryo-plates. *Plant Genetic Resource: Characterization Utilization* 10: 14–19.

Yamamoto, S. I., Rafique, T., Arizaga, M. V., Fukui, K., Gutierrez, E. J. C., Martinez, C. R. C., Watanabe, K. and Niino, T. 2015. The aluminum cryo-plate increases efficiency of cryopreservation protocols for potato shoot tips. *American Journal of Potato Research* 92(2): 250–257.

Yi, J., Lee, G., Chung, J., Lee, Y., Kwak, J. and Lee, S. 2015. Morphological and genetic stability of dormant apple winter buds after cryopreservation. *Korean Journal of Plant Resources* 28(6): 697–703.

Yi, J. Y., Lee, G. A., Lee, S. Y., Chung, J. W. and Shin, S. 2013. Cryopreservation of winter-dormant apple buds using two-step freezing. *Plant Breeding and Biotechnology* 1(3): 283–289.

Yong, S. H., Park, D., Yang, W. H., Seol, Y., Choi, E., Jeong, M. J., Suh, G. U., Lee, C. H. and Choi, M. S. 2019. Cryopreservation of sievers wormwood (*Artemisia sieversiana* Ehrh. Ex Willd.) seeds by vitrification and encapsulation. *Forest Science and Technology* 15(4): 180–186.

Zhang, D., Ren, L., Chen, G. et al. 2015. ROS-induced oxidative stress and apoptosis-like event directly affect the cell viability of cryopreserved embryogenic callus in Agapanthus praecox. *Plant Cell Reports* 34: 1499–1513.

Zhang, J. M., Han, L., Lu, X. X., Volk, G. M., Xin, X., Yin, G. K., He, J. J., Wang, L. and Chen, X. L. 2017. Cryopreservation of Jerusalem artichoke cultivars using an improved droplet-vitrification method. *Plant Cell Tissue and Organ Culture* 128(3): 577–587.

14

Role of Biotechnology in the Improvement of Cole Crops

Shweta Sharma
CSKHPKV
Shoolini University

Priya Bhargava and Bharti Shree
CSKHPKV

CONTENTS

14.1 Introduction

Cole crops are the economically important temperate vegetables including crops like cauliflower (*Brassicaoleracea* var. *botrytis*), cabbage (*B. oleracea* var. *capitata*), broccoli or sprouting broccoli (*B. oleracea* var. *italica*), knol-khol (*B. oleracea* var. *gongylodes*), brussels sprout (*B. oleracea* var. *gemmifera*) and kale (*B. oleracea* var. *acephala*) which belong to Cruciferae or the mustard/*Brassica* family, therefore also known as *brassica* vegetables. All these crops have been derived from *B. oleracea* L. var. *sylvestris* (wild cabbage), cliff cabbage or coleworts, from which the name 'cole' has probably been derived. Cole crops have been derived from the word 'caulis' which means stalk or stem of the plant (Dhaliwal 2017). These vegetables are a fantastic source of vitamin C; proteins; fibre; soluble sugars; minerals including potassium, manganese, calcium, iron, magnesium and health-promoting phytochemicals such as luteolin, myricetin, quercetin, glucosinolates, carotenoids (especially broccoli and kale), anthocyanins (red cabbage) (Yang 2018), flavanols (Ahmed and Ali 2013) and phenolic compounds which show a role in disease prevention (Uher et al. 2017). Cole crops contain sulphur-rich compounds (glucosinolates and S-methyl cysteine sulphoxide) which are responsible for the characteristic flavour associated with cole crops and are the potent inducer of the endogenous antioxidant defence (Amany et al. 2013). Sulphoraphane belongs to glucosinolate and is known to possess anti-cancerous properties and protection against diabetes as well as cardiovascular and neurodegenerative diseases (Kim and Park 2016). Unfortunately, the productivity and quality of these crops are affected by a plethora of biotic (such as black rot of cabbage) and abiotic (such as salt, drought) factors. Different conventional and biotechnological approaches are being utilised for the improvement of cole crops, although conventional methods are time-consuming and possess cross-incompatibility barriers, especially on crossing with wild relatives. These bottlenecks can be overcome by tissue culture techniques, modern breeding approaches such as marker-assisted breeding and transgenic approaches such as *Agrobacterium*-mediated gene transfer and through genome editing techniques, which offer a new opportunity for genetic improvement of the cole crops (Kumar and Srivastava 2016). For successful crop improvement, introduction of new genetic materials into cultivated ones is necessary either by single gene transfer through genetic engineering or by multiple gene transfer through tissue culture techniques like pollen culture to overcome the pre-zygotic barrier and embryo culture and ovule or ovary culture to overcome pre-zygotic

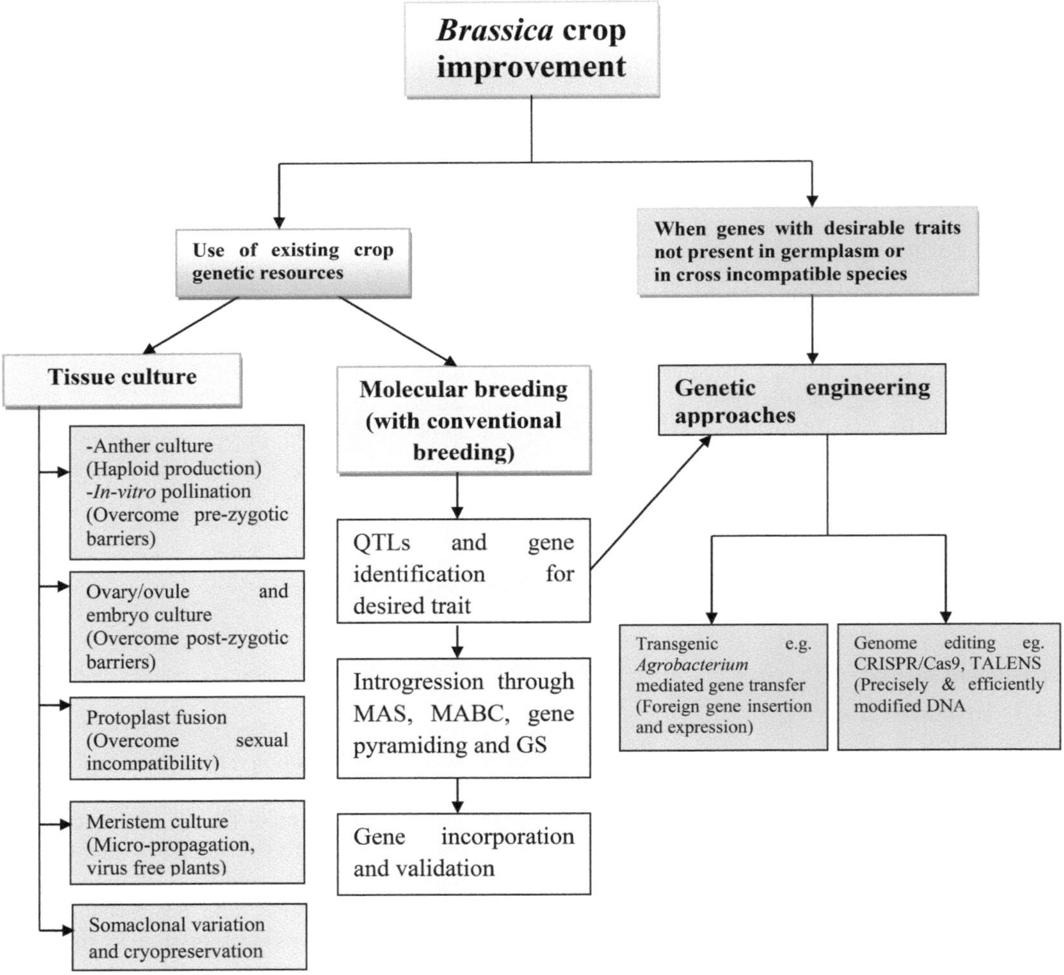

FIGURE 14.1 A schematic presentation of biotechnological strategies for the development of varieties/cultivars with enhanced quality and resistance to biotic and abiotic stresses in cole/*Brassica* vegetables. CRISPR, clustered regularly interspersed short palindromic repeat; GS, genomic selection; MAS, marker-associated selection; MABC, marker-assisted backcrossing; QTL, quantitative trait loci; TALENs, transcription activator-like effector nuclease.

and post-zygotic barriers present between sexually hybridizing plants resulting in *brassica* vegetables with desired traits (Gerszberg 2018) (Figure 14.1). Therefore the objective of this chapter is to explore the role of various tissue culture techniques, application of molecular breeding, genetic transformation, and genome editing technologies like clustered regularly interspersed short palindromic repeat (CRISPR)/Cas9 and to assess the future potential of these approaches in the genetic improvement of cole crops.

14.2 Tissue Culture and Cole Crops

Tissue culture is an *in vitro* cultivation technique of plant cell, tissue or organ under the aseptic condition on suitable culture media. Tissue culture techniques take advantage of totipotency defined as the ability of a cell to develop into the whole plant. It was first observed by Haberlandt, known as the father of plant tissue culture, in 1902. Murashige and Skoog (MS) medium (1962) is the first and most commonly used medium in tissue culture, but variants of MS medium and B5 medium (Gamborg et al. 1968) are most suitable for *Brassica* crop development. The optimum pH and temperature for proper plant growth and hormonal activity should be a pH 5.4–5.8 and temperature 18–25°C. The tissue culture techniques like anther culture, ovule culture, embryo culture, protoplast fusion, meristem culture and somaclonal variation can be applied for the development of haploids and doubled haploids, somatic hybrids, selection for disease and pest resistance, salinity resistance and metal toxicity resistance, selection of drought resistance, micropropagation, creation of variability and so forth in crucifers (Singh 2009).

14.2.1 Anther/Microspore Culture

Using the anther/microspore culture technique, whole plant is regenerated from anther, pollen or microspore in liquid or semi-solid agar medium by the formation of pollen embryos or callus to develop into haploid/double haploid (DH) plantlets. This is done for production of isogenic or homozygous lines or converted into fertile DH plantlets by using colchicine (Gosal et al. 2018). Guha and Maheshwari (1964) were the first to develop anther culture in *Datura* for haploid embryo production and this was first reported in cabbage by Kameya and Hinata (1970). Liu et al. (2017a) developed the DH cabbage line DH134 by using microspore culture for hybrid line formation. Li et al. (2020) also developed four elite DH lines (D29, D70, D120 and D162) by microspore culture to develop resistant hybrid lines for cabbage *Fusarium* wilt and head-splitting. These protocols can be effective to hasten production of hybrid varieties in cole crops.

14.2.2 Ovule and Ovary Culture

Ovule and ovary culture is an alternative approach to the regeneration of plants by culturing the ovary/ovule under culture medium. It was first applied in the development of interspecific hybrids between *B. campestris* and *B. oleracea* by Inomata (1977). This technique can be used when pollen culture is unable to produce haploid, DH, and hybrid plants because ovule/ovary culture produces only one or a few embryo sacs per ovary and the rate of success depends on the developmental stage of the ovary/ovule (Gosal et al. 2018). Sharma et al. (2017) developed black rot–resistant F_1 and BC_1 cauliflower hybrids by rescuing embryo through the direct ovule culture developed from interspecific hybridisation between susceptible *B. oleracea* var. *botrytis* (Pusa Sharad) × resistant *B. carinata* (NPC-9).

14.2.3 Embryo Culture

Embryo culture was first used by Hannig (1904), who cultured the mature embryo of crucifers. This technique is mainly used in embryo rescue during the formation of hybrids where endosperm development sometimes becomes arrested and finally, the embryo is aborted (Loyola-Vargas and Ochoa-Alejo 2018). Zhang et al. (2016a) developed an interspecific hybrid between cytoplasmic male sterile line 4E286 of *B. oleracea* var. *alboglabra* and the HCT3-inbred line of *B. rapa* L. var. *purpurea* by using embryo culture and backcrossing.

14.2.4 Protoplast Culture and Fusion

Protoplast fusion is an efficient alternative to conventional cross-breeding programmes as it bypasses fertilisation and produces plants containing new characteristics by fusion of protoplasts (Ragavendran and Natarajan 2017). By using protoplast culture, intraspecific or interspecific hybrids can also be generated in sexually incompatible plant species (Loyola-Vargas and Ochoa-Alejo 2018). The first time regeneration in *B. oleracea* was obtained by Vatsya and Bhaskaran (1982) utilising protoplast culture. Tonguc et al. (2003) reported the black rot resistance in breeding line "11B-1-12" obtained by protoplast fusion between *B. oleracea* (susceptible) and *B. carinata* (resistant). Wang et al. (2016) reported the production of 28 putative introgression lines (ILs) from somatic hybridisation between cauliflower (Korso) and black mustard (G1/1) based on their morphological characters, curd and flower traits, clubroot and black rot resistance.

14.2.5 Meristem Culture

Meristem culture is applied in vegetatively propagating plants involving the regeneration of the whole plant from actively growing parts such as shoot tips, root tips and auxiliary buds under nutrient medium. Micropropagation, virus-free production and germplasm conservation by cryopreservation are the main applications of this technique. Shoot regeneration and callus formation in broccoli (Solan green head) was reported by culturing of hypocotyls, cotyledons, leaf and petiole explants (Kumar and Srivastava 2015).

14.3 Recent Trends in Molecular Breeding of Cole Crops

Molecular breeding is the genetic manipulation at the DNA level to improve interesting traits by the use of approaches like marker-assisted selection (MAS), marker-assisted recurrent selection (MARS), marker-assisted backcrossing (MABC), gene pyramiding and genomic selection (GS) (Ribaut et al. 2010). In this approach, the selection relies on molecular marker closely linked with the gene of interest and not with gene itself by showing polymorphism, and it can be detected either by polymerase chain reaction (PCR) or by Southern blotting (Jiang 2013). It gives us opportunities to monitor the incorporation of the desirable allele of the donor source to the recipient and improves the efficiency and speed of selection during breeding (Singh 2009). Therefore, there is the need to identify and characterise suitable molecular markers such as random amplified polymorphic DNA (RAPD), amplified fragment length polymorphism (AFLP), inter-simple sequence repeats (ISSRs), simple sequence repeats (SSRs), single nucleotide polymorphism (SNP) and so forth (Cooper et al. 2014). These markers can be used to transfer the gene into a new cultivar or to inspect multigenic inheritance (Mishra and Singh 2015). Due to the development of both molecular/DNA markers and genetic maps of a crop, MAS has to turn out to be achievable for both traits governed by major genes as well as by quantitative trait loci (QTLs). For MAS, appropriate genetic material is an important factor and it includes wild/cultivated germplasm resources, natural/artificial mutants and other populations such as F_2 progenies, F_2-derived F_3 progenies, DH, recombinant inbred lines (RILs), near-isogenic lines (NILs), backcross progenies (BCs) and multiparent advanced generation inter-cross (MAGIC). Molecular markers are being utilised for resistance to biotic (e.g., black leg, black rot, clubroot, diamondback moth) and abiotic stresses (e.g., salt, drought), yield and quality improvement of cole crops (Table 14.1).

14.3.1 Application of Molecular Breeding for Resistance to Biotic and Abiotic Stresses

Matsumoto et al. (2012) introduced three clubroot resistance genes, namely *CRa, CRk* and *CRc*, into Chinese cabbage through MAS. Kalia et al. (2017b) developed sequence characterized amplified region (SCAR) markers, ScOPO-04_{833} and ScPKPS-11_{635}, linked to the black rot resistance and are the first genetic markers found linked to the locus Xca1Bo in cauliflower. A salt-tolerant candidate gene 'Bra003640' was identified in their QTL region in *Brassicae* by Lang et al. (2017). Seven candidate genes

TABLE 14.1

Molecular Marker Studies Conducted for Resistance to Biotic and Abiotic Stresses and to Improve Agronomic and Quality Traits in Cole Crops

Source of Resistance	Mapping Population	Marker Type	No. of QTLs/ Genes	Chromosome/ Linkage Group	Reference
Biotic Stress					
Black rot-resistant DH line 'Reiho P01' (*Brassica oleracea* subsp. *capitata*)	F_3	CAPS and SRAP	2	LG2 and LG9	Doullah et al. (2011)
Downy mildew-resistant 'BR-2' cultivar	F_2	RAPD and ISSR	1	-	Singh et al. (2012)
Downy mildew-resistant 'S4 line' derived from broccoli accession number OL87125	F_2	RAPD, AFLP, STS, SSR	1	C8	Carlier et al. (2012)
Five clubroot-resistant lines, T136-8, K13, K10, C9, and RC22 in Chinese cabbage	DH	SCAR	3	-	Matsumoto et al. (2012)
Black rot-resistant line 'Early Fuji' (*B. oleracea* L. var *capitata*)	F_2	SNP	3	C2, C4 and C5	Kifuzi et al. (2013)
B. oleracea var. *botrytis* L. BR 161 genotype (black rot)	F_2	RAPD and ISSR	1	3	Saha et al. (2014)
DBM resistance [resistant glossy leaf cabbage (748) with a susceptible smooth cabbage line (747)]	F_3	SSR	8	-	Ramchiary et al. (2015)
B. oleracea var. *capitata* C1234 inbred line (black rot resistant)	F_2	SNP-based CAPS markers	1 major and 3 minor	1, 3 and 6	Lee et al. (2015)
Five TuMV-resistant Chinese cabbage lines ('80122', '80124', 'BP058', '80186' and '2079')	Inbred lines	CAPS and KASP	1	A04	Li et al. (2016)
Resistant *B. oleracea* var. *botrytis* accession BR-207 (black rot)	F_3	RAPD and ISSR	1	3	Saha et al. (2016)
Downy mildew-resistant line 'T12–19'of *B. rapa* L.	DH	SNP	6	A04, A06 and A08	Yu et al. (2016)
Twenty-seven (black rot) different *B. oleracea* inbred lines	Inbred lines	SSR	-	C01, C03, C06 and C08	Afrin et al. (2018)
B. rapa resistant line 'P143'(black rot)	DH	AFLP and SSR	1	A03, A06 and A09	Artemyeva et al. (2018)
Cross between cauliflower LSQUO Pusa Sharad RSQUO and Ethiopian mustard NPC-9 (black rot resistance)	BC_1	SSR		LG B-7	Kalia et al. (2018)
Cross between L29 (resistant) and L16 (susceptible) in cabbage (black leg)	BC_1	SNP	6		Ferdous et al. (2019)
B. oleracea var. *botrytis* L. (BolTBDH) by crossing DH broccoli line 'Early Big' and a DH rapid cycling of Chinese kale line'TO1000DH3'(black rot)	DH	SSR and RAPD	3 single QTLs and 4 multi QTLs	6, 8 and 9 (single QTLs) and 1,3,4 and 5 (multi QTLs)	Iglesias-Bernabé et al. (2019)
Club root-resistant 'Darmor-*bzh*' nearly isogenic line of genotype 'Darmor' (*B. napus*)	DH	SNP	8	C02, C03, C07, C09, A01 and A10	Wagner et al. (2019)
Clubroot Resistance in CR38 (resistant)×CS22 (susceptible) in pak choi	F2	SNP	7	A07, A08	Zhu et al. (2019)
Abiotic Stress					

(Continued)

TABLE 14.1
(Continued)

Source of Resistance	Mapping Population	Marker Type	No. of QTLs/ Genes	Chromosome/ Linkage Group	Reference
B. oleracea var. *capitata* 96–100 inbred line (head-splitting resistance)	DH	SSR and InDel	3	3, 4 and 9	Su et al. (2015)
Temperature-regulated curd induction *B. oleracea* var. *botrytis* lines	Commercial lines	SNP	18	O1, O2, O3, O4, O6, O8, and O9	Matschegewski et al. (2015)
Heat-tolerant breeding lines 'USVL138' (*B. oleracea* var. *italica*)	DH	SNP	5	C02, C03, C05, C07 and C09	Branham et al. (2017)
Salt tolerance in *B. napus*	$F_{2:3}$	SSR and AFLP	1 (*Bra003640*)	-	Lang et al. (2017)
B. oleracea var. *capitata* BN1 variety (high temperature and high humidity tolerant)	Inbred lines, F_1 and F_2	SNPs	1	-	Song et al. (2020)
Agronomic and Quality Traits					
Male sterility in the line 79-399-3 of spring cabbage for hybrid seed production	F_2	EST-SSR and SSR	1 *CDMs399-3)*	-	Chen et al. (2013)
Whole genome mapping for agronomic traits in cabbage	DH	-	12	8 chromosomes	Lv et al. (2016)
Cross between 24-5 (inbred line, oblate head) × 01-88 (inbred line, round head) (*B. oleracea* L. var. *capitata*) for head shape	DH	InDel and SSR	14	2, 3, 4, 5, 6, 7 and 8	Zhang et al. (2016b)
'4305' × 'ZN198' lines of *B. oleracea* var. *botrytis* for curd-related traits and molecular breeding	DH	SLAF	1776	C01, C07 and C08	Zhao et al. (2016)
'Huangxiaoza' and 'Bqq094-11' inbred line of *B. rapa* L. syn. *B. campestris* for morphological and agronomic traits	F_2	RAPD and SNP	15	A03, A05, A06, A07, A09 and A10	Huang et al. (2017)
Introgression of the Or gene from donor line EC625883 in cauliflower	BC_1F_1	SCAR	-	-	Kalia et al. (2017a)
Cuticular wax biosynthesis in cabbage	F_2	SSR	1 (*Cgl1*)	8	Liu et al. (2017b)
97 F1 of HiQ × RIL-144 lines of *B. napus*	DH	SSR	2	C01 and C09	Rahman et al. (2018)
BN623 (early flowering) and BN3848 (late flowering) inbred line of *B. oleracea L.* var. *capitata*	F_1 and F_2	InDel	1	9	Abuyusuf et al. (2019)
Wax less mutant in cabbage	F2	SSR	1 (*BoGL-3*)	8	Dong et al. (2019)
(Chinese kale × broccoli) –head quality	DH	SNP	4	-	Stansell et al. (2019)
B. oleracea L. var. *italica* inbred lines 86101 (P1) and 90196 (P2) for plant and leaf characteristics	DH	SSR	4(plant height) 1(petiole length) 2(leaf width)	1, 6 and 3	Huang et al. (2020)
'IL4305' × 'ZN198' in cauliflower for curd architecture	DH	SNPs	2 (stalk length) 2 (curd solidity)	6	Zhao et al. (2020)

Abbreviations: AFLP, amplified fragment length polymorphism; CAPS, cleaved amplified polymorphic sequences; DBM, diamondback moth; DH, double haploid; EST-SSR, expressed sequence tag-simple sequence repeats; InDel, insertion-deletions; ISSR, inter-simple sequence repeat; KASP, kompetitive allele-specific polymerase chain reaction; QTL, quantitative trait loci; RAPD, random amplified polymorphic DNA; SCAR, sequence characterized amplified region; SLAF, specific locus amplified fragment; SNP, single nucleotide polymorphism; SRAP, sequence-related amplified polymorphism; SSR, simple sequence repeat; STS, sequence-tagged sites.

(*BraA07002249, BraA07002412, BraA07002494, BraA08002451, BraA08002452, BraA08002455* and *BraA08002471*) located on A07 and A08 chromosome were detected in the CR38 inbred line of pak-choi for clubroot resistance which can be used in MAS breeding programme by using their associated SNP markers (Zhu et al. 2019). Li et al. (2020) developed five elite hybrid lines ($DH_{70}\times IL_{15}$, $DH_{120}\times IL_{15}$, $DH_{120}\times IL_{16}$, $DH_{162}\times IL_{2348}$ and $DH_{162}\times IL_{108}$) resistant to cabbage *Fusarium* wilt and head-splitting by using MAS and microspore culture.

14.3.2 Application of Molecular Breeding for the Improvement of Quality and Agronomic Traits

The first wax synthesis *'CGL1'* gene was mapped in *B. oleracea* by Liu et al. (2017b) through fine-mapping technique and can be used in improving the cabbage breeding programme. Kalia et al. (2017a) introgressed '*Or* gene (β-carotene-rich)' from the line (1227) to three mid-early maturing open-pollinated variety Pusa Sharad (DC 309) and the parents of Pusa Hybrid-2 (CC-35 and DC 18-19) by using SCAR and SSR markers. This '*Or* gene' turned the curd colour of the cauliflower from normal white to orange 'Pusa Kesari VitA-1' (first β-carotene biofortified cauliflower cultivar). Xiao et al. (2019) created a line (SC96–100), which exhibits both agronomic traits similar to the 96–100 line and a similar compatibility index (>5.0), that was effectively utilised in the development of the 06–88 hybrid. This study provides a new paradigm of knowing the genetic basis of self-compatibility and facilitating cabbage breeding in male-sterile lines using self-compatible lines. Two DH populations were used to identify QTL, and 20 QTLs were detected with the logarithm of odds (LOD) values (2.61–8.38) and percentage of the phenotypic variance (7.69–25.10%).

14.4 Genetic Engineering Approaches for the Improvement of Cole Crops

This is a process where foreign genes are introduced and expressed in a host plant organism. This transfer and manipulation of gene technology are also known as recombinant DNA technology or genetic engineering. With the advancement in technology tools like genetic engineering and genome editing, the introduction of novel and agronomically beneficial genes has been made possible in different vegetables. So far, many genes of agronomical importance like pest and insect resistance, herbicide resistance, flavours, aroma, size, colour, taste, oil content improvement, nutrient enhancement and so forth have been identified and transferred in cruciferous vegetables (Table 14.2). These transgenic approaches also have the potential to develop biofortified vegetables.

14.4.1 Genetic Transformation

Genetic transformation includes two methods to transfer foreign genes into plants: *Agrobacterium*-mediated gene transfer or the indirect method (vector mediated) and direct gene transfer (vectorless) method. The former method is based on *A. tumefaciens*, which is a soil-borne, gram-negative bacterium that is responsible for crown gall disease in many dicot plants or plant species mostly used for genetic transformation in crucifers. Direct methods for the gene transformation are liposome encapsulation, microinjection, electroporation of protoplasts, particle bombardment, physiochemical uptake of DNA and so forth. Transgenic broccoli was developed using *A. tumefaciens* to integrate the cryIAa gene, which exhibited diamondback moth (DBM) resistance in broccoli (Kumar et al. 2018). Similarly, a transcription factor MYB29 was overexpressed to enhance the glucosinolate biosynthesis in *B. oleracea* (Zuluaga et al. 2019).

14.4.2 Genome Editing Techniques

Genome editing (also called gene editing) technologies can either knockout or knockdown a gene as per the desired objective to improve a trait. Gene knockdown simply means to reduce the expression of genes by targeting the mRNA (degraded or blocked translation) transcript of a gene instead of the DNA

TABLE 14.2

Genetic Engineering Techniques Used for Improvement of Traits in Cole Crops

Plant Species/ Cultivar	Technique Used to Improve Trait	Gene(s) Transfer/ Editing	Trait Improvement	Reference(s)
Transgenic cabbage	*Agrobacterium*-mediated transformation	Bt cry1Ba3	Resistance to DBM	Yi et al. (2011)
Transgenic broccoli	*Agrobacterium*-mediated transformation	Superoxide Dismutase (Rsr SOD)	Resistance to downy mildew	Jiang et al. (2012)
Transgenic cabbage	*Agrobacterium*-mediated transformation	Bt cry1Ia8 and cry1Ba3	Resistance to DBM	Yi et al. (2013)
Transgenic broccoli	Antisense RNA technology and *Agrobacterium*-mediated transformation	GIGANTEA (BoGI)	Delayed flowering, leaf senescence, and post-harvest yellowing retardation	Thiruvengadam et al. (2015)
Transgenic broccoli	*Agrobacterium*-mediated transformation	Endochitinase (Tch)	Resistance to *Botrytis cinerea* and *Rhizoctonia solani*	Yu et al. (2015)
Chinese cabbage (*Brassica rapa* L. subsp. *pekinensis*)	*Agrobacterium*-mediated transformation	Arabidopsis AtWRKY75	Bacterial soft rot resistance	Choi et al. (2016)
Transgenic cabbage	*Agrobacterium*-mediated transformation	Cry1Ia8	Lepidopteran resistance	Yi et al. (2016)
Chinese kale (*B. oleracea* var. *alboglabra*)	*Agrobacterium*-mediated transformation	AtEDT1/HDG11	Abiotic (drought and osmotic) stress tolerance	Zhu et al. (2016)
Chinese cabbage (*B. campestris* ssp. *chinensis*)	*Agrobacterium*-mediated transformation	Sporamin	Resistance against DBM	Cui et al. (2017)
Chinese cabbage (*B. campestris* L. ssp. *pekinensis*)	*Agrobacterium*-mediated gene transformation	Cowpea trypsin inhibitor (CpTI), SCK	Tolerance to *Pieris rapae*	Ma et al. (2017)
Transgenic broccoli	*Agrobacterium*-mediated transformation	cryIAa	Resistance against DBM	Kumar et al. (2018)
Chinese kale	CRISPR/Cas9-mediated genome editing	BaPDS1 and BaPDS2	Clear albino phenotype	Sun et al. (2018)
Transgenic broccoli	*Agrobacterium*-mediated transformation	BoERF1	Resistance to salt stress and *Sclerotinia* stem rot	Jiang et al. (2019)
Cabbage	Gene editing (CRISPR/Cas9 system)	Phytoene desaturase (BoPDS)	Albino phenotype	Ma et al. (2019)
Cabbage cv. Pride of India	*Agrobacterium*-mediated transformation	cryIAa	Resistance to DBM	Gambhir et al. (2020)
Chinese cabbage (*B. rapa* ssp. *pekinesis*)	Plant mediated RNAi and *Agrobacterium*-mediated transformation	COPB2	Resistance to mite	Shin et al. (2020)

Abbreviation: CRISPR, clustered regularly interspersed short palindromic repeat; DBM, diamondback moth; RNAi, RNA interference.

and not completely silencing the gene. Genome editing is one of the recent advancements in the field of biotechnology to modify the crops as per the desirable trait and facilitates efficient, precise and targeted modifications at genomic loci (Zhang et al. 2019). This editing technique relies on special endonucleases called engineered endonucleases (EENs). These endonucleases cut the DNA in a specific manner due to the occurrence of a sequence-specific DNA binding domain. After recognition or identification of the specific DNA sequence, these nucleases can very precisely cut the targeted genes (Xiong et al. 2015).

This editing technique is a mighty technology developed for the precise and site-specific addition, modification or deletion of the gene of interest from the genome. Genome editing is utilised in the improvement of the basic understanding of gene functions of plants. This editing technique includes CRISPR/Cas, transcription activator-like effector nucleases (TALENs) and zinc-finger nucleases (ZFNs) systems. The engineered CRISPR system (role in adaptive immunity in bacteria) is composed of two parts, a single small guide RNA and Cas9 nuclease, which are simple and effective compared with ZFNs and TALENs. For Cas9 to cleave the DNA, it must first recognize a DNA sequence known as the protospacer adjacent motif (PAM) sequence defined by guide RNA. After binding to the target DNA sequence, Cas9 will cause a double-stranded break at the desired site. Error-prone non-homologous end joining (NHEJ) repair of the site by the cell will cause indel mutations. Because each cell will undergo a different editing event, screening is necessary to isolate a cell line that has frameshift mutations resulting in a significant change in the translated protein, leading to gene knockout. Techniques like RNA interference (RNAi) exploiting the endogenous system for miRNA-induced gene silencing to artificially inhibit gene expression via transcriptional regulation have also been used for crop improvement. One example of the efficient knockout was seen in cabbage where the CRISPR/Cas9 system, along with an endogenous tRNA-processing system, was used to improve cabbage traits by overcoming self-incompatibility and prolonged vernalization (Ma et al. 2019). Therefore, this system can be utilised to improve different agronomic and nutritional traits and to solve the various problems in plant breeding.

14.5 Conclusion and Future Prospects

Great advent in the field of tissue culture, molecular biotechnology, genetic engineering techniques has offered an opportunity for the genetic improvement of cole crops. Efficient procedures for plant regeneration and transformation might be a way to obtain plants with desirable traits in less time compared with traditional approaches. Different tissue culture techniques like anther culture, protoplast fusion, embryo culture and genetic engineering techniques can overcome the cross-incompatibility barriers associated with *Brassica* crops. Plant tissue culture represents the potential areas of application at present and has a significant role in its association with transgenic plants in the future. They are the means to conserve the genetic resources through cryopreservation for conventional and non-conventional breeding programmes (Hussain et al. 2012). Modern approaches along with conventional breeding are resilient approaches that utilise wild species/relatives to surpass various biotic and abiotic stresses for the development of new varieties. Molecular breeding tools offer an opportunity to facilitate the development of linkage maps and to enhance accuracy and efficiency of selection for various economically essential traits in a breeding programme (Gantait et al. 2019; Osei et al. 2018). In the future, it will be essential to focus on the improvement of nutritional traits, resistance to biotic and abiotic stresses and herbicide resistance through recent biotechnological tools. So far, fewer efforts have been made towards the improvement of cole crops through advanced technologies like transgenic and genome editing (e.g., CRISPR/Cas9) which are necessary to exploit the introduction of genes encoding resistance to pest and diseases. Although few developments have taken place, issues like biosafety and environmental safety associated with transgenic crops need to be addressed, and there is a need to utilise the full potential of these modern biotechnological approaches for the genetic improvement of cole crops.

Acknowledgements

Authors are highly thankful to CSK HPKV, Palampur for the infrastructure and support required for making data available.

Conflict of Interest

No conflict of interest was reported by the authors.

REFERENCES

Abuyusuf, Md, Ujjal Kumar Nath, Hoy-Taek Kim, Md Rafiqul Islam, Jong-In Park, and Ill-Sup Nou. "Molecular markers based on sequence variation in BoFLC1. C9 for characterizing early-and late-flowering cabbage genotypes." *BMC Genetics* 20, no. 1 (2019): 42.

Afrin, Khandker Shazia, Md Abdur Rahim, Jong-In Park et al. "Screening of cabbage (*Brassica oleracea* L.) germplasm for resistance to black rot." *Plant Breeding and Biotechnology* 6, no. 1 (2018): 30–43.

Ahmed, Fouad A., and Rehab FM Ali. "Bioactive compounds and antioxidant activity of fresh and processed white cauliflower." *BioMed Research International* 9 (2013): 367819.

Amany, A. S., Faten F. Mohammed, and A. Samiha. "Nutritional impact of cauliflower and broccoli against development of early vascular lesions induced by animal fat diet (biochemical and immunohistochemical studies)." *Life Science Journal* 10, no. 4 (2013): 2496–2509.

Artemyeva, A. M., A. N. Ignatov, A. I. Volkova, M. N. Kocherina, N. V. Konopleva, and Yu V. Chesnokov. "Physiological and genetic components of black rot resistance in double haploid lines of *Brassica rapa* L." *Agricultural Biology (Sel'skokhozyaistvennaya Biologiya)* 53, no. 1 (2018): 157–169.

Branham, Sandra E., Zachary J. Stansell, David M. Couillard, and Mark W. Farnham. "Quantitative trait loci mapping of heat tolerance in broccoli (*Brassica oleracea* var. italica) using genotyping-by-sequencing." *Theoretical and Applied Genetics* 130, no. 3 (2017): 529–538.

Carlier, Jorge D., Claudia A. Alabaça, Paula S. Coelho, António A. Monteiro, and José M. Leitão. "The downy mildew resistance locus Pp523 is located on chromosome C8 of *Brassica oleracea* L." *Plant Breeding* 131, no. 1 (2012): 170–175.

Chen, C. H. E. N., Mu Zhuang, Zhi-yuan FANG et al. "A co-dominant marker BoE332 applied to marker-assisted selection of homozygous male-sterile plants in cabbage (*Brassica oleracea* var. *capitata* L.)." *Journal of Integrative Agriculture* 12, no. 4 (2013): 596–602.

Choi, Changhyun, Sangryeol Park, Ilpyung Ahn, Shinchul Bae, and Duk-Ju Hwang. "Generation of Chinese cabbage resistant to bacterial soft rot by heterologous expression of Arabidopsis WRKY75." *Plant Biotechnology Reports* 10, no. 5 (2016): 301–307.

Cooper, Mark, Carlos D.Messina, DeanPodlichet al. "Predicting the future of plant breeding: complementing empirical evaluation with genetic prediction." *Crop and Pasture Science* 65, no. 4 (2014): 311–336.

Cui, Jie, Ming Li, Lin Qiu, Jiashu Cao, and Li Huang. "Stable expression of exogenous imported sporamin in transgenic Chinese cabbage enhances resistance against insects." *Plant Growth Regulation* 81, no. 3 (2017): 543–552.

Dhaliwal, M. S. "Chapter 6: Cole Crops." In *Handbook of vegetable crops*, 3rd ed., pp. 148–176. Kalyani Publishers, New Delhi, India, 2017.

Dong, Xin, Jialei Ji, Limei Yang et al. "Fine-mapping and transcriptome analysis of BoGL-3, a wax-less gene in cabbage (*Brassica oleracea* L. var. *capitata*)." *Molecular Genetics and Genomics* 294, no. 5 (2019): 1231–1239.

Doullah, M. A. U., G. M. Mohsin, K. Ishikawa, H. Hori, and K. Okazaki. "Construction of a linkage map and QTL analysis for black rot resistance in *Brassica oleracea* L." *International Journal of Natural Sciences* 1, no. 1 (2011): 1–6.

Ferdous, Mostari Jahan, Mohammad Rashed Hossain, Jong-In Park et al. "Inheritance pattern and molecular markers for resistance to blackleg disease in cabbage." *Plants* 8, no. 12 (2019): 583.

Gambhir, Geetika, Pankaj Kumar, Gaurav Aggarwal, D. K. Srivastava, and Ajay Kumar Thakur. "Expression of cry1Aa gene in cabbage imparts resistance against diamondback moth (*Plutella xylostella*)." *Biologia Futura* (2020): 1–9.

Gamborg, O. L., R. A. Miller, and K. Ojima. "Nutrient requirements of suspension cultures of soybean root cells." *Experimental Cell Research* 50 (1968): 151–158.

Gantait, Saikat, Sutanu Sarkar, and Sandeep Kumar Verma. "Marker-assisted selection for abiotic stress tolerance in crop plants." In *Molecular plant abiotic stress: biology and biotechnology*, pp. 335–368. John Wiley & Sons, New York, 2019.

Gerszberg, Aneta. "Tissue culture and genetic transformation of cabbage (*Brassica oleracea* var. *capitata*): an overview." *Planta*248, no. 5 (2018): 1037–1048.

Gosal, Satbir Singh, and Shabir Hussain Wani. "Cell and tissue culture approaches in relation to crop improvement." In *Biotechnologies of crop improvement*, Volume 1, pp. 1–55. Springer, Cham, Switzerland, 2018.

Guha, Sipra, and S. C. Maheshwari. "*In vitro* production of embryos from anthers of Datura." *Nature* 204, no. 4957 (1964): 497–497.

Haberlandt, Gottlieb. "Kulturversuche mit isolierten Pflanzenzellen. Sitzungsber. Akad. Wiss. Wien. *Math.-Naturwiss.*" *Kl. Abt. J.* 111 (1902): 69–92.

Hannig, E. "Zur Physiologie pflanzlicher Embryonen. I. Ueber die Cultur von Cruciferen-Embryonen ausserhalb des Embryasacks." *Botanische Zeitung* 62 (1904): 45–80.

Huang, Jingjing, Jifeng Sun, Eryan Liu et al. "Mapping of QTLs detected in a broccoli double diploid population for plant height and leaf characteristics." *BMC Genetics* (2020). https://doi.org/10.21203/rs.3.rs-25053/v1

Huang, Li, Yafei Yang, Fang Zhang, and Jiashu Cao. "A genome-wide SNP-based genetic map and QTL mapping for agronomic traits in Chinese cabbage." *Scientific Reports* 7 (2017): 46305.

Hussain, Altaf, Iqbal Ahmed Qarshi, Hummera Nazir, and Ikram Ullah. "Plant tissue culture: current status and opportunities." *Recent Advances in Plant In Vitro Culture* (2012): 1–28.

Iglesias-Bernabé, Laura, Pari Madloo, Víctor Manuel Rodríguez, Marta Francisco, and Pilar Soengas. "Dissecting quantitative resistance to *Xanthomonas campestris* pv. *campestris* in leaves of *Brassica oleracea* by QTL analysis." *Scientific Reports* 9, no. 1 (2019): 1–11.

Inomata, Nobumichi. "Production of interspecific hybrids between *Brassica campestris* and *Brassica oleracea* by culture *in vitro* of excised ovaries: I. Effects of yeast extract and casein hydrolysate on the development of excised ovaries." *Japanese Journal of Breeding* 27, no. 4 (1977): 295–304.

Jiang, Guo-Liang. "Molecular markers and marker-assisted breeding in plants." *Plant Breeding from Laboratories to Fields* (2013): 45–83.

Jiang, Ming, Li-Xiang Miao, and Caiming He. "Overexpression of an oil radish superoxide dismutase gene in broccoli confers resistance to downy mildew." *Plant Molecular Biology Reporter* 30, no. 4 (2012): 966–972.

Jiang, Ming, Zi-Hong Ye, Hui-Juan Zhang, and Li-Xiang Miao. "Broccoli plants over-expressing an ERF transcription factor gene BoERF1 facilitates both salt stress and sclerotinia stem rot resistance." *Journal of Plant Growth Regulation* 38, no. 1 (2019): 1–13.

Kalia, Pritam, B. B. Sharma, and D. Singh. "Marker-assisted introgression of black-rot resistance gene Xca1bc into cauliflower through interspecific hybridization." In *IV international symposium on molecular markers in horticulture* 1203 (2018):51–58.

Kalia, Pritam, P. Muthukumar, and S. Soi. "Marker-assisted introgression of the Or gene for enhancing β-carotene content in Indian cauliflower." In *IV international symposium on molecular markers in horticulture* 1203 (2017a): 121–128.

Kalia, Pritam, Partha Saha, and Soham Ray. "Development of RAPD and ISSR derived SCAR markers linked to Xca1Bo gene conferring resistance to black rot disease in cauliflower (*Brassica oleracea* var. *botrytis* L.)." *Euphytica* 213, no. 10 (2017b): 232.

Kameya, Toshiaki, and Kokichi Hinata. "Induction of haploid plants from pollen grains of Brassica." *Japanese Journal of Breeding* 20, no. 2 (1970): 82–87.

Kifuzi, Yasuko, Hideaki Hanzawa, Yuuichi Terasawa, and Takeshi Nishio. "QTL analysis of black rot resistance in cabbage using newly developed EST-SNP markers." *Euphytica* 190, no. 2 (2013): 289–295.

Kim, Jae Kwang, and Sang Un Park. 2016. Current potential health benefits of sulforaphane. *EXCLI Journal* 15 (2016): 571–577.

Kumar, Pankaj, and D. K. Srivastava. "High frequency organogenesis in hypocotyl, cotyledon, leaf and petiole explants of broccoli (*Brassica oleracea* L. var. *italica*), an important vegetable crop." *Physiology and Molecular Biology of Plants* 21, no. 2 (2015): 279–285.

Kumar, Pankaj, and Dinesh Kumar Srivastava. "Biotechnological advancement in genetic improvement of broccoli (*Brassica oleracea* L. var. *italica*), an important vegetable crop." *Biotechnology Letters* 38, no. 7 (2016): 1049–1063.

Kumar, Pankaj, Geetika Gambhir, Ayesh Gaur, Krishan C. Sharma, Ajay K. Thakur, and Dinesh K. Srivastava. "Development of transgenic broccoli with cryIAa gene for resistance against diamondback moth (*Plutella xylostella*)." *3 Biotech* 8, no. 7 (2018): 299.

Lang, Lina, Aixia Xu, Juan Ding et al. "Quantitative trait locus mapping of salt tolerance and identification of salt-tolerant genes in *Brassica napus* L." *Frontiers in Plant Science* 8 (2017): 1000.

Lee, Jonghoon, Nur Kholilatul Izzah, Murukarthick Jayakodi et al. "Genome-wide SNP identification and QTL mapping for black rot resistance in cabbage." *BMC plant Biology* 15, no. 1 (2015): 1–11.

Li, Guo-Liang, Wei Qian, Shu-Jiang Zhang et al. "Development of gene-based markers for the Turnip mosaic virus resistance gene retr02 in *Brassica rapa*." *Plant Breeding* 135, no. 4 (2016): 466–470.

Li, Qiang, Yantong Shi, Ying Wang et al. "Breeding of cabbage lines resistant to both head splitting and fusarium wilt via an isolated microspore culture system and marker-assisted selection." *Euphytica* 216, no. 2 (2020): 34.

Liu, Xing, Fengqing Han, Congcong Kon get al. "Rapid introgression of the Fusarium wilt resistance gene into an elite cabbage line through the combined application of a microspore culture, genome background analysis, and disease resistance-specific marker assisted foreground selection." *Frontiers in Plant Science* 8 (2017a): 354.

Liu, Zezhou, Zhiyuan Fang, Mu Zhuang et al. "Fine-mapping and analysis of Cgl1, a gene conferring glossy trait in cabbage (*Brassica oleracea* L. var. *capitata*)." *Frontiers in Plant Science* 8 (2017b): 239.

Loyola-Vargas, Victor M., and Neftalí Ochoa-Alejo. "An introduction to plant tissue culture: advances and perspectives." In *Plant cell culture protocols*, pp. 3–13. Humana Press, New York, NY, 2018.

Lv, Honghao, Qingbiao Wang, Xing Liu et al. "Whole-genome mapping reveals novel QTL clusters associated with main agronomic traits of cabbage (*Brassica oleracea* var. *capitata* L.)." *Frontiers in Plant Science* 7 (2016): 989.

Ma, Cunfa, Mengci Liu, Qinfei Li, Jun Si, Xuesong Ren, and Hongyuan Song. "Efficient BoPDS gene editing in cabbage by the CRISPR/Cas9 system." *Horticultural Plant Journal* 5, no. 4 (2019): 164–169.

Ma, Xiaoli, Zhen Zhu, Yane Li, Guangdong Yang, and Yanxi Pei. "Expressing a modified cowpea trypsin inhibitor gene to increase insect tolerance against Pieris rapae in Chinese cabbage." *Horticulture, Environment, and Biotechnology* 58, no. 2 (2017): 195–202.

Matschegewski, Claudia, Holger Zetzsche, Yaser Hasan et al. "Genetic variation of temperature-regulated curd induction in cauliflower: elucidation of floral transition by genome-wide association mapping and gene expression analysis." *Frontiers in Plant Science* 6 (2015): 720.

Matsumoto, Etsuo, Hiroki Ueno, Daisuke Aruga, Koji Sakamoto, and Nobuaki Hayashida. "Accumulation of three clubroot resistance genes through marker-assisted selection in Chinese cabbage (*Brassica rapa ssp. pekinensis*)." *Journal of the Japanese Society for Horticultural Science* 81, no. 2 (2012): 184–190.

Mishra, G. P., and R. K. Singh. "Molecular breeding approach for crop improvement of quality traits." *Journal of Biotechnology and Crop Science* 4, no. 5 (2015): 4–21.

Murashige, Toshio, and Folke Skoog. "A revised medium for rapid growth and bio assays with tobacco tissue cultures." *Physiologia Plantarum* 15, no. 3 (1962): 473–497.

Osei, Michael K., Ruth Prempeh, Joseph Adjebeng-Danquah et al. "Marker-assisted selection (MAS): a fast-track tool in tomato breeding." In *Recent advances in tomato breeding and production*, pp. 93–113. IntechOpen, London, 2018.

Ragavendran, Chinnasamy, and Devarajan Natarajan. "Role of plant tissue culture for improving the food security in India: A review update." In *Sustainable agriculture towards food security*, pp. 231–262. Springer, Singapore, 2017.

Rahman, Habibur, Rick A. Bennett, and Berisso Kebede. "Molecular mapping of QTL alleles of *Brassica oleracea* affecting days to flowering and photosensitivity in spring *Brassica napus*." *PLoS One* 13, no. 1 (2018): e0189723.

Ramchiary, Nirala, Wenxing Pang, Van Dan Nguyen et al. "Quantitative trait loci mapping of partial resistance to Diamondback moth in cabbage (*Brassica oleracea* L)." *Theoretical and Applied Genetics* 128, no. 6 (2015): 1209–1218.

Ribaut, J. M., M. C. De Vicente, and X. Delannay. "Molecular breeding in developing countries: challenges and perspectives." *Current Opinion in Plant Biology* 13, no. 2 (2010): 213–218.

Saha, Partha, Pritam Kalia, Humira Sonah, and Tilak R. Sharma. "Molecular mapping of black rot resistance locus X ca1bo on chromosome 3 in Indian cauliflower (*Brassica oleracea* var. *botrytis* L.)." *Plant Breeding* 133, no. 2 (2014): 268–274.

Saha, Partha, Pritam Kalia, Munish Sharma, and Dinesh Singh. "New source of black rot disease resistance in *Brassica oleracea* and genetic analysis of resistance." *Euphytica* 207, no. 1 (2016): 35–48.

Sharma, Brij B., Pritam Kalia, Dinesh Singh, and Tilak R. Sharma. "Introgression of black rot resistance from *Brassica carinata* to cauliflower (*Brassica oleracea botrytis* group) through embryo rescue." *Frontiers in Plant Science* 8 (2017): 1255.

Shin, Yun Hee, Si Hyeock Lee, and Young-Doo Park. "Development of mite (*Tetranychus urticae*)-resistant transgenic Chinese cabbage using plant-mediated RNA interference." *Horticulture, Environment, and Biotechnology* 61, no. 2 (2020): 305–315.

Singh, B. D. *Plant breeding: principles and methods*. Kalyani Publishers, New Delhi, India, 2009.

Singh, S., S. R. Sharma, P. Kalia et al. "Molecular mapping of the downy mildew resistance gene Ppa3 in cauliflower (*Brassica oleracea* var. *botrytis* L.)." *The Journal of Horticultural Science and Biotechnology* 87, no. 2 (2012): 137–143.

Song, Hayoung, Myungjin Lee, Byung-Ho Hwang, Ching-Tack Han, Jong-In Park, and Yoonkang Hur. "Development and application of a PCR-based molecular marker for the identification of high temperature tolerant cabbage (*Brassica oleracea* var. *capitata*) genotypes." *Agronomy* 10, no. 1 (2020): 116.

Stansell, Zachary, Mark Farnham, and Thomas Björkman. "Complex horticultural quality traits in broccoli are illuminated by evaluation of the immortal BolTBDH mapping population." *Frontiers in Plant Science*10 (2019): 1104.

Su, Yanbin, Yumei Liu, Zhansheng Li et al. "QTL analysis of head splitting resistance in cabbage (*Brassica oleracea* L. var. *capitata*) using SSR and InDel makers based on whole-genome re-sequencing." *PLoS One* 10, no. 9 (2015): e0138073.

Sun, Bo, Aihong Zheng, Min Jiang et al. "CRISPR/Cas9-mediated mutagenesis of homologous genes in Chinese kale." *Scientific Reports* 8, no. 1 (2018): 1–10.

Thiruvengadam, Muthu, Ching-Fang Shih, and Chang-Hsien Yang. "Expression of an antisense *Brassica oleracea* GIGANTEA (BoGI) gene in transgenic broccoli causes delayed flowering, leaf senescence, and post-harvest yellowing retardation." *Plant Molecular Biology Reporter* 33, no. 5 (2015): 1499–1509.

Tonguc, Muhammet, Elizabeth D. Earle, and Phillip D. Griffiths. "Segregation distortion of *Brassica carinata* derived black rot resistance in *Brassica oleracea*." *Euphytica* 134, no. 3 (2003): 269–276.

Uher, Anton, Ivanam Mezeyová, Alžbeta Hegedűsová, and Miroslav Šlosár. "Impact of nutrition on the quality and quantity of cauliflower florets." *Potravinárstvo: Slovak Journal of Food Sciences* 11, no. 1 (2017): 113–119.

Vatsya, Bhartendu, and S. Bhaskaran. "Plant regeneration from cotyledonary protoplasts of cauliflower (*Brassica oleracea* L. var. botrytis L.)." *Protoplasma* 113, no. 2 (1982): 161–163.

Wagner, Geoffrey, Anne Laperche, Christine Lariagon et al. "Resolution of quantitative resistance to clubroot into QTL-specific metabolic modules." *Journal of Experimental Botany* 70, no. 19 (2019): 5375–5390.

Wang, Gui-Xiang, Jing Lv, Jie Zhang et al. "Genetic and epigenetic alterations of *Brassica nigra* introgression lines from somatic hybridization: a resource for cauliflower improvement." *Frontiers in Plant Science* 7 (2016): 1258.

Xiao, Zhiliang, Fengqing Han, Yang Hu et al. "Overcoming cabbage crossing incompatibility by the development and application of self-compatibility-QTL-specific markers and genome-wide background analysis." *Frontiers in Plant Science* 10 (2019): 189.

Xiong, Jin-Song, Jing Ding, and Yi Li. "Genome-editing technologies and their potential application in horticultural crop breeding." *Horticulture Research* 2, no. 1 (2015): 1–10.

Yang, Dong Kwon. "Cabbage (*Brassica oleracea* var. *capitata*) protects against H_2O_2-induced oxidative stress by preventing mitochondrial dysfunction in H9c2 cardiomyoblasts." *Evidence-Based Complementary and Alternative Medicine* (2018): 2179021.

Yi, Dengxia, C. U. I. Lei, Yu-Mei Liu et al. "Transformation of cabbage (*Brassica oleracea* L. var. *capitata*) with Bt cry1Ba3 gene for control of diamondback moth." *Agricultural Sciences in China* 10, no. 11 (2011): 1693–1700.

Yi, Dengxia, Lei Cui, Li Wang et al. "Pyramiding of Bt cry1Ia8 and cry1Ba3 genes into cabbage (*Brassica oleracea* L. var. *capitata*) confers effective control against diamondback moth." *Plant Cell, Tissue and Organ Culture* 115, no. 3 (2013): 419–428.

Yi, Dengxia, Weijie Yang, Jun Tang et al. "High resistance of transgenic cabbage plants with a synthetic cry1Ia8 gene from *Bacillus thuringiensis* against two lepidopteran species under field conditions." *Pest Management Science* 72, no. 2 (2016): 315–321.

Yu, Shuancang, Tongbing Su, Shenghua Zhi et al. "Construction of a sequence-based bin map and mapping of QTLs for downy mildew resistance at four developmental stages in Chinese cabbage (*Brassica rapa* L. *ssp. pekinensis*)." *Molecular Breeding* 36, no. 4 (2016): 44.

Yu, Ya, Lei Zhang, Wei-ran Lian, et al. "Enhanced resistance to *Botrytis cinerea* and *Rhizoctonia solani* in transgenic broccoli with a *Trichoderma viride* endochitinase gene." *Journal of Integrative Agriculture* 14, no. 3 (2015): 430–437.

Zhang, Xiaohui, Tongjin Liu, Xixiang Li et al. "Interspecific hybridization, polyploidization, and backcross of *Brassica oleracea* var. *alboglabra* with *B. rapa* var. *purpurea* morphologically recapitulate the evolution of Brassica vegetables." *Scientific Reports* 6 (2016a): 18618.

Zhang, Xiaoli, Yanbin Su, Yumei Liu et al. "Genetic analysis and QTL mapping of traits related to head shape in cabbage (*Brassica oleracea* var. *capitata* L.)." *Scientia Horticulturae* 207 (2016b): 82–88.

Zhang, Yi, Karen Massel, Ian D. Godwin, and Caixia Gao. "Correction to: Applications and potential of genome editing in crop improvement." *Genome Biology* 20, no. 1 (2019): 1–1.

Zhao, Zhen-Qing, Xiao-Guang Sheng, Hui-Fang Yu et al. "Identification of QTLs associated with curd architecture in cauliflower." *BMC Plant Biology* 20 (2020): 1–8.

Zhao, Zhenqing, Honghui Gu, Xiaoguang Sheng et al. "Genome-wide single-nucleotide polymorphisms discovery and high-density genetic map construction in cauliflower using specific-locus amplified fragment sequencing." *Frontiers in Plant Science* 7 (2016): 334.

Zhu, Hongfang, Wen Zhai, Xiaofeng Li, and Yuying Zhu. "Two QTLs controlling clubroot resistance identified from bulked segregant sequencing in Pakchoi (*Brassica campestris ssp. chinensis Makino*)." *Scientific Reports* 9, no. 1 (2019): 1–9.

Zhu, Zhangsheng, Binmei Sun, Xiaoxia Xu et al. "Overexpression of AtEDT1/HDG11 in Chinese kale (*Brassica oleracea* var. *alboglabra*) enhances drought and osmotic stress tolerance." *Frontiers in Plant Science* 7 (2016): 1285.

Zuluaga, Diana L., Neil S. Graham, Annett Klinder et al. "Overexpression of the MYB29 transcription factor affects aliphatic glucosinolate synthesis in *Brassica oleracea*." *Plant Molecular Biology* 101, no. 1–2 (2019): 65–79.

15

Application of Nanotechnology in Crop Improvement: An Overview

Pritom Biswas and Nitish Kumar
Central University of South Bihar

CONTENTS

15.1 Introduction

Recent projections have demonstrated that global demand for food production will double by 205 (Tilman et al., 2011) However, a constant decrease in farmland and scarce resources of water along with the impact of climate change coupled with the ineffective use of fertilisers and pesticides worsens the biotic and abiotic stresses on crop plants, resulting in low production. Drought and salinity stress are the reason for billions of dollars in crop damages yearly (Suzuki et al., 2014). Subsequently, increasing the production of food is a vital problem. Nanotechnology has an immense perspective to create a valuable effect on numerous areas, including agriculture. Recently, nanotechnology is demanding research at both industrial as well as academic points (Dasgupta et al., 2015; Parisi et al., 2015). The huge contact area and various sites of binding that come with the extremely small size make nanoparticles (NPs) make them excellent nanocarriers for bioactive molecules such as plasmid DNA (Demirer et al., 2019; Kwak et al., 2019) and double-stranded (ds) RNA (Mitter et al., 2017). Demirer et al. (2019) and Kwak et al. (2019) stated that single-walled carbon nanotubes (SWCNTs) have the potential for delivering the genetic material into the nucleus and chloroplasts, respectively. The main advantage of SWCNTs is that they do not need *Agrobacterium* infection. They are also not dependent on a gene gun or biolistic bombardment, which is of little efficacy and harms the crops. Not only can these NPs act as transporters for biological molecules, but they can also be encapsulated or loaded with conventional pesticides and fertilisers as well as active components to accomplish their targeted and controlled release (Avellan et al., 2019; Kottegoda et al., 2017). Thus, it can be stated that NPs have the potential for the control transfer of nutrients to definite plant tissues or organelles and can reduce the amount of nutrients required.

The applications of NPs in nanoencapsulation and as nano-guards for agro-chemicals are described by Adisa et al. (2019), Nuruzzaman et al. (2016) and Raliya et al. (2018).

In addition to being carrier molecules, NPs also have a number of other exclusive properties including NPs can directly enhance the growth of plants, can increase the photosynthesis rate of plants and can increase the resistance of plants to abiotic and biotic stresses. These are the optoelectronic, physiochemical and catalytic properties of NPs. Dutta et al. (2006) and Walkey et al. (2015) through their experiments proved that CeO_2 NPs can scavenge the reactive oxygen species (ROS) because NPs have the huge number of surface oxygen vacancies which in turn help to substitute among two oxidation states (Ce^{4+} and Ce^{3+}). Various research work can be done in the future to increase the biotic and abiotic stress response of crop plants and thus their existence by exploiting this antioxidant-enzyme-mimicking activity.

15.2 Application of Nanoparticles in Sustainable Agriculture

In 2009, the Royal Society defined sustainable intensification as 'concept related to a production system aiming to increase the yield without adverse environmental impact while cultivating the same agricultural area'. This provides the foundation for the evaluation and range of the finest grouping of methods to the production of agriculture products, keeping in mind the effect of several factors, including the present biophysical situation as well as cultural, economic and social conditions (Garnett & Godfray, 2012). In this perspective, new NPs, whether based on lipid, polymeric and inorganic NPs, have been created to enhance agricultural production. These NPs may be developed in various ways (e.g., oxidoreduction ionic gelation, emulsification, polymerisation, etc.). They may be applied for different functions from manufacture of smart nanosystems for the nutrient immobilisation to their smart discharge in the topsoil system. They not only improve the uptake of nutrients by plants, but they also minimise leaching. By decreasing the nitrogen transfer to groundwater, they also mitigate eutrophication (Liu & Lal, 2015). Apart from that, NPs could also be exploited to attain various other results like improving the function and structure of agro-chemicals by enhancing solubility, or to enhance tolerance against photodecomposition and hydrolysis and/or by giving a large controlled and specific release towards target organisms (Grillo et al., 2016; Mishra & Singh, 2015; Nuruzzaman et al., 2016).

15.2.1 Nanobiotechnology for Improvement of Plant Growth and Yield

Enhanced growth of the plant and production of agricultural products can be attributed to the smart delivery developed through the nanobiotechnology detection system. The NPs can affect plant growth based on various factors, like the concentration of NPs and their composition as well as their chemical properties, plant species and size (Ma et al., 2010). The application of nano-SiO_2 in little amounts has been described to enhance the seed germination of tomato plants (Siddiqui & Al-Whaibi, 2014). According to Suriyaprabha et al. (2012), nano-SiO_2 improved the germination of seed in maize by enhancing the availability of nutrients, pH of the growth medium and altering conductivity. Nano-SiO_2 not only enhanced the germination of squash seeds, but it also activated the antioxidant system under salinity (Siddiqui et al., 2015).

NPs of gold, copper, palladium and silica all have a remarkable influence on the germination of lettuce seeds (Shah & Belozerova, 2009). Apart from germination, nano-SiO_2 has been reported to improve other growth parameters as well. Siddiqui et al. (2015) reported that gaseous exchange is enhanced by nano-SiO_2 and it amplifies photosynthetic features, such as transpiration rate, stomatal conductance, the potential activity of PSII photochemical efficiency, electron transport and photochemical quench. The zinc NPs (ZnONPs) have also been studied and are valuable for the development of plant growth. At low amounts, ZnONPs have been reported to enhance the germination of wheat and onion seed (Raskar & Laware, 2014). According to Raliya & Tarafdar (2013), ZnONPs have been found to enhance the growth of root and shoot, protein and pigment concentration, microbes of rhizosphere and activity of an enzyme in *Cyamopsis tetragonoloba*. When added as the supplementation of an MS nutrient medium, ZnONPs stimulate the proline synthesis and antioxidant enzymes (Helaly et al., 2014). This results in the enhancement of plants to abiotic and biotic stress and amplifies the process of embryo development from a somatic cell, organogenesis of explant and shoot development. Plant growth in *Crocus sativus* (Rezvani et al., 2012), *Arabidopsis thaliana* (Syu et al., 2014), *Brassica juncea* (Sharma et al., 2012) and *Boswellia ovalifoliolata* (Savithramma et al., 2012) have been found to be enhanced by the application of silver NPs (AgNPs). Likewise, gold NPs (AuNPs) also have been reported to positively impact the growth and development of plants (Siddiqi & Husen, 2016).

15.2.2 Intervention of Nanoparticles in Metabolic Pathways

Secondary metabolites are the phytochemicals or natural products that have therapeutic properties. The secondary metabolites, as per the reports of various research papers, are useful for humans and can be measured as phytomedicines. During unfavourable conditions, secondary metabolites have a pivotal role, paving the way for plant survival. They also protect the plants against the attack of insects and pests, abiotic and biotic stresses and mechanical injury. Secondary metabolites comprise different chemical components like phenolics, alkaloids and terpenoids (Kabera et al., 2014). In plants, high variability of bioactive metabolites occur which are extremely significant as agro-chemicals, nutraceuticals and pharmaceuticals. The secondary metabolite content in some medicinal plants is generally low. Nanotechnological approaches are thought to have a positive impact on enhancing the important therapeutic compound production in plants. Though there are a number of observations that have described the impact of NPs on the growth and development of plants, very few reports are available that address the application of NPs and their role in the enhancement of secondary metabolite production in plants (Nair et al., 2010).

NPs can be used as new operative elicitors in plant biology to facilitate the production of secondary metabolites. NPs and their role have been reported by several researchers (Aditya et al., 2010; Ghanati & Bakhtiarian, 2014; Ghasemi et al., 2015; Sharafi et al., 2013). Ghasemi et al. (2015) has extensively described the potential role of NPs as elicitors to improve the expression level of those genes which are accountable for secondary metabolite production. Isoflavonoids and flavonoids are the furthermost prevalent groups of secondary metabolites in plants. The enhanced secondary metabolite production (flavonoids and phenols) in grams, under *in vitro* conditions, was documented on introduction to TiO_2 NPs (Mohammed AL-oubaidi & Kasid, 2015). Heiras-Palazuelos et al. (2013) studied that one of the rich sources of secondary metabolites are legumes.

Hence, it can be recommended that NPs are suitable candidates for elicitation studies for *in vitro* production of secondary metabolites. It was concluded in the research conducted by Isah (2019) that the stress hormone jasmonate (JA), in addition to enhancing several responses of plant defence, also amplifies defensive secondary metabolite biosynthesis. NPs also show a pivotal role in the regulation of the gene expression for the production of JA in treated cells. Enhanced production of JA can be attributed to the elevated production of hyperforin and hypericin. Nutrient media with various hormonal mixtures were prepared, qualitative analysis using iodine fumes was performed and growth was studied. From these studies, it can be resolved that the NPs performed as an elicitor for secondary metabolites. The *in vitro* secondary metabolite production through various approaches is extensively exploited. Still, there remains numerous problems with the synthetic production of these secondary metabolites because of trouble in understanding and decoding their biosynthetic pathway. Therefore, *in vitro* approaches (e.g., plant cell/tissue culture and free cell suspensions in bioreactors) are favourable approaches for obtaining the products in their natural form.

15.2.3 Nanoparticle-Based Plant Gene Transfer

The prosperous use of nanotechnology in pharmacology under *in vitro* conditions has created a great deal of attention in agri-nanotechnology as well. Several micro- and macromolecules are essential for the protection of plants, for their enhancement and for nutrient utilisation. Nanotechnology not only ensures the precise use of fertilisers and pesticides, but also their site-specific transfer for enhanced production in agriculture (Nair et al., 2010). Traditional approaches in agriculture are susceptible to restrictions. Nanotechnology has delivered answers for these restrictions through species-independent passive

TABLE 15.1
Nanoparticle-Mediated Genetic Transformation of Plant Species

NP	Cargo	Plant Species	Outcomes	References
Gold-plated mesoporous silica NPs	GFP- or mCherry expressing plasmid DNA	Tobacco, maize and onion	Outstanding enhancement in cargo transformation during co-bombardment with 0.6-mm AuNPs	((Martin-Ortigosa et al., 2012)
Gold-functionalised mesoporous silica NPs	Protein and GFP plasmid co-delivery	Onion	High expression of the fluorescent protein and marker genes	(Martion-Ortigosa et al., 2012)
Magnetic gold NPs	b-Glucuronidase and fluorescein isothiocyanate	Mustard and carrot	Effective b-glucuronidase gene expression	(Singh et al., 2016)
Mesoporous silica NPs	*cryIAb* gene	Tomato	Successful expression of *cryIAb* gene	(Hajiahmadi et al., 2019)
Chitosan and g-polyglutamic	GA	*Phaseolus vulgaris*	Germination of seed occurs within 24 h after treatment; increased root development and leaf area	(Pereira et al., 2017)
Silica NPs	SA	*Arabidopsis*	Expression of the PR-1 protein to support plant protection against biotic stress	(Yi et al., 2015)
Silica NPs	-	*Pisum sativum*	Protection seedlings against phytotoxicity of Cr(VI)	(Tripathi et al., 2012)

Abbreviations: AuNP, gold nanoparticles; GA, gibberellic acid; GFP, green fluorescent protein NPs, nanoparticles SA, salicylic acid.

transfer of the DNA and proteins. Nanotechnology is an outstanding mode for nuclease-dependent editing of genome cargo. It also delivers important NP delivery for the enhanced genetic engineering of plants (Table 15.1). NPs have the capability to penetrate the plant cell wall without any outside energy, enhancing the delivery of biomolecules. Moreover, these particles possess tuneable physiochemical properties for several cargo conjugations as well as show extensive use on a range of hosts.

Silica NPs are surface-functionalised nanomaterial. They can deliver drug and genetic material into the animal's cells/tissues. However, this application is restricted in plant cells because of the occurrence of cell wall (Bharali et al., 2005). Torney et al. (2007) worked using a Mesoporous silica nanoparticle (MSN) method that had 3-nm pores, through which chemical molecules and genetic materials could be transported into targeted plant cells. Advanced development is expected to pave an approach for new opportunities in the transfer of genetic material, chemical molecules and protein and to a specific target, by pore expansion and multi-functionalisation of the MSNs.

Biopolymers like proteins, nucleic acids and polysaccharides are mostly produced by living organisms. To achieve a remarkable result on the application of proteins, nucleic acids and polysaccharides as drug or gene transfer transporters, various factors like regulating charge, surface morphology, particle size and the release speed of molecules are to be organised (Nitta & Numata, 2013). Nano-encapsulation promises safer handling and efficient use of pesticides along with eco-protection through less exposure to the environment. The impact of various NPs and their uptake capability may differ between species of plants mainly on metabolic functions and the plant growth. It is strongly believed that a plant's genetic engineering can provide additional enhancement of NP-mediated plant genetic engineering. Nair et al. (2010) studied the use of NPs in plants, studied their impact and concluded that they could deliver an imminent opinion for the harmless usage of this progressive technology in agricultural production.

Chitosan is considered a valuable transporter for the controlled transfer of agro-chemical biomolecules and genetic materials due to its biodegradability, absorption abilities, biocompatibility and non-toxicity. The matrix of chitosan in the encapsulation of agro-chemical biomolecules and genetic materials acts in a defensive way for the biomolecules and protects them from damage as well as from environmental factors.

15.2.4 Plant-Nanoparticle Interaction

Absorption, translocation and accumulation of NPs is greatly influenced by their shape, dimensions and molecular structure, as well as stability and functionalisation. These properties are found to be controlled in many methods by the kind of plant species as well as the spot that assists the internalisation of NPs. In the plant cells, the chief site of contact with the exterior atmosphere is the cell wall, which limits its entrance to all distant elements, with NPs. In a plant cell wall, there are certain functionalised active sites along with other components. These contain functional groups like amine, sulfhydryl, hydroxyl, phosphate, imidazole and carboxylate (Vinopal et al., 2007). Complex biomolecules such as proteins, carbohydrates and cellulose are formed when these functional groups fit together and these biomolecules, in turn, facilitate selective absorption of NPs. Due to its semipermeable property, differential or selective NP translocation and absorption occur through the cell wall. This allows the small-sized NPs to move through and enter into the plant system. Thus, the screening property of the plant is due to the pore size of the cell wall, the diameter of which varies from 5–20 nm (Fleischer et al., 1999). Therefore, those NPs with a diameter equal to that of the pore size of the cell wall can easily penetrate through the cell wall and reach the plasma membrane. Ovečka et al. (2005) and Navarro et al. (2008) also reported that during reproduction, new pores are formed in the cell walls and existing pores get expanded under the influence of NPs, which subsequently improves the absorption of NPs. After the NPs enter inside the cell wall, it undergoes endocytosis-mediated internalisation, wherein the NPs are enclosed by the plasma membrane in a structure similar to a cavity. NPs can enter the plant cell via other routes like stomata, stigmas, cortexes and cuticles. NPs can also enter via root tips, lateral root junction and by wounds.

15.2.5 Nanofertiliser

Nanofertiliser is a neoteric concept and happens to be one of the recent advancements in the agricultural sector. The more surface area to volume ratio and very small size make nanofertiliser highly effective compared with regular fertilisers. Various kinds of nanofertilisers have been manufactured and applied to various crops. Various researchers, after extensive studies, have found nanomaterials to show a positive effect on plant biomass, shoot and root elongation and the content of chlorophyll and germination of seed at a definite amount. For example, Mochizuki et al. (2009) reported that fertilisers in nanoform are found to dissolve quicker compared with their traditional counterparts. They further clarified that the thermodynamics law is followed by nanofertilisers which have a high entropy due to colloid suspension state compared with their conventional forms (Tables 15.2 and 15.3).

TABLE 15.2

Comparative Study of Conventional Fertilisers Applications and Nanotechnology-Based Formulations

Properties	Nanofertilisers	Conventional Fertilisers
Dispersion and solubility of mineral micronutrients	High bioavailability and solubility to plants due to small particle size	Low bioavailability and solubility to plants due to large particle size
Controlled-release modes	Release amount and release design of nutrients for water-soluble fertilisers might be precisely controlled through encapsulation in envelope forms	Surplus release of fertilisers may give toxicity and abolish soil ecological balance
Nutrient release duration	Nanofertilisers can extend effective nutrient supply duration of fertilisers into soil	Used by the plants at the time of delivery, the rest is converted into insoluble salts in the soil
Fertiliser nutrients loss rate	Reduce fertilizer nutrients loss rate into soil by leaching and/or leaking	High fertilizer nutrient loss rate by drift, rain off and leaching
Efficiency of nutrient uptake	Might enhance efficiency of fertiliser and ratio of uptake of the soil nutrients in crop production and save fertiliser resource	Bulk composite is not available for roots and reduce efficiency

TABLE 15.3
Use of Nanoparticles (NPs) as Nanofertiliser on Different Crops

Sl. No.	Nanoparticles	Crop	Positive Impact on Crop	References
1	Silver NPs	*Crocus sativus*	Elongation of root	(Rezvani et al., 2012)
		Phaseolus vulgaris	Enhanced growth and development	(Zea & Salama, 2012)
		Boswellia ovalifoliolata	Enhanced growth and development	(Savithramma et al., 2012)
		Hordeum vulgare	Elongation of root	(Gruyer et al., 2014)
		Brassica juncea	Enhanced growth and development	(Sharma et al., 2012)
2	Gold NPs	*Gloriosa superba*	Enhanced growth and development	(Gopinath et al., 2014)
		B. juncea	Enhanced growth and development	(Arora et al., 2012)
3	Carbon nanotubes	*Cicer arietinum*	Growth and development improved	(Tripathi et al., 2012)
		Solanum lycopersicum	Seed germination improved	(Khodakovskaya et al., 2009)
4	Carbon-coated Fe-NPs	*Cucurbita pepo*	Content of chlorophyll enhanced	(Khodakovskaya et al., 2009)
5	Iron NPs			
		Vigna unguiculata	Content of chlorophyll enhanced	(Liu et al., 2005)
6	Copper NPs	*Lactuca sativa*	Enhanced growth and development	(Shah & Belozerova, 2009)
		Elodea densa	Rate of photosynthesis improved	(Nekrasova et al., 2011)
		Triticum aestivum	Enhanced growth and development	Lee et al. (2008)
		Phaseolus radiates	Enhanced growth and development	Lee et al. (2008)
7	Manganese NPs	*V. radiate*	Enhanced growth and development	Ghafariyan et al. (2013)
8	Molybdenum NPs	*C. arietinum*	Enhanced the formation of nodule	Ghafariyan et al. (2013)
9	Silicon NPs	*Lycopersicum Esculentum*	Enhanced growth and development	(Siddiqui et al., 2015)
		Cucurbita pepo	Enhanced growth and development	Siddiqui et al., (2015)
		Larix olgensis	Enhanced growth and development	(Bao-shan et al., 2004)
		Zea mays	Enhanced growth and development	Suriyaprabha et al. (2012)
10	Single walled carbon Nanotubes	*Cucumis sativus*	Elongation of root enhanced	(Cañas et al., 2008)
		Allium cepa	Elongation of root enhanced	Canas et al. (2008)
11	Titanium NPs	*V. radiata L.*	Enhanced growth and development	(Raliya et al., 2015)
		Lemna minuta	Enhanced growth and development	(Song et al., 2012)
12	Zinc complexed chitosan NPs	*T. aestivum*	Improvement of zinc utilisation	(Dapkekar et al., 2018)
13	Zinc NPs	*G. max*	Germination of seed increased	(Sedghi et al., 2013)
		Allium cepa	Germination of seed increased	(Raskar & Laware, 2014)
		T. aestivum	Enhanced growth and development	(Ramesh et al., 2014)
		C. sativus	Enhanced growth and development	(Zhao et al., 2014)
			Enhanced growth and development	(de la Rosa et al., 2013)

15.2.6 Nanoparticles in Crop Pathology

Various NPs have been exploited for their more inhibitory action in contradiction of certain pathogens of crops, both *in vitro* and in field conditions.

Polymeric-based nanomaterials: Various polymers (e.g., artificial polymers and natural bio-polymers) have been applied for pesticide delivery by encapsulating the dynamic components

(Nuruzzaman et al. 2016). These encapsulating materials, to deliver the active ingredients, can make many types of arrangements, viz. nanogel, nanosphere, micelle and nanocapsule.

Synthetic polymers: These are amphiphilic block copolymers, constituting hydrophilic and hydrophobic parts. These can be formed by linking together two or higher diverse chains of polymers. The different types of copolymer structures that can be synthesised (depending on the number of blocks used) include AB (biblock), ABA (triblock), AnBn (multiblock), AB3 (hetero-arm star), A2B (Y structure) and ABC (miktoarm) (Ruzette & Leibler, 2005). Some of the examples are poly (ε-caprolactone) and poly (d, l-lactide) (Kothamasu et al., 2012).

15.2.7 Biopolymers

Biopolymers include chitin, starch, cellulose, DNA and RNA, peptides and chitosan proteins. Biopolymers are those molecules that result from living sources such as animals, microbes, plants and waste of agriculture and fossils. However, using biological monomer units such as natural oils and fats, nucleotides, amino acids, and sugars and their chemical synthesis is also possible (Perlatti et al., 2013; Sanchez-Vazquez et al., 2013). They have been reported to be a sustainable source of eco-friendly, biodegradable low-cost polymeric materials, and nontoxic. Perlatti et al. (2013) reported about the exploration of their biocompatibility in agriculture.

15.2.8 Nanosensor

The advent of nanosensors has played an important part in the progress of agriculture. Nanosensors are trusted and proven facilitators in the sector of agriculture for real-time crop monitoring and monitoring of field conditions, along with plant disease, pest attack, environmental stressors and crop growth (Chen & Yada, 2011) (Table 15.4).

TABLE 15.4

Application of Nanotechnology in Sensing

NPs	Target Compound/Species	Detection limit	References
Titanium oxide (TiO_2)	Atrazine	0.1 ppt	Yu et al. (2010)
Carbon	Detection of herbicide	-	Luo et al. (2014)
Gold (Au)	Acetamiprid	5 nM	Shi et al. (2013)
	Activity of urease and urea	5 μM and 1.8 U/L	Deng et al. (2016)
	Pantoea stewartii subsp.	7.8 × 103 cfu/mL	Zhao (2014a,b)
	Detection of herbicide	-	Boro et al. (2011)
	Detection of organophosphates	-	Kang et al. (2010)
Gold nanorods	Cymbidium mosaic virus	48 pg/mL	Lin (2014)
	Odontoglossum ringspot virus	42 pg/mL	
Multi-walled carbon nanotubes (MWCNT)	Glufosinate and glyphosate	0.35 ng/mL and 0.19 ng/mL	Prasad et al. (2014)
Graphene oxide	Nitrate	10^{-5} M	Pan et al. (2016)
Graphene	Detection of herbicide	-	Zhao et al. (2011)
ZnO-chitosan nanocomposite membrane	*Trichoderma harzianum*	1.0×10^{-19} mol/L	Siddiquee et al. (2014)
Quantum dots	DNA	3.55×10^{-9} M	Bakhori et al. (2013)
Silver (Ag)	Detection of herbicide	-	Dubas and Pimpan (2008)

15.3 Regulatory Guidance

The strict need for efficient safety guidance is of tremendous importance because nanomaterials are present in a wide variety of products. Several workshops on the application of nanotechnology concerning the agricultural sector have been conducted by various international organisations and the conclusions were compiled in reports and were made accessible online, like Food and Agriculture Organization of the United Nations (FAO)/World Health Organization (WHO) (Joint Meeting on Pesticide Residues, 2014; Kah, 2015). As per the reports of FAO/WHO (2013), Australian Pesticides and Veterinary Medicines Authority (AVPMA 2014)), various activities regarding the legislation issued/adapted by various governing bodies regarding the nano-agro-chemicals vary considerably (Kah, 2015). There is currently no specific body fully dedicated to or with the jurisdiction to strictly impose on the production of nanomaterials or their use. Only nonbinding guidance for the industry is available for most regions in the world, and it is likely to vary in different regions. Here, in this section, the available guidance and regulation is briefly discussed.

15.3.1 International Efforts

The International Organization for Standardization (ISO) and Organization for Economic Cooperation and Development (OECD) are the main guiding and regulating bodies at the international level. They generally aim and focus on deciding the regulatory frameworks among nanotechnology research and development as well as communication and standardisation of the regulatory guidance (Sadeghi et al., 2017).

15.4 Conclusion and Future Aspects

The Indian economy is highly dependent on agriculture, but the growth rate of agriculture is of immense concern and it needs a serious boost for improvement of productivity. It can be asserted that interdisciplinary methodologies could increase agricultural productivity by incorporation of new evolving departments such as biotechnology and chemical science integrated with nanotechnology. This is predicted to challenge the prevailing living blockages that are presently providing hindrance for further growth. Nanotechnology can be the vital cog in the wheel for the improvement of productivity and vigour. With the advent of nanotechnology, a large quantity of application of agro-chemicals has been reduced, while keeping the quantity of yield steady. This has been made possible due to well-developed delivery mechanisms. The study of plant-NP interaction has greatly modified the gene and protein profiling of plants at the cellular level, thereby altering the biological pathways and enhancing the growth and development of plants. Therefore, it is highly recommended to conduct further research at the molecular levels about the phenomena of absorption and translocation of NPs and their aftermath. The potential benefits of nanotechnology have to be balanced in such a way that they do not harm the environmental components such as air, water and soil.

REFERENCES

Adisa, I. O., Pullagurala, V. L. R., Peralta-Videa, J. R., Dimkpa, C. O., Elmer, W. H., Gardea-Torresdey, J. L., & White, J. C. (2019). Recent advances in nano-enabled fertilizers and pesticides: A critical review of mechanisms of action. *Environmental Science: Nano*, *6*(7), 2002–2030. https://doi.org/10.1039/c9en00265k

Aditya, N. P., Patankar, S., Madhusudhan, B., Murthy, R. S. R., & Souto, E. B. (2010). Arthemeter-loaded lipid nanoparticles produced by modified thin-film hydration: Pharmacokinetics, toxicological and in vivo anti-malarial activity. *European Journal of Pharmaceutical Sciences*, *40*(5), 448–455. https://doi.org/10.1016/j.ejps.2010.05.007

APVMA (2014). A draft report from the Australian Pesticides and Veterinary Medicines Authority. Regulatory considerations for nanopesticides and neterinary nanomedecines. Available online at: http://apvma.gov.au/sites/default/files/docs/report-draft-regulatory-considerations-nanopesticides-veterinary-nanomedicines.pdf [Accessed February 21, 2021]

Arora, S., Sharma, P., Kumar, S., Nayan, R., Khanna, P. K., & Zaidi, M. G. H. (2012). Gold-nanoparticle induced enhancement in growth and seed yield of Brassica juncea. *Plant Growth Regulation*, *66*(3), 303–310. https://doi.org/10.1007/s10725-011-9649-z

Avellan, A., Yun, J., Zhang, Y., Spielman-Sun, E., Unrine, J. M., Thieme, J., Li, J., Lombi, E., Bland, G., & Lowry, G. V. (2019). Nanoparticle size and coating chemistry control foliar uptake pathways, translocation, and leaf-to-rhizosphere transport in wheat [Research article]. *ACS Nano*, *13*(5), 5291–5305. https://doi.org/10.1021/acsnano.8b09781

Bakhori, N. M., Yusof, N. A., Abdullah, A. H., & Hussein, M. Z. (2013). Development of a fluorescence resonance energy transfer (fret)-based DNA biosensor for detection of synthetic oligonucleotide of *Ganoderma boninense*. *Biosensors*, *3*(4), 419–428.

Bao-Shan, L., Chun-hui, L., Li-jun, F., Shu-chun, Q., & Min, Y. (2004). Effect of TMS (nanostructured silicon dioxide) on growth of Changbai larch seedlings. *Journal of Forestry Research*, *15*(2), 138–140.

Bharali, D. J., Klejbor, I., Stachowiak, E. K., Dutta, P., Roy, I., Kaur, N., Bergey, E. J., Prasad, P. N., & Stachowiak, M. K. (2005). Organically modified silica nanoparticles: A nonviral vector for in vivo gene delivery and expression in the brain. *Proceedings of the National Academy of Sciences of the United States of America*, *102*(32), 11539–11544. https://doi.org/10.1073/pnas.0504926102

Boro, R. C., Kaushal, J., Nangia, Y., Wangoo, N., Bhasin, A., & Suri, C. R. (2011). Gold nanoparticles catalyzed chemiluminescence immunoassay for detection of herbicide 2, 4-dichlorophenoxyacetic acid. *Analyst*, *136*(10), 2125–2130.

Cañas, J. E., Long, M., Nations, S., Vadan, R., Dai, L., Luo, M., Ambikapathi, R., Lee, E. H., & Olszyk, D. (2008). Effects of functionalized and nonfunctionalized single-walled carbon nanotubes on root elongation of select crop species. *Environmental Toxicology and Chemistry*, *27*(9), 1922–1931. https://doi.org/10.1897/08-117.1

Chen, H., & Yada, R. (2011). Nanotechnologies in agriculture: New tools for sustainable development. In *Trends in Food Science and Technology* (Vol. 22, Issue 11, pp. 585–594). https://doi.org/10.1016/j.tifs.2011.09.004

Dapkekar, A., Deshpande, P., Oak, M. D., Paknikar, K. M., & Rajwade, J. M. (2018). Zinc use efficiency is enhanced in wheat through nanofertilization. *Scientific Reports*, *8*(1), 1–7. https://doi.org/10.1038/s41598-018-25247-5

Dasgupta, N., Ranjan, S., Mundekkad, D., Ramalingam, C., Shanker, R., & Kumar, A. (2015). Nanotechnology in agro-food: From field to plate. In *Food Research International* (Vol. 69). Elsevier Ltd. https://doi.org/10.1016/j.foodres.2015.01.005

de la Rosa, G., López-Moreno, M. L., de Haro, D., Botez, C. E., Peralta-Videa, J. R., & Gardea-Torresdey, J. L. (2013). Effects of ZnO nanoparticles in alfalfa, tomato, and cucumber at the germination stage: Root development and X-ray absorption spectroscopy studies. *Pure and Applied Chemistry*, *85*(12), 2161–2174. https://doi.org/10.1351/PAC-CON-12-09-05

Demirer, G. S., Zhang, H., Matos, J. L., Goh, N. S., Cunningham, F. J., Sung, Y., Chang, R., Aditham, A. J., Chio, L., Cho, M. J., Staskawicz, B., & Landry, M. P. (2019). High aspect ratio nanomaterials enable delivery of functional genetic material without DNA integration in mature plants. *Nature Nanotechnology*, *14*(5), 456–464. https://doi.org/10.1038/s41565-019-0382-5

Deng, H. H., Hong, G. L., Lin, F. L., Liu, A. L., Xia, X. H., & Chen, W. (2016). Colorimetric detection of urea, urease, and urease inhibitor based on the peroxidase-like activity of gold nanoparticles. *Analytica Chimica Acta*, *915*, 74–80.

Dubas, S. T., & Pimpan, V. (2008). Humic acid assisted synthesis of silver nanoparticles and its application to herbicide detection. *Materials Letters*, *62*(17–18), 2661–2663.

Dutta, P., Pal, S., Seehra, M. S., Shi, Y., Eyring, E. M., & Ernst, R. D. (2006). Concentration of Ce 3+ and oxygen vacancies in cerium oxide nanoparticles. *Chemistry of Materials*, *18*(21), 5144–5146. https://doi.org/10.1021/cm061580n

FAO/WHO (2013). Food and Agriculture Organization of the United Nations and World Health Organization. State of the art on the initiatives and activities relevant to risk assessment and risk management of nanotechnologies in the food and agriculture sectors, in FAO/WHO Technical Paper. Available online at: http://www.fao.org/docrep/018/i3281e/i3281e.pdf [Accessed February 21, 2021]

Fleischer, A., O'Neill, M. A., & Ehwald, R. (1999). The pore size of non-graminaceous plant cell walls is rapidly decreased by borate ester cross-linking of the pectic polysaccharide rhamnogalacturonan II. *Plant Physiology*, *121*(3), 829–838. https://doi.org/10.1104/pp.121.3.829

Garnett, T., & Godfray, H. C. J. (2012). *Sustainable intensification in agriculture. Navigating a course through competing food system priorities.* https://hdl.handle.net/10568/52201

Ghafariyan, M. H., Malakouti, M. J., Dadpour, M. R., Stroeve, P., & Mahmoudi, M. (2013). Effects of magnetite nanoparticles on soybean chlorophyll. *Environmental Science & Technology*, *47*(18), 10645–10652. https://doi.org/10.1021/es402249b

Ghanati, F., & Bakhtiarian, S. (2014). Effect of methyl jasmonate and silver nanoparticles on production of secondary metabolites by Calendula Officinalis L (Asteraceae). *Tropical Journal of Pharmaceutical Research*, *13*(11), 1783–1789. https://doi.org/10.4314/tjpr.v13i11.2

Ghasemi, B., Hosseini, R., & Dehghan Nayeri, F. (2015). Effects of cobalt nanoparticles on artemisinin production and gene expression in Artemisia annua. *Turkish Journal of Botany*, *39*(5), 769–777. https://doi.org/10.3906/bot-1410-9

Gopinath, K., Gowri, S., Karthika, V., & Arumugam, A. (2014). Green synthesis of gold nanoparticles from fruit extract of Terminalia arjuna, for the enhanced seed germination activity of Gloriosa superba. *Journal of Nanostructure in Chemistry*, *4*(3). https://doi.org/10.1007/s40097-014-0115-0

Grillo, R., Abhilash, P. C., & Fraceto, L. F. (2016). Nanotechnology applied to bio-encapsulation of pesticides. *Journal of Nanoscience and Nanotechnology*, *16*(1), 1231–1234. https://doi.org/10.1166/jnn.2016.12332

Gruyer, N., Dorais, M., Bastien, C., Dassylva, N., & Triffault-Bouchet, G. (2014). Interaction between silver nanoparticles and plant growth. *Acta Horticulturae*, *1037*, 795–800. https://doi.org/10.17660/ActaHortic.2014.1037.105

Hajiahmadi, Z., Movahedi, A., Wei, H., Li, D., Orooji, Y., Ruan, H., & Zhuge, Q. (2019). Strategies to increase on-target and reduce off-target effects of the CRISPR/Cas9 system in plants. In *International Journal of Molecular Sciences* (Vol. 20, Issue 15). MDPI AG. https://doi.org/10.3390/ijms20153719

Heiras-Palazuelos, M. J., Ochoa-Lugo, M. I., Gutiérrez-Dorado, R., López-Valenzuela, J. A., Mora-Rochín, S., Milán-Carrillo, J., Garzón-Tiznado, J. A., & Reyes-Moreno, C. (2013). Technological properties, antioxidant activity and total phenolic and flavonoid content of pigmented chickpea (Cicer arietinum L.) cultivars. *International Journal of Food Sciences and Nutrition*, *64*(1), 69–76. https://doi.org/10.3109/09637486.2012.694854

Helaly, M. N., El-Metwally, M. A., El-Hoseiny, H., Omar, S. A., & El-Sheery, N. I. (2014). Effect of nanoparticles on biological contamination of *in vitro* cultures and organogenic regeneration of banana. *Australian Journal of Crop Science*, *8*(4), 612–624.

Isah, T. (2019). Stress and defense responses in plant secondary metabolites production. Biological research, 52. Available online at: https://biolres.biomedcentral.com/articles/10.1186/s40659-019-0246-3

Joint Meeting on Pesticide Residues. (2014). *Food and Agriculture Organization of the United Nations World Health Organization Joint FAO/WHO Expert Committee on Food Additives Fiftieth meeting.* 2(41), 1–11. Available online at: www.who.int/pcs/jecfa/jecfa.htm

Kabera, J. N., Semana, E., Mussa, A. R., & He, X. (2014). Plant secondary metabolites: Biosynthesis, classification, function and pharmacological properties. *Journal of Pharmacy and Pharmacology*, *2*(January), 377–392.

Kah, M. (2015). Nanopesticides and nanofertilizers : Emerging contaminants or opportunities for risk mitigation? *Frontiers in Chemistry*, *3*(November), 1–6. https://doi.org/10.3389/fchem.2015.00064

Kang, T. F., Wang, F., Lu, L. P., Zhang, Y., & Liu, T. S. (2010). Methyl parathion sensors based on gold nanoparticles and Nafion film modified glassy carbon electrodes. *Sensors and Actuators B: Chemical*, *145*(1), 104–109.

Khodakovskaya, M., Dervishi, E., Mahmood, M., Xu, Y., Li, Z., Watanabe, F., & Biris, A. S. (2009). Carbon nanotubes are able to penetrate plant seed coat and dramatically affect seed germination and plant growth. *ACS Nano*, *3*(10), 3221–3227. https://doi.org/10.1021/nn900887m

Kothamasu, P., Kanumur, H., Ravur, N., Maddu, C., Parasuramrajam, R., & Thangavel, S. (2012). Nanocapsules: The weapons for novel drug delivery systems. *BioImpacts*, *2*(2), 71–81. https://doi.org/10.5681/bi.2012.011

Kottegoda, N., Sandaruwan, C., Priyadarshana, G., Siriwardhana, A., Rathnayake, U. A., Berugoda Arachchige, D. M., Kumarasinghe, A. R., Dahanayake, D., Karunaratne, V., & Amaratunga, G. A. J. (2017). Urea-hydroxyapatite nanohybrids for slow release of nitrogen. *ACS Nano*, *11*(2), 1214–1221. https://doi.org/10.1021/acsnano.6b07781

Kwak, S. Y., Lew, T. T. S., Sweeney, C. J., Koman, V. B., Wong, M. H., Bohmert-Tatarev, K., Snell, K. D., Seo, J. S., Chua, N. H., & Strano, M. S. (2019). Chloroplast-selective gene delivery and expression in planta using chitosan-complexed single-walled carbon nanotube carriers. *Nature Nanotechnology*, *14*(5), 447–455. https://doi.org/10.1038/s41565-019-0375-4

Lee, W. M., An, Y. J., Yoon, H., & Kweon, H. S. (2008). Toxicity and bioavailability of copper nanoparticles to the terrestrial plants mung bean (*Phaseolus radiatus*) and wheat (*Triticum aestivum*): Plant agar test for water-insoluble nanoparticles. *Environmental Toxicology and Chemistry: An International Journal*, *27*(9), 1915–1921. https://doi.org/10.1897/07-481.1

Lin, H. Y., Huang, C. H., Lu, S. H., Kuo, I. T., & Chau, L. K. (2014). Direct detection of orchid viruses using nanorod-based fiber optic particle plasmon resonance immunosensor. *Biosensors and Bioelectronics*, *51*, 371–378.

Liu, J., Curry, J. A., Rossow, W. B., Key, J. R., & Wang, X. (2005). Comparison of surface radiative flux data sets over the Arctic Ocean. *Journal of Geophysical Research C: Oceans*, *110*(2), 1–13. https://doi.org/10.1029/2004JC002381

Liu, R., & Lal, R. (2015). Potentials of engineered nanoparticles as fertilizers for increasing agronomic productions. *Science of the Total Environment*, *514*(2015), 131–139. https://doi.org/10.1016/j.scitotenv.2015.01.104

Luo, M., Liu, D., Zhao, L., Han, J., Liang, Y., Wang, P., et al. (2014). A novel magnetic ionic liquid modified carbon nanotube for the simultaneous determination of aryloxyphenoxypropionate herbicides and their metabolites in water. *Analytica Chimica Acta*, *852*, 88–96. https://doi.org/10.1016/j.aca.2014.09.024

Ma, X., Geiser-Lee, J., Deng, Y., & Kolmakov, A. (2010). Interactions between engineered nanoparticles (ENPs) and plants: Phytotoxicity, uptake and accumulation. *Science of the Total Environment*, *408*(16), 3053–3061. https://doi.org/10.1016/j.scitotenv.2010.03.031

Martin-Ortigosa, S., Valenstein, J. S., Lin, V. S. Y., Trewyn, B. G., & Wang, K. (2012). Gold functionalized mesoporous silica nanoparticle mediated protein and DNA codelivery to plant cells via the biolistic method. *Advanced Functional Materials*, *22*(17), 3576–3582. https://doi.org/10.1002/adfm.201200359

Mishra, S., & Singh, H. B. (2015). Biosynthesized silver nanoparticles as a nanoweapon against phytopathogens: Exploring their scope and potential in agriculture. *Applied Microbiology and Biotechnology*, *99*(3), 1097–1107. https://doi.org/10.1007/s00253-014-6296-0

Mitter, N., Worrall, E. A., Robinson, K. E., Li, P., Jain, R. G., Taochy, C., Fletcher, S. J., Carroll, B. J., Lu, G. Q., & Xu, Z. P. (2017). Clay nanosheets for topical delivery of RNAi for sustained protection against plant viruses. *Nature Plants*, *3*(January). https://doi.org/10.1038/nplants.2016.207

Mochizuki, H., Gautam, P., Sinha, S., & Kumar, S. (2009). Increasing fertilizer and pesticide use efficiency by nanotechnology in desert afforestation, arid agriculture. *Journal of Arid Land Studies*, *19*(1), 129–132.

Mohammed AL-oubaidi, H., & Kasid, N. (2015). Increasing phenolyic and flavoniods compoundes of cicer Arietinum L. from embryo explant using titanum dioxide nanoparticle *in vitro*. *World Journal of Pharmaceutical Research*, *4*(11), 1791–1799. http://www.wjpr.net/dashboard/abstract_id/4142

Nair, R., Varghese, S. H., Nair, B. G., Maekawa, T., Yoshida, Y., & Kumar, D. S. (2010). Nanoparticulate material delivery to plants. *Plant Science*, *179*(3), 154–163. https://doi.org/10.1016/j.plantsci.2010.04.012

Navarro, E., Piccapietra, F., Wagner, B., Marconi, F., Kaegi, R., Odzak, N., Sigg, L., & Behra, R. (2008). Toxicity of silver nanoparticles to *Chlamydomonas reinhardtii*. *Environmental Science and Technology*, *42*(23), 8959–8964. https://doi.org/10.1021/es801785m

Nekrasova, G. F., Ushakova, O. S., Ermakov, A. E., Uimin, M. A., & Byzov, I. V. (2011). Effects of copper(II) ions and copper oxide nanoparticles on Elodea densa Planch. *Russian Journal of Ecology*, *42*(6), 458–463. https://doi.org/10.1134/S1067413611060117

Nitta, S. K., & Numata, K. (2013). Biopolymer-based nanoparticles for drug/gene delivery and tissue engineering. *International Journal of Molecular Sciences*, *14*(1), 1629–1654. https://doi.org/10.3390/ijms14011629

Nuruzzaman, M., Rahman, M. M., Liu, Y., & Naidu, R. (2016). Nanoencapsulation, nano-guard for pesticides: A new window for safe application. *Journal of Agricultural and Food Chemistry, 64*(7), 1447–1483. https://doi.org/10.1021/acs.jafc.5b05214

Ovečka, M., Lang, I., Baluška, F., Ismail, A., Illeš, P., & Lichtscheidl, I. K. (2005). Endocytosis and vesicle trafficking during tip growth of root hairs. *Protoplasma, 226*(1–2), 39–54. https://doi.org/10.1007/s00709-005-0103-9

Pan, P., Miao, Z., Yanhua, L., Linan, Z., Haiyan, R., Pan, K., et al. (2016). Preparation and evaluation of a stable solid state ion selective electrode of polypyrrole/electrochemically reduced graphene/glassy carbon substrate for soil nitrate sensing. *International Journal of Electrochemical Science, 11*, 4779–4793.

Parisi, C., Vigani, M., & Rodríguez-Cerezo, E. (2015). Agricultural nanotechnologies: What are the current possibilities? *Nano Today, 10*(2), 124–127. https://doi.org/10.1016/j.nantod.2014.09.009

Pereira, A. E. S., Silva, P. M., Oliveira, J. L., Oliveira, H. C., & Fraceto, L. F. (2017). Chitosan nanoparticles as carrier systems for the plant growth hormone gibberellic acid. In *Colloids and Surfaces B: Biointerfaces* (Vol. 150). Elsevier B.V. https://doi.org/10.1016/j.colsurfb.2016.11.027

Perlatti, B., Souza Bergo, P. L. de, Fernandes da Silva, M. F. das G., Batista, J., & Rossi, M. (2013). Polymeric nanoparticle-based insecticides: A controlled release purpose for agrochemicals. *Insecticides – Development of Safer and More Effective Technologies.* https://doi.org/10.5772/53355

Prasad, B. B., Jauhari, D., & Tiwari, M. P. (2014). Doubly imprinted polymer nanofilmmodified electrochemical sensor for ultra-trace simultaneous analysis of glyphosate and glufosinate. *Biosensors and Bioelectronics, 59*, 81–88.

Raliya, R., Biswas, P., & Tarafdar, J. C. (2015). TiO2 nanoparticle biosynthesis and its physiological effect on mung bean (Vigna radiata L.). *Biotechnology Reports, 5*(1), 22–26. https://doi.org/10.1016/j.btre.2014.10.009

Raliya, R., Saharan, V., Dimkpa, C., & Biswas, P. (2018). Nanofertilizer for precision and sustainable agriculture: Current state and future perspectives [Review article]. *Journal of Agricultural and Food Chemistry, 66*(26), 6487–6503. https://doi.org/10.1021/acs.jafc.7b02178

Raliya, R., & Tarafdar, J. C. (2013). ZnO Nanoparticle biosynthesis and its effect on phosphorous-mobilizing enzyme secretion and gum contents in clusterbean (*Cyamopsis tetragonoloba* L.). *Agricultural Research, 2*(1), 48–57. https://doi.org/10.1007/s40003-012-0049-z

Ramesh, M., Palanisamy, K., & Kumar Sharma, N. (2014). Effects of bulk & nano-titanium dioxide and zinc oxide on physio-morphological changes in *Triticum aestivum* Linn. *Journal of Global Biosciences ISSN, 3*(2), 2320–1355.

Raskar, S. V., & Laware, S. L. (2014). Effect of zinc oxide nanoparticles on cytology and seed germination in onion. *International Journal of Current Microbiology and Applied Sciences, 3*(2), 467–473. http://www.ijcmas.com/vol-3-2/S.V.Raskar and S.L.Laware.pdf

Rezvani, S., Dehkordi, G. J., Rahman, M. S., Fouladivanda, F., Habibi, M., & Eghtebasi, S. (2012). A conceptual study on the country of origin effect on consumer purchase intention. *Asian Social Science, 8*(12), 205–215. https://doi.org/10.5539/ass.v8n12p205

Ruzette, A. V., & Leibler, L. (2005). Block copolymers in tomorrow's plastics. In *Nature Materials* (Vol. 4, Issue 1, pp. 19–31). Nature Publishing Group. https://doi.org/10.1038/nmat1295

Sadeghi, R., Rodriguez, R. J., Yao, Y., & Kokini, J. L. (2017). Advances in nanotechnology as they pertain to food and agriculture: Benefits and risks. *Annual Review of Food Science and Technology, 8*(January), 467–492. https://doi.org/10.1146/annurev-food-041715-033338

Sanchez-Vazquez, S. A., Hailes, H. C., & Evans, J. R. G. (2013). Hydrophobic polymers from food waste: Resources and synthesis. *Polymer Reviews, 53*(4), 627–694. https://doi.org/10.1080/15583724.2013.834933

Savithramma, N., Ankanna, S., & Bhumi, G. (2012). Effect of nanoparticles on seed germination and seedling growth of *Boswellia ovalifoliolata* – an endemic and endangered medicinal tree Taxon. *Nano Vision, 2*(1), 1–68.

Sedghi, M., Hadi, M., & Toluie, S. G. (2013). Effect of nano zinc oxide on the germination parameters of soybean seeds under drought stress. *Annals of West University of Timisoara, Series of Biology, XVI*(2), 73–78.

Shah, V., & Belozerova, I. (2009). Influence of metal nanoparticles on the soil microbial community and germination of lettuce seeds. *Water, Air, and Soil Pollution, 197*(1–4), 143–148. https://doi.org/10.1007/s11270-008-9797-6

Sharafi, E., Nekoei, S. M. K., Fotokian, M. H., Davoodi, D., Mirzaei, H. H., & Hasanloo, T. (2013). Improvement of hypericin and hyperforin production using zinc and iron nano-oxides as elicitors in cell suspension culture of St John's wort (*Hypericum perforatum* L.). *Journal of Medicinal Plants and By-Products*, *2*(2), 177–184.

Sharma, P., Bhatt, D., Zaidi, M. G. H., Saradhi, P. P., Khanna, P. K., & Arora, S. (2012). Silver nanoparticle-mediated enhancement in growth and antioxidant status of *Brassica juncea*. *Applied Biochemistry and Biotechnology*, *167*(8), 2225–2233. https://doi.org/10.1007/s12010-012-9759-8

Shi, H., Zhao, G., Liu, M., Fan, L., & Cao, T. (2013). Aptamer-based colorimetric sensing of acetamiprid in soil samples: Sensitivity, selectivity and mechanism. *Journal of hazardous materials*, *260*, 754–761.

Siddiqi, K. S., & Husen, A. (2016). Engineered gold nanoparticles and plant adaptation potential. *Nanoscale Research Letters*, *11*(1). https://doi.org/10.1186/s11671-016-1607-2

Siddiqui, M. H., & Al-Whaibi, M. H. (2014). Role of nano-SiO2 in germination of tomato (Lycopersicum esculentum seeds Mill.). *Saudi Journal of Biological Sciences*, *21*(1), 13–17. https://doi.org/10.1016/j.sjbs.2013.04.005

Siddiqui, M. H., Al-Whaibi, M. H., & Mohammad, F. (2015). Nanotechnology and plant sciences: Nanoparticles and their impact on plants. *Nanotechnology and Plant Sciences: Nanoparticles and Their Impact on Plants*, 1–303. https://doi.org/10.1007/978-3-319-14502-0

Siddiquee, S., Rovina, K., Yusof, N. A., Rodrigues, K. F., & Suryani, S. (2014). Nanoparticle enhanced electrochemical biosensor with DNA immobilization and hybridization of *Trichoderma harzianum* gene. *Sensing and Bio-Sensing Research*, *2*, 16–22.

Singh, B., Rani, M., Singh, J., Moudgil, L., Sharma, P., Kumar, S., Saini, G. S. S., Tripathi, S. K., Singh, G., & Kaura, A. (2016). Identifying the preferred interaction mode of naringin with gold nanoparticles through experimental, DFT and TDDFT techniques: Insights into their sensing and biological applications. *RSC Advances*, *6*(83), 79470–79484. https://doi.org/10.1039/c6ra12076h

Song, G., Gao, Y., Wu, H., Hou, W., Zhang, C., & Ma, H. (2012). Physiological effect of anatase TiO2 nanoparticles on Lemna minor. *Environmental Toxicology and Chemistry*, *31*(9), 2147–2152. https://doi.org/10.1002/etc.1933

Suriyaprabha, R., Karunakaran, G., Yuvakkumar, R., Prabu, P., Rajendran, V., & Kannan, N. (2012). Growth and physiological responses of maize (*Zea mays* L.) to porous silica nanoparticles in soil. *Journal of Nanoparticle Research*, *14*(12). https://doi.org/10.1007/S11051-012-1294-6

Suzuki, N., Rivero, R. M., Shulaev, V., Blumwald, E., & Mittler, R. (2014). Abiotic and biotic stress combinations. *New Phytologist*, *203*(1), 32–43. https://doi.org/10.1111/nph.12797

Syu, Y. yu, Hung, J. H., Chen, J. C., & Chuang, H. wen. (2014). Impacts of size and shape of silver nanoparticles on Arabidopsis plant growth and gene expression. *Plant Physiology and Biochemistry*, *83*, 57–64. https://doi.org/10.1016/j.plaphy.2014.07.010

Tilman, D., Balzer, C., Hill, J., & Befort, B. L. (2011). Global food demand and the sustainable intensification of agriculture. *Proceedings of the National Academy of Sciences of the United States of America*, *108*(50), 20260–20264. https://doi.org/10.1073/pnas.1116437108

Torney, F., Trewyn, B. G., Lin, V. S. Y., & Wang, K. (2007). Mesoporous silica nanoparticles deliver DNA and chemicals into plants. *Nature Nanotechnology*, *2*(5), 295–300. https://doi.org/10.1038/nnano.2007.108

Tripathi, D. K., Singh, V. P., Kumar, D., & Chauhan, D. K. (2012). Impact of exogenous silicon addition on chromium uptake, growth, mineral elements, oxidative stress, antioxidant capacity, and leaf and root structures in rice seedlings exposed to hexavalent chromium. *Acta Physiologiae Plantarum*, *34*(1), 279–289. https://doi.org/10.1007/s11738-011-0826-5

Vinopal, S., Ruml, T., & Kotrba, P. (2007). Biosorption of Cd2+ and Zn2+ by cell surface-engineered Saccharomyces cerevisiae. *International Biodeterioration and Biodegradation*, *60*(2), 96–102. https://doi.org/10.1016/j.ibiod.2006.12.007

Walkey, C., Das, S., Seal, S., Erlichman, J., Heckman, K., Ghibelli, L., Traversa, E., McGinnis, J. F., & Self, W. T. (2015). Catalytic properties and biomedical applications of cerium oxide nanoparticles. *Environmental Science: Nano*, *2*(1), 33–53. https://doi.org/10.1039/c4en00138a

Yi, Z., Hussain, H. I., Feng, C., Sun, D., She, F., Rookes, J. E., Cahill, D. M., & Kong, L. (2015). Functionalized mesoporous silica nanoparticles with redox-responsive short-chain gatekeepers for agrochemical delivery. *ACS Applied Materials and Interfaces*, *7*(18), 9937–9946. https://doi.org/10.1021/acsami.5b02131

Yu, Z., Zhao, G., Liu, M., Lei, Y., & Li, M. (2010). Fabrication of a novel atrazine biosensor and its subpart-per-trillion levels sensitive performance. *Environmental Science & Technology*, *44*(20), 7878–7883. https://doi.org/10.1021/es101573s

Zea, L., & Salama, H. M. H. (2012). Effects of silver nanoparticles in some crop plants, common bean (*Phaseolus vulgaris* L.) and corn. *International Research Journal of Biotechnology*, *3*(10), 190–197. http://www.interesjournals.org/IRJOB/Pdf/2012/December/Salama.pdf

Zhao, G., Song, S., Wang, C., Wu, Q., & Wang, Z. (2011). Determination of triazine herbicides in environmental water samples by high-performance liquid chromatography using graphenecoated magnetic nanoparticles as adsorbent. *Analytica Chimica Acta*, *708*(1–2), 155–159.

Zhao, L., Peralta-Videa, J. R., Rico, C. M., Hernandez-Viezcas, J. A., Sun, Y., Niu, G., Servin, A., Nunez, J. E., Duarte-Gardea, M., & Gardea-Torresdey, J. L. (2014a). CeO_2 and ZnO nanoparticles change the nutritional qualities of cucumber (*Cucumis sativus*). *Journal of Agricultural and Food Chemistry*, *62*(13), 2752–2759. https://doi.org/10.1021/jf405476u

Zhao, Y., Liu, L., Kong, D., Kuang, H., Wang, L., & Xu, C. (2014b). Dual amplified electrochemical immunosensor for highly sensitive detection of Pantoea stewartii sbusp. Stewartii. *ACS Applied Materials & Interfaces*, *6*(23), 21178–21183.

16

Biosensors in Food Industry: An Exposition of Novel Food Safety Approach

Satish Kumar
Dr. Yashwant Singh Parmar University of Horticulture and Forestry

Vikas Kumar
Punjab Agricultural University

Priyanka Suthar and Rajni Saini
Lovely Professional University

Taru Negi
Govind Ballabh Pant University of Agriculture and Technology

CONTENTS

16.1 Introduction

Food contamination and adulterations are frequent food safety issues in most of the food production facilities and the finished food products. The World Health Organization (WHO) reports suggest that one-fourth of children below 5 years age are experiencing stunting due to the consumption of contaminated foods. These contaminants are hazardous and responsible for various foodborne illnesses and affect the quality of food as well. The maintenance of safety and quality during and after the processing is an onerous plight during manufacturing the food in households and in food industries. The rapid growth and transition of food industries in the recent years to meet the demands of specific population groups according to their changing lifestyles has been observed. The occurrence of hazardous food contaminants may take place at any stage of the production process and these contaminants may be either intentional or non-intentional. The assessment of safety and quality of food during production and manufacturing is important to define food quality at appropriate points during the production and to generate the status of food quality immediately. To minimise the deteriorative effects of foodborne pathogens and contaminants in raw materials and processed foods, the food scientists and food industries are trying their level best for the development of rapid, sensitive, cost-effective and reliable strategies, technologies and analytical techniques to confirm the presence of contaminants as soon as possible (Ibrišimović et al. 2015). One such development in the food industry for rapid detection of food contamination is with the help of biosensors.

By definition, a biosensor is a transducer device with an integrated receptor which has the potential to give selective analytical information (quantitative or semi-quantitative) by using biological recognition elements. The overall functioning of these biosensors basically depends on two parts, a biological component known as a bioreceptor and a transducer. The purpose of the transducer is to transform the response of the bioreceptor into an analytical signal. The advantages of using biosensors over traditional methods like spectrophotometry and chromatography includes less labour, low cost, real-time monitoring and specificity for the detection of particular food pathogens. Typically, biosensors must be autonomous and should not be affected by any change in physical parameters such as pH and temperature. Along with this independent factor, other factors 'must be' qualities which include specificity and the reusability of biosensors which will make their use intended in the field of food safety and security. In food industries, biosensors are used mainly for two purposes: the detection of adulterants or additives or for the estimation or detection of spoiling contaminants (both chemical and microbial).

In this chapter, we will discuss the current developments in biosensing techniques and their potential applications in the food industry for detecting the hazardous contaminants. Biosensors are well-established substituents to the traditional methods and are used to ensure the quality and safety of food with many other additional advantages like better process control, rapid testing, high precision and so forth. The recent progress in biosensors ensures its reliability in both industrial and research fields.

16.2 Overview of the Traditional and Modern Food Safety Approach

There are various advantages to using biosensors over traditional technology in food processing as the conventional methods are more time-consuming, complex, costly and need trained experts. As sample analysis is a cumbersome and time-consuming process and relies on the type of sample technique and size of the sample, it provides additional hurdles for rapid detection. Sometimes during the operating processes compounds develop in the food which may be toxic and may result in the production of incompetent food. Hence, there is an immediate need to develop methods for quick assessment of such compounds and to eliminate them from the food chain. Some traditional methods which are conventionally used lack applicability on-site, possess complex processes and are expensive and time-consuming. Monitoring of the microbial quality of food during production may limit itself to tedious plating methods which are time-consuming and laborious, and results may sometimes be obtained when the manufacturing process is over and the product is out for delivery. Hence, the development of biosensors for the assessment of food quality is a long-term realisation of both quality testing and microbial food stability, which could benefit by the application of sensing technology. Applications of biosensors in various food products are tabulated in Table 16.1.

TABLE 16.1

An Exposition on Different Types of Spoilage and the Biosensor Used or Their Detection in Different Food Industries

Food Industry	Spoilage Microbe	Biosensor for Detection	Working Principle	Detection Limit	References
Dairy	*Listeria cells*	Electrochemical impedance spectroscopy	Resistance to electron transfer due to immobilised endolysin	1.1 × 104 CFU in pure culture and 10^{-5} CFU in milk	Tolba et al. (2012)
	Melamine (from plastic)	Wave-fibre optic immune-sensor	Light excitation from single-multimode optic fibre couple and collecting generated fluorescence from optic fibre probe	5.14 µg/L	Hao et al. (2014)
	Bacillus cereus	Gold nanoparticle–modified pencil graphite electrode based on DNA	The rise in charge transfer resistance (RCT) of the biosensor for the detection of *B. cereus* resulted from the single-stranded DNA hybridisation with target DNA; an atomic force hybridisation (AFM) has been used for confirmation of hybridisation	9.4×10^{-12} mol L^{-1}	Izadi et al. (2016)
Fruits and vegetables	Hormone concentrations	Potentiometric naphthalene acetic acid sensor	A potentiometric naphthalenic acid (NAA) sensor on graphene nanosheets was developed by incorporating lipid films (stablilised) where auxin-binding protein 1 (ABP1) is immobilised, which provided selective detection of phytohormone	Limit of 10 Nm	Bratakou et al. (2016)
	Carbamate pesticides(carbaryl and methomyl)	GC/multiwalled carbon nanotube (MWCNT)/PANI/(Acetylcholinesterase) AChE sensor: electrochemical AChE-based biosensor was constructed on MWCNTs with core shell–modified glassy carbon (GC) electrode and polyaniline (PANI)	Used for the detection of carbaryl and methomyl pesticides using chronoamperometry	Limits of carbaryl and methomyl are 1.4 and 0.95 µmol L^{-1}, respectively	Cesarino et al. (2012)
	Pesticide; formetanate hydrochloride (FMT)	Electrochemical enzymatic biosensor	Biosensor is based on FMT's capacity to restrict the laccase catalytic reaction, which occurs when phenolic substrates are present	$9.5 \times 10^{-8} \pm 9.5 \times 10^{-10}$ M	Ribeiro et al. (2014)

(Continued)

TABLE 16.1
(Continued)

Food Industry	Spoilage Microbe	Biosensor for Detection	Working Principle	Detection Limit	References
Meat and poultry	Concentration of microbial load, pH, amount of inosine, inosine 5′-monophosphate and hypoxanthine	Redox potential is determined for pork sample by using potentiometric-based e-tongue	Six electrodes are made by C, Zn, CU, Pb, Ag and Au and a reference electrode to produce an e-tongue; the pork is stored under refrigerated conditions and spoilage is studied		Gil et al. (2011)
	B. subtilis	Cell-based biosensor	Using DNA microarray analysis several promoters were identified which are upregulated when the bacterial cells are exposed to volatile compounds released from the spoiled meat	Headspace gas for fresh (<103 CFU/g) and for spoiled (>106 CFU/g)	Daszczuk et al. (2014)
	Xanthine concentration (in fish meat)	Amperometric-based biosensors	Xanthine oxidase is immobilised on polymeric-mediated multi-walled carbon nanotubes, which have nanocomposite film, and measurement is done by electrochemical impedance spectroscopy	0.12 µM	Dervisevic et al. (2015)
Beverage (alcoholic/ non-alcoholic)	Ethanol (fermented beverage products)	Enzyme biosensor (alcohol oxidase–based biosensors)	Alcohol oxidase enzyme (catalyst) is used to detect ethanol, which is coupled with PANI, which has the optical properties for a visual sensor; hence, the detection of alcohol can be done by the naked eye as there is a change in colour to blue from green	0.001%	Kuswandi et al. (2014)
	L-lactate in apple juice	Based on current produced (amperometric)	L-lactate is immobilised on a screen-printed electrode (biochips), and on oxidation of L-lactate, pyruvate is produced along with H_2O_2; the produced H_2O_2 oxidises the surface of the electrode resulting in the generation of current that is proportional to the concentration of L-lactate		Przybyt et al. (2010)

In the current and future, development of biosensors may include high reproducibility and sensitivity, rapid performance and specificity. The biosensors can be used to detect any compound of interest, which can be toxin, bacterial antigen, adulterant or the precursor of spoilage or microbial contaminants.

16.3 Classification of Biosensors

The notion of biosensors was first reported in the mid-90s by Clark and Lyons and it is being used diversely in numerous fields of science and technology, viz, immune sensors, tissue-based biosensors, thermal piezoelectric biosensors, enzyme-based biosensors and DNA biosensors. Immobilisation procedures have been long exploited in the field of enzyme biosensing. Plants and animals were primarily utilized for making sensors to determine arginine, an amino acid in various substrates.

To be known, antibodies are distinctive in nature and show high affinity against respective antigens. Keeping this fact in mind, the immunosensors were entrenched as these antibodies bind specifically with toxins or pathogens or with the components of the host immune system to generate a response. The use of DNA-based sensors crafted on their property of recognising the complimentary strand when nucleic acid of a single-strand molecule is present in the sensor gets in phase with the sample. The stable hydrogen bond formation within the two nucleic acid molecules is the reason for these interactions. The use of the magnetoresistance effect for detection of nano- and magnetic microparticles in the microfluidic channel in case of magnetic biosensors provides greater size and sensitivity. Nanomaterial-based biosensors are depicted in Figure 16.1. Piezoelectric biosensors like quartz crystal microbalance and the surface acoustic wave devices are based on changes in the resonating frequency of the piezoelectric crystal due to the variation in mass of the of crystal. Optical biosensors use a number of components as a light source for the generation of a light beam with a definite frequency focused straightaway onto

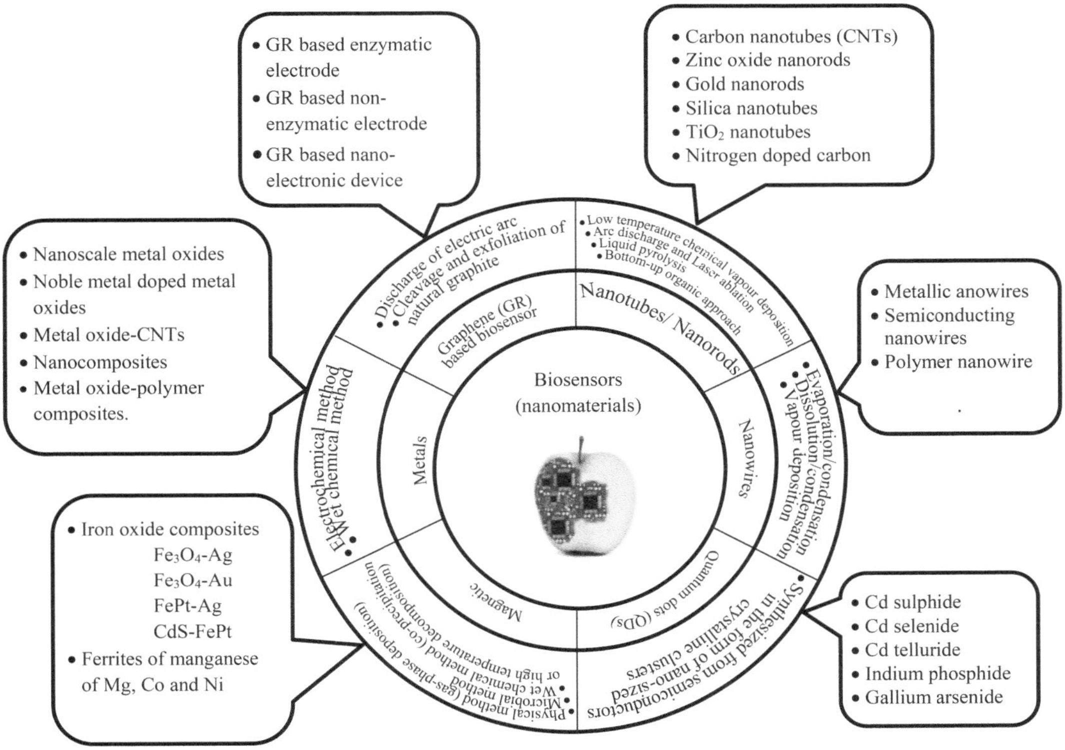

FIGURE 16.1 Nanomaterials used for fabrication of biosensors, their production technology and diversity.

the modulating agent, which is further attached with the photo detector for modified sensing. Brief classification, working principles and the most commercial applications of biosensors are tabulated and discussed in Table 16.2. Generally due to the specificity and sensitivity of enzymes, they have been used widely among these biological elements. Human, animal and plant cells are more difficult to manipulate compared with microbial cells, which simplify the fabrication process and have better stability and viability leading to a boost in the biosensor's performance.

Microbes are host to a wide range of enzymes that are responsible for sensing various chemicals, so these enzymes can be applied as biosensors to sense chemicals that are generated as a part of a reaction. Because of the recent advances in genetic and molecular biological techniques, substrate-specific microbes with desired metabolisms can be obtained by tweaking DNA. A transducer must be efficiently associated with microbes for an effective transformation of biochemical signal to physical signal; this is achieved by immobilising micro-organisms on transducers.

TABLE 16.2

Classification of Biosensors and Their Applications

		Working	Application	Reference
Electrochemical biosensor	Amperometric	Based on the current recorded (due to oxidation or reduction of metabolic product) and concentration of target compound	Detection of glucose, vitamin B_{12}, *Salmonella pullorum*, benzophenone	Su et al. (2011); Li et al. (2012); Liu et al. (2013); Ovalle et al. (2015); Hassan et al. (2016)
Electrochemical biosensor	Potentiometric	Device measures the electrical potential between the working electrode and the reference electrode	Detection of cadmium, *Escherichia coli*, toxins and fumaric acid	Santini et al. (2012); Li et al. (2016); Zhang et al. (2018); Shaibani et al. (2018)
Electrochemical biosensor	Conductimetric	Measures the conductivity of solution	Detection of heavy metal ions, aflatoxin B_1, alkaline phosphatase activity (APA) analysis, *E. coli*, *Salmonella* and O157:H7	Su et al. (2011); Soldatkin et al. (2013); Adley and Ryan (2015); Kumar and Neelam (2016)
Electrochemical biosensor	Microbial fuel cell	Electronic pathways between electrodes and bacteria	Estimation of biochemical oxygen demand, metals, metabolites like ethanol, glucose, hormones, lactate, L-lysine, vitamin B_1 and urate and in wastewater treatment in a potato processing unit	Abrevaya et al. (2014); Pacheco et al. (2017); Gutiérrez et al. (2017); Gonchar et al. (2017); Ismail and Radeef (2018)
Optical microbial sensors	Fluorescent microbial biosensor	The fluorescent microbial biosensor is based on the fuse of an inducible promoter and reporter gene and further fluorescent protein is encoded; this emits detectable fluorescence in a genetically modified organism (GMO)	Detection of pathogens (*S. typhimurium*, *Salmonella* and *E. coli*)	Bhardwaj et al. (2017)
Optical microbial sensors	Bioluminescent microbial biosensor	Bioluminescent microbial sensors measure bioluminescence emitted by living cells (wild or GMO) as a response when there is a change in the concentration of analytes	Heavy metals estimation like mercury	Brányiková et al. (2020)
Optical microbial sensors	Colorimetric microbial biosensor	Bacterium hydrolysed methyl parathion into chromophoric products (p-nitrophenol) which can be detected by the colorimetric method	Detection of fish spoilage compounds like dimethylamine, trimethylamine, cadaverine and putrescine	Morsy et al. (2016)

16.3.1 Electrochemical Biosensor

In electrochemical biosensors, the chemical reaction between target analyte and immobilised biomolecules causes the generation of electrons and ions. This leads to the variation in the electrical properties of the solution, which serves as the basic principle of electrochemical biosensors. There are four types of electrochemical transducers, i.e., amperometric, potentiometric, conductimetric and impedimetric. The electrochemical processes are low-cost technologies with quick detection and high sensitivity; portable, low-power-consuming and compatible with modern microfabrication (Diculescu et al. 2016). The interference with electrochemical reaction of the species instead of the analyte is a major reported drawback of electrochemical biosensors (Hassan et al. 2016). E-nose and e-tongue technology is one technique which is developed by applying various amperometric and potentiometric chemical sensors that help to analyse various sensory parameters like fruity, sweet, burnt, bitter, sour, artificial, caramel, body and intensity along with determination of polyphenols (Kraujalytė et al. 2015).

16.3.1.1 Amperometric Microbial Biosensor

The working of amperometric biosensors is based on the given potential between the references electrode and working electrode, and current generated relates with target analyte concentration. Amperometric biosensors have been widely utilised for various applications in the health, fermentation industry and environmental applications. These biosensors are helpful in the detection of glucose, which is important in quality control during the fermentation process and is of great interest for treating diabetes; it comprises nearly 85% of the whole biosensor market (Hassan et al. 2016). The two pairs of synthetic colours used in food products, sunset-yellow and brilliant blue or sunset-yellow and tartrazine (SY-BB OR SY-TT), were estimated by using a boron-doped diamond electrode with multiple pulse amperometric and coupling flow injection analysis (FIA) with a single-line flow injection system.

16.3.1.2 Potentiometric Microbial Biosensor

Potentiometric microbial biosensors possess an immobilised microbial layer, which is coated on the gas-sensing electrode or ion-selective electrode. The analyte is consumed by a micro-organism, leading to changes in potential, which is due to depletion or accumulation of ions. Zhang et al. (2018) developed a potentiometric biosensor for fast detection of cadmium content in rice samples. Shaibani et al. (2018) used a potentiometer biosensor for fast detection of *Escherichia coli* from a complex food sample like orange juice. Nanofiber-light addressable potentiometric sensors (NF-LAPSs) are potable and able to detect bacteria in less than 60 minutes. Li et al. (2016) successfully developed immunoassay-based potentiometric sensors for estimation of toxins from peanut-like B_1 aflatoxin.

16.3.1.3 Conductimetric Biosensor

Conductimetric biosensors examine the conductance property of solution to estimate the rate of reaction (Parmin et al. 2019). The conductimetric sensors are cheap thin films and are extremely sensitive as they combine or immobilise the biological component for detection. Various intentional and unintentional food ingredients like sucrose, urea, glucose, trichloroethylene, lactate, heavy metals, cyanide, ethanol, pesticides, creatinine, formaldehyde and nitrite have been detected by using conductimetric biosensors. Micro-organisms like *E. coli* 0157: H7 and *Salmonella* have been successfully detected using a conductimetric biosensor with limits of detection (LOD) of 81 CFU/mL and within 10 minutes (Kumar and Neelam 2016). Poor specificity and complex conditions of experiments like low dissolution of ingredients and maintaining the concentration of buffer solution are regarded as the major research areas in the field of immunosensors (Melo et al. 2016).

16.3.1.4 Microbial Fuel Cell Type Biosensor

An anaerobic biofilm is present on the anode in an anodic chamber, which causes the extraction of electrons by the digestion of organic carbon. The electrogenic biofilm on the anode passes the electrons to the cathode by an external load; anodes have microbes that are able to survive the change in the electrochemical gradient that can induce electric power. The electricity is induced due to metabolic activity of microbes. Microbial fuel cell biosensors have advantages like long-term stability, frequent response and portability. The yeast-based microbial sensors, by using genetically modified or engineered microbes, have wide spectrum diagnostic applications for the estimation of metabolites like ethanol, glucose, hormones, lactate, L-lysine, vitamin B_1 and urate. Food applications are ethanol estimation in alcoholic beverages by *Saccharomyces cerevisiae*, L -arginine in wines and juices by *Hansenula polymorpha*, L -Lactic acid in milk by *S. cerevisiae* and cyanide in fruit brandies by *S. cerevisiae*. Recently, probiotics like bifidobacteria and lactic acid bacteria (LAB) have been used for biosensors which are selective for the determination of bioagents and chemicals. The electrodes of the biosensors can be modified to improve results in the performance of microbial fuel cells in the treatment of wastewater and generation of electricity by using various methods like sterilisation/autoclave, sonicator or nano-sized materials like oxides of iron and gold (Kumar et al. 2018).

16.4 Optical Microbial Biosensor

The construction of optical biosensors is based on optical properties such as bioluminescence and chemiluminescence, UV-Vis absorption, fluorescence and reflectance. Bacterial detection is directly possible through these biosensors, which are also able to determine the small changes in refractive indices. These sensors are able to detect the small change in refractive indices, when binding of cells with receptors are immobilised on receptors. These biosensors have been used to monitor the product quality during production and to detect any potential contamination in the food material.

16.4.1 Bioluminescence Biosensor

When living organisms produce and emit light is called bioluminescence, which has been observed in insects, fish, bacteria, fungi, a variety of marine invertebrates and plants. The variation in the bioluminescence functions ranges from predation, defence and communication to metabolism, as a terminal oxidase. The generalised process of bioluminescent light emission includes the internal biochemical reaction that leads to the generation of excited state product that emits photons on relaxation. Biosensors based on bioluminescence are widely applicable in the estimation of heavy metal ions, phosphorous, heavy metals, genotoxicants, chlorophenols and naphthalene.

16.4.2 Fluorescence Biosensor

These biosensors are made from two components, a receptor and a transducer. The receptor component captures the target ligand and the latter one converts the event of ligand binding into measurable signals such as chemiluminescent, fluorescent, magnetic, colorimetric and electrochemical responses. Due to the high selectivity and sensitivity of fluorescence detection, it is the most widely used method currently in biomolecular imaging and spatial resolution, with a sufficient temporal and low cost for use.

16.4.2.1 Microbial Metabolism-Based Biosensors

Micro-organisms are able to transduce their metabolic redox reaction to quantify electric signals by using the oxidoreductase reaction and a mediator. The presence of polysaccharides and proteins in the outer layer of micro-organisms helps in bioassay development for bacterial detection. In the case of fluorescent immunoassay, for the labelling of immunoglobulin fluorochrome, molecules are used, which absorb the light of shorter wavelengths but emit consecutively higher range light that is further detected by using fluorescent microscopy. Fluorescein isothiocyanate (FITC), the original fluorescein molecule with a functional group of isothiocyanate by replacing hydrogen atom; bovine serum albumin (BSA), a cow serum albumin protein; and rhodamine isothiocyanate (RITC), is an organic fluorescence dye, are used for anthrax detection as it is considered as the most useful fluorochrome.

16.4.2.2 Green Fluorescence Protein (GFP)-Based Biosensor

Numerous organisms contain green fluorescent proteins (GFPs). It is exclusively found in jellyfish *Aequorea aequorea* (also known as *A. forskalea and A. victoria*). *Aequorea* GFP was first to be cloned, expressed and further used for tracer studies. These types are easy to engineer, are user friendly, manipulate and transfer into cells and can be further used for assessment of compounds like toluene, arsenite, galactosides and other related compounds. This system is used because of its advantageous features which include low metabolic toxicity, long-lasting signals, no interfering background fluorescence and most importantly no requirement of adding an exogenous factor. A most interesting and important feature of GFP based on the formation of chromophore resulted from a rare process of peptide autocatalytic posttranslational cyclisation from its own backbone structure. In the initial phase of autocatalytic cyclisation of GFP, it was considered to be unique, but the recently conducted research indicates that a family of enzymes (histidine ammonia lyase [HAL] and phenylalanine ammonia lyase [PAL]) also constitutes formations of a posttranslational ring that autocatalytically occurs when protein backbone attacks itself.

16.4.2.3 O_2-Sensitive Fluorescent Material-Based Biosensor

For the determination of biochemical oxygen demand (BOD), which represents volume of O_2 consumed by microorganisms during decomposition of the organic matter under conditions at a specific temperature fibre-optical sensors for microbial determination have been recently reported. These biosensors are constituted either from a sol-gel matrix along with a quenching indicator based on oxygen fluorescence with a linear range of 4–200 mg/L or a layer of material made from oxygen-sensitive fluorescence developed by immobilisation of seawater micro-organisms in polyvinyl alcohols or *Pseudomonas putida* membrane (immobilised) attached through a sensor principled on optical fibres for the detection of oxygen (limit of 0.5 mg/L) from ASR Co. Ltd.

16.4.3 Calorimetric Biosensor

Calorimetric biosensors are also known as thermometric or thermal biosensors.They measure the change in temperature of solutions containing analyte which is then followed by the action of the enzyme and the results are interpreted as a concentration of analyte in the solution. The solution of analyte is passed through an enzyme immobilised in a small packed bed column, the temperature detection is followed just before and after the entry and exit of the solution into the column via thermistors separately used according to the purpose. These biosensors are extensively used in the detection of pesticides and other enzymatic reactions.In the food quality analysis field, such biosensor usage to detect metabolites has been defined.

16.5 Other Types of Microbial Biosensors

16.5.1 Sensors Based on Baroxymeter for the Detection of Pressure Change

Another type of biosensor is a baroxymeter which acts on the fluctuation in the respiration of micro-organisms for the water toxicity measurement known as bacterial respirometry. The oxygen utilisation rate by the metabolising micro-organisms can be monitored by electrolytically, monometrically or by directly measuring the oxygen dissolved. The micro-organism's oxygen uptake leads to a pressure drop in the headspace of a vessel (closed) when the instrument is in contact with the sample.

16.5.2 Sensors Based on Infrared Analyzer for the Detection of the Microbial Respiration Product CO_2

A new method to measure inhibitory effects in wastewater treatment plants is based on a continuous measurement of the microbial respiration product (CO_2). Heavy metals and non-metallic wastes (phenols, ammonia, cyanide, arsenic, mercury, cadmium and copper) may have toxic effect or inhibit a biological treatment process. Microbial sensors can detect the toxic compounds present in a sample by a lowering of CO_2 concentration due to inhibition of microbial activity respiration. At the start, the microbes are given a nutritious solution to obtain baseline measurements for CO_2 production (from microbial respiration) as it acts as a substrate for the activated sludge microbes. After the baseline is stabilised, the test sample (toxic substance) replaces the nutritious solution. The restricting effect of the substance on the micro-organism's respiration is estimated by continuously measuring the CO_2 concentration; it is known as the mean percentage reduction in concentration of CO_2.

16.6 Future Trends and Conclusion

Food contamination has remained an important impediment of traditional analytical techniques for efficient results and the least consequences. The need to find tools for various food applications like food quality monitoring has successfully stimulated developments in the testing methodologies. Considering the needs of environmental and product safety, biosensors have been a substitute with versatile applications ranging from monitoring the raw material to quality assurance in processing to safeguarding the healthy food supply to the consumers. Most generalised applications of biosensors have been in the fields of food adulteration, contamination, toxicity and overall food testing. The continuous amplification of technology in food industries for product verification, raw material monitoring, checking the freshness of product, etc. demands sensitive technologies with real-time output. The specificity, storage, operational and environmental stability help to choose the biological material and also the detection of analyte, which includes microbes, chemical compounds, nucleic acids, hormones and any other parameter, such as smell and taste, by biosensors. Biosensors developed so far are based on DNA, enzymes, receptors, antibodies, organelles, micro-organisms and the living tissues of animals and plants. Major attributions for a good biosensing system implies reliability, specificity, portability, sensitivity, real-time analysis and simplicity of operation. The impeccable specific excellent sensitivity of laser-based optical detection and biological recognition probes are the recent developments that make biosensing capable of detection and differentiation of big/chemical constituents of complex systems and to have explicit identification and accurate quantification. Lab-on-chip interfacing with biosensor systems is increasingly in demand. Despite the complex functioning, the handling of biosensors should be made simple so that anyone can use it without the help of qualified persons. Biosensor research should be encouraged and supported by governments by investing in the infrastructure and research and development activities in order to achieve the goal of healthy food delivery.

REFERENCES

Abrevaya, Ximena C., Natalia J. Sacco, Maria C. Bonetto, Astrid Hilding-Ohlsson, and Eduardo Cortón. "Analytical applications of microbial fuel cells. Part I: Biochemical oxygen demand." *Biosensors and Bioelectronics* 63 (2015): 580–590.

Adley, C. C., and M. P. Ryan. "Conductometric biosensors for high throughput screening of pathogens in food." In *High Throughput Screening for Food Safety Assessment*, pp. 315–326. Woodhead Publishing, 2015.

Bhardwaj, Neha, Sanjeev K. Bhardwaj, Manoj K. Nayak, Jyotsana Mehta, Ki-Hyun Kim, and Akash Deep. "Fluorescent nanobiosensors for the targeted detection of foodborne bacteria." *TrAC Trends in Analytical Chemistry* 97 (2017): 120–135.

Brányiková, Irena, Simona Lucáková, Gabriela Kuncová, Josef Trögl, Václav Synek, Jan Rohovec, and Tomáš Navrátil. "Estimation of Hg (II) in soil samples by bioluminescent bacterial bioreporter *E. coli* ARL1, and the effect of humic acids and metal ions on the biosensor performance." *Sensors* 20, no. 11 (2020): 3138.

Bratakou, Spyridoula, Georgia-Paraskevi Nikoleli, Christina G. Siontorou, Stephanos Karapetis, Dimitrios P. Nikolelis, and Nikolaos Tzamtzis. "Electrochemical biosensor for naphthalene acetic acid in fruits and vegetables based on lipid films with incorporated auxin-binding protein receptor using graphene electrodes." *Electroanalysis* 28, no. 9 (2016): 2171–2177.

Cesarino, Ivana, Fernando C. Moraes, Marcos R. V. Lanza, and Sergio A. S. Machado. "Electrochemical detection of carbamate pesticides in fruit and vegetables with a biosensor based on acetylcholinesterase immobilized on a composite of polyaniline–carbon nanotubes." *Food Chemistry* 135, no. 3 (2012): 873–879.

Daszczuk, Alicja, Yonathan Dessalegne, Ismaêl Drenth, Elbrich Hendriks, Emeraldo Jo, Tom van Lente, Arjan Oldebesten et al. "*Bacillus subtilis* biosensor engineered to assess meat spoilage." *ACS Synthetic Biology* 3, no. 12 (2014): 999–1002.

Dervisevic M., E. Cevik, and M. Şenel. Development of glucose biosensor based on reconstitution of glucose oxidase onto polymeric redox mediator coated pencil graphite electrodes. *Enzyme and Microbial Technology* 68 (2015): 69–76.

Dias, Luís G., Andreia Fernandes, Ana C. A. Veloso, Adélio A. S. C. Machado, José A. Pereira, and António M. Peres. "Single-cultivar extra virgin olive oil classification using a potentiometric electronic tongue." *Food Chemistry* 160 (2014): 321–329.

Diculescu, Victor Constantin, Ana-Maria Chiorcea-Paquim, and Ana Maria Oliveira-Brett. "Applications of a DNA-electrochemical biosensor." *TrAC Trends in Analytical Chemistry* 79 (2016): 23–36.

Gil, Luis, José M. Barat, Diana Baigts, Ramón Martínez-Máñez, Juan Soto, Eduardo Garcia-Breijo, M-Concepción Aristoy, Fidel Toldrá, and Eduard Llobet. "Monitoring of physical–chemical and microbiological changes in fresh pork meat under cold storage by means of a potentiometric electronic tongue." *Food Chemistry* 126, no. 3 (2011): 1261–1268.

Gonchar, Mykhailo, Oleh Smutok, Maria Karkovska, Nataliya Stasyuk, and Galina Gayda. "Yeast-based biosensors for clinical diagnostics and food control." In *Biotechnology of Yeasts and Filamentous Fungi*, pp. 391–412. Springer, Cham, 2017.

Gutiérrez, Juan Carlos, Francisco Amaro, and Ana Martín-González. "Microbial biosensors for metal (loid) s." In *Microbial Ecotoxicology*, pp. 313–336. Springer, Cham, 2017.

Hao, Xiu-Juan, Xiao-Hong Zhou, Yan Zhang, Lan-Hua Liu, Feng Long, Lei Song, and Han-Chang Shi. "Melamine detection in dairy products by using a reusable evanescent wave fiber-optic biosensor." *Sensors and Actuators B: Chemical* 204 (2014): 682–687.

Hassan, Sedky H. A., Steven W. Van Ginkel, Mohamed A. M. Hussein, Romany Abskharon, and Sang-Eun Oh. "Toxicity assessment using different bioassays and microbial biosensors." *Environment International* 92 (2016): 106–118.

Ibrišimović, Nadira, Mirza Ibrišimović, Aldina Kesić, and Fritz Pittner. "Microbial biosensor: a new trend in the detection of bacterial contamination." *Monatshefte für Chemie-Chemical Monthly* 146, no. 8 (2015): 1363–1370.

Ismail, Zainab Ziad, and Ahmed Yasir Radeef. "Biotreatment of actual potato chips processing wastewater with electricity generation in microbial fuel cell." *Journal of Engineering* 24, no. 12 (2018): 26–34.

Izadi, Zahra, Mahmoud Sheikh-Zeinoddin, Ali A. Ensafi, and Sabihe Soleimanian-Zad. "Fabrication of an electrochemical DNA-based biosensor for *Bacillus cereus* detection in milk and infant formula." *Biosensors and Bioelectronics* 80 (2016): 582–589.

Kraujalytė, Vilma, Petras Rimantas Venskutonis, Audrius Pukalskas, Laima Česonienė, and Remigijus Daubaras. "Antioxidant properties, phenolic composition and potentiometric sensor array evaluation of commercial and new blueberry (*Vaccinium corymbosum*) and bog blueberry (*Vaccinium uliginosum*) genotypes." *Food Chemistry* 188 (2015): 583–590.

Kumar, Harish, and Rani Neelam. "Enzyme-based electrochemical biosensors for food safety: a review." *Nanobiosensors in Disease Diagnosis* 5 (2016): 29–39.

Kumar, Ravinder, Lakhveer Singh, A. W. Zularisam, and Faisal I. Hai. "Microbial fuel cell is emerging as a versatile technology: a review on its possible applications, challenges and strategies to improve the performances." *International Journal of Energy Research* 42, no. 2 (2018): 369–394.

Kuswandi, Bambang, Titi Irmawati, Moch Amrun Hidayat, and Musa Ahmad. "A simple visual ethanol biosensor based on alcohol oxidase immobilized onto polyaniline film for halal verification of fermented beverage samples." *Sensors* 14, no. 2 (2014): 2135–2149.

Li, Haidong, Huaimin Guan, Hong Dai, Yuejin Tong, Xianen Zhao, Wenjing Qi, Saadat Majeed, and Guobao Xu. "An amperometric sensor for the determination of benzophenone in food packaging materials based on the electropolymerized molecularly imprinted poly-o-phenylenediamine film." *Talanta* 99 (2012): 811–815.

Li, Qunfang, Shuzhen Lv, Minghua Lu, Zhenzhen Lin, and Dianping Tang. "Potentiometric competitive immunoassay for determination of aflatoxin B 1 in food by using antibody-labeled gold nanoparticles." *Microchimica Acta* 183, no. 10 (2016): 2815–2822.

Liu, Guoyan, Chunyan Chai, and Bing Yao. "Rapid evaluation of *Salmonella pullorum* contamination in chicken based on a portable amperometric sensor." *Journal of Biosensors and Bioelectronics* 4, no. 137 (2013): 2.

Melo, Airis Maria Araújo, Dalila L. Alexandre, Roselayne F. Furtado, Maria F. Borges, Evânia Altina T. Figueiredo, Atanu Biswas, Huai N. Cheng, and Carlúcio R. Alves. "Electrochemical immunosensors for *Salmonella* detection in food." *Applied Microbiology and Biotechnology* 100, no. 12 (2016): 5301–5312.

Morsy, Mohamed K., Kinga Zor, Nathalie Kostesha, Tommy Sonne Alstrøm, Arto Heiskanen, Hassan El-Tanahi, Ashraf Sharoba et al. "Development and validation of a colorimetric sensor array for fish spoilage monitoring." *Food Control* 60 (2016): 346–352.

Ovalle, M., E. Arroyo, M. Stoytcheva, R. Zlatev, L. Enriquez, and A. Olivas. "An amperometric microbial biosensor for the determination of vitamin B 12." *Analytical Methods* 7, no. 19 (2015): 8185–8189.

Pacheco, J. G., M. F. Barroso, H. P. A. Nouws, S. Morais, and C. Delerue-Matos. "Biosensors." In *Current Developments in Biotechnology and Bioengineering*, pp. 627–648. Elsevier, 2017.

Parmin, N. A., U. Hashim, Subash C. B. Gopinath, and M. N. A. Uda. "Biosensor recognizes the receptor molecules." In *Nanobiosensors for Biomolecular Targeting*, pp. 195–210. Elsevier, 2019.

Przybyt, Małgorzata, Jan Iciek, Agnieszka Papiewska, and Joanna Biernasiak. "Application of biosensors in early detection of contamination with lactic acid bacteria during apple juice and concentrate production." *Journal of Food Engineering* 99, no. 4 (2010): 485–490.

Ribeiro, Francisco Wirley Paulino, Maria Fátima Barroso, Simone Morais, Subramanian Viswanathan, Pedro de Lima-Neto, Adriana N. Correia, Maria Beatriz Prior Pinto Oliveira, and Cristina Delerue-Matos. "Simple laccase-based biosensor for fermetanate hydrochloride quantification in fruits." *Bioelectrochemistry* 95 (2014): 7–14.

Santini, Alberto Oppermann, Helena Redigolo Pezza, and Leonardo Pezza. "Development of a sensitive potentiometric sensor for determination of fumaric acid in powdered food products." *Food Chemistry* 134, no. 1 (2012): 483–487.

Shaibani, Parmiss Mojir, Hashem Etayash, Keren Jiang, Amirreza Sohrabi, Mahtab Hassanpourfard, Selvaraj Naicker, Mohtada Sadrzadeh, and Thomas Thundat. "Portable nanofiber-light addressable potentiometric sensor for rapid *Escherichia coli* detection in orange juice." *ACS Sensors* 3, no. 4 (2018): 815–822.

Soldatkin, O. O., O. S. Burdak, T. A. Sergeyeva, V. M. Arkhypova, S. V. Dzyadevych, and A. P. Soldatkin. "Acetylcholinesterase-based conductometric biosensor for determination of aflatoxin B1." *Sensors and Actuators B: Chemical* 188 (2013): 999–1003.

Su, Liang, Wenzhao Jia, Changjun Hou, and Yu Lei. "Microbial biosensors: a review." *Biosensors and Bioelectronics* 26, no. 5 (2011): 1788–1799.

Tolba, Mona, Minhaz Uddin Ahmed, Chaker Tlili, Fritz Eichenseher, Martin J. Loessner, and Mohammed Zourob. "A bacteriophage endolysin-based electrochemical impedance biosensor for the rapid detection of Listeria cells." *Analyst* 137, no. 24 (2012): 5749–5756.

Zhang, Wen, Yiwei Xu, and Xiaobo Zou. "Rapid determination of cadmium in rice using an all-solid RGO-enhanced light addressable potentiometric sensor." *Food Chemistry* 261 (2018): 1–7.

17

Advances in Microbes Use in Agricultural Biotechnology

Devki, Deepesh Neelam and Jebi Sudan
JECRC University

Deepak Kumar
University of Rajasthan

CONTENTS

17.1 Introduction

Plant growth is the spatial and temporal cell division and differentiation. This is efficiently regulated by the interaction between plant and associated micro-organisms. The micro-organism utilises nutrients from the host which in turn produces phytohormones or other compounds that are key determinants of plant growth, development, yield and immunity. Increasing the population and limiting the agriculture area demands novel strategies for crop quality and production improvement. Among other approaches, the microbial use in crop improvement has a vast area of opportunities. The microbiome, which is an important fraction of the soil and has various interactions with plants, is a field to explore. Microbes are associated with plants and help them with growth, production and survival in climatic conditions. Microbes such as *Azospirillum, Alcaligenes, Arthrobacter, Acinetobacter, Bacillus, Paenibacillus, Burkholderia, Enterobacter, Erwinia, Flavobacterium, Methylobacterium, Pseudomonas, Rhizobium* and *Serratia* can interact with plants by their roots (rhizospheric microbe). Some microbes reside inside the plant tissue and may be found within the cells or intracellular spaces. These microbes enter the plant via the wounds on plant tissues and can remain localised there or spread throughout the plant without causing them any harm. This chapter highlights the present scenario of microbe involvement in plant growth and development and the need for the development of microbial inoculants that are used along with fertiliser to maintain sustainable agriculture.

17.2 The Interaction of Microbes with Plants

The human population is increasing very fast, which causes challenges in agricultural science to develop different new technologies for food security and sustainable development under deteriorating global climate changes, shrinkages in land area, industrialisation, urbanisation and the use of chemicals in agriculture which mainly affects crop production in all nations (Hamilton et al. 2016; Vimal et al. 2017). To overcome these hurdles in the agricultural field, microbes increase the crop yield productivity with fewer uses of the resources from the soil. In stress, microbes play a very important role in plants because they are associated with different parts of the plant, i.e., roots, stem, leaves, fruits, etc. either symbiotically or free living. Plants interact with environmental stresses directly in different ways like salinity, drought, flooding, freezing, radiation effects, nutrient imbalance, heavy metal accumulation, high temperature and so forth. Prolonged stress causes several deformities in the plants, i.e., epinasty, abscissions, ion toxicity, desiccation, defoliation, senescence, electrolyte leakage, hormonal imbalance and destabilisation in the membrane proteins that limit plant productivity (Xu et al. 2016).

Endophytes are the microbes that reside in healthy plant tissues without causing any damage to the host; they were discovered in 1866. They colonise ranges such as the rhizosphere (wounds caused by abrasion, pathogenic damage, lateral root formation and micropores), phyllosphere (hydathodes, stomata and lenticels), laimosphere (stems), carposphere (fruits), spermosphere (seeds) and anthosphere (flowers), and are transmitted via seeds and vertical transmission (Lata et al. 2018; Brader et al. 2017; Shahzad et al. 2018). Seed-borne endophytes help in plant growth and defence, mitigate the stress and pass to the next generation in seedling form by vertical transmission. Endophytes synthesise the antistress biochemical substances, which help to cope with various stress conditions of the plant. Some of the compounds synthesised are indole acetic acid (IAA), cytokines (CKs), gibberellins (GAs), essential vitamins and phosphate solubilisation for promoting plant growth which helps during nutrient stress of the plants (Shahzad et al. 2016; Shade et al. 2017). During drought stress, endophytes increase the level of solute accumulation in tissues, which affects decreased transpiration rate and reduces water conductance in leaves or by the formation of the thick cuticle. Osmotic stress responses result in triggering the secondary signals by releasing abscisic acid (ABA), ethylene, reactive oxygen species (ROS) and intra-secondary messengers, i.e., phospholipids that regulate the ascent of sap in the shoot system of the plant (Lata et al. 2018).

Plant growth-promoting rhizobacteria (PGPRs) help with productivity and crop yield by secreting various exopolysaccharides (EPS), phytohormones, 1-amino-cyclopropane-1-carboxylate (ACC), siderophores, hydrogen cyanide (HCN), different antibiotics and organic volatile compound production which promotes the growth of the plant in stress conditions (Zimmer et al. 2016). PGPRs interact with the N_2 fixers, which help improve the nitrogen concentration in the soil. Arbuscular mycorrhizal fungi (AMF) have a special role in the soil by increasing the antioxidant system, helping phosphate solubilisation and enhancing the osmotic and stress tolerance capacity, while, along with PGPRs, they restore micronutrients in the degraded soil profile, working as an antagonist in biocontrol of the root pathogens, improved phytoremediation, enhancing the nutrient uptake and so forth (Zhipeng et al. 2016; Rashid et al. 2016).

17.3 The Role of Microbes in Biopesticides

In India, every year there is a 30% loss of crop yield caused by arthropod groups and from diseases and weeds which destroy grains. For a long time, damage imposed on crops by pests lead to a serious problem in agricultural fields and practices. For controlling the pests, diverse measures have been personalised for the primary planning that worked to eliminate the pests by using chemical pesticides in developing countries. Due to the harmful effects of chemical pesticides, biopesticides are being developed that are suitable for removing different species of pests in agriculture fields (Nawaz et al. 2016; Gouda et al. 2017).

Biopesticides are the chemicals derived from natural substances such as animals, plants and bacteria (Kachhawa 2017). Generally, these are categorised as microbial pesticides, plant-included-protectants and biochemical pesticides, which can be used for pest management by controlling insects and disease-causing pathogens due to which they are non-hazardous to the environment and health of humans (Singh 2019). In the agricultural sector, the purposes of biopesticides are to control insects, diseases, weeds and nematodes and to improve plant physiology and crop productivity (Gouda et al. 2017; Mishra et al. 2020). A microbial pesticide includes the bacterium, fungus, virus or protozoans that provide a defence against different types of pests like flies, beetle larvae, caterpillars, fungus, bacterial diseases, soilborne pathogens and so forth. Some of the genera studied so far as microbial pesticides include '*Beauveria bassiana*,

Metarhizium anisopliae, Nomuraea rileyi, Paecilomyces farinosus and *Verticillium lecanii*'. *Azotobacter* species reported the secretion of chemical compounds such as phytohormones, vitamin B complex, and other bioactive compounds that work as biopesticides (Mahanty et al. 2017). For the sustainable development in the agricultural field, some of the bacterial biopesticides such as *Bacillus, Agrobacterium,* and *Pseudomonas* on the other hand fungal biopesticides such as *Ampelomyces, Fusarium, Gliocladium, Trichoderma, Beauveria, Metarhizium, Paecilomyces, Penicillim, Verticillium* and *Pythium* have great efficacy (Umesha et al. 2018; Lengai and Muthomi 2018).

Virus-based biopesticides include baculoviruses (BVs) andnuclear polyhedrosis viruses (NPVs). NPVs include '*Helicoverpa armigera, Amsacta moorei, Agrotis ipsilon, A. segetum, Anadividia peponis, Trichoplusia orichalcea, Adisura atkinsoni, Plutella xylostella, Corcyra cephalonica, Mythimna separata* and *Phthorimaea operculella*'. Granuloviruses (GVs) are reoviruses, and cytoplasmic polyhedrosis viruses (CPVs) include nodaviruses, picorna-like viruses, tetraviruses and so forth. They are used for controlling gypsy moths, pine sawflies, Douglas fir tussock moths and pine caterpillars and are actively used in agricultural land to control cabbage moths, corn earworms, cotton leafworms and bollworms, celery loppers and tobacco budworms (Dhir 2017; Ruiu 2018; Mishra et al. 2020).

Plant-incorporated protectants (PIPs) are substances developed by genetic incorporation in plants, which are effective against pest and insects by targeting insects or providing resistance. For insect management, different plant-derived products from custard apple, neem, tobacco, pyrethrum and so forth are used. Across the globe, some of the chemical substances like limonene, pyrethrum/pyrethrins, rotenone, sabadilla and ryania are utilised widely to control fleas, aphids and mites; ants; roaches; ticks; beetles; caterpillars and thrips; squash bugs; harlequin bugs and so forth (Nawaz et al. 2016). Biochemical pesticides are extracts of either plants or animals which consist of secondary metabolites such as pheromones, leave extracts, juices, latex and so forth that are effectively used in controlling pests. Extracts obtained from *Madhuca longifolia, Derris scandals,* and *Eupholsia antignomum* and pineapple have a nauseating smell effective against insects. An amalgamation of Kohomba, cinnamon and *Croton lexifenio* keeps away worms when it is mixed with water and sprinkled on crops (Kumari et al. 2016). Biochemical pesticides function by interfering with mating in the breeding cycle which helps in reducing the population of insects.

Bacillus thuringiensis (*Bt*) is one of the microbial insecticides widely used as commercial sources of a bacterial group. The insecticidal properties of this bacterium are the biosynthesis of the crystal protein toxins (Cry and Cyt) which release parasporal bodies during the sporulation phase and virulence factors during the vegetative phase of growth (VIP). These toxins paralyze the mid-gut by binding specific receptors that trigger the pore-forming process, altering the permeability of the epithelial membrane of the intestine as a result of disruptions leading to insect killing within 24–48 hours, which prevents further feeding. *Bt* toxin is used for controlling caterpillars/larvae moths, mosquitoes and black flies that feed on cabbage, potatoes and other crops. Commercial preparations of *B. thuringiensis* contain a mixture of spores, Cry protein and an inert carrier that controls the diamondback moths, *Helicoverpa* and *Trichoplusia*, on cotton, pigeon pea, tomato and so forth. (Umesha et al. 2018). Cry proteins cause osmotic cell lysis that forms the pores in ion channels in the membrane. Also, cell death in insect cells is performed by an adenylyl cyclase/protein kinase (PKA) signalling pathway promoted by Cry toxin monomers. Similarly, *Beauveria bassiana, B. tabaci* and *Metarhizium anisopliae* produce toxic metabolites known as beauvericin and destruxin and so forth, which cause the death of adult *Aedes aegypti* and *A. albopictus* mosquitoes.

Entomopathogenic nematodes are soft-bodied, non-segmented roundworms that present naturally in soils and are obligate or facultative parasites of insects. They locate their host in response to carbon dioxide, vibration and other chemical cues. Two families, i.e., Heterorhabditidae and Steinernematidae, are used in pest control as biological insecticides. The juvenile stage of nematodes penetrates the host insect via outer natural openings (mouth, spiracles, anus and inter-segmental membranes of the cuticle) and finally release into the hemocoel. *Heterorhabditis* and *Steinernema* both have mutual relationships with *Photorhabdus* and *Xenorhabdus* bacteria, respectively. The juvenile stage releases symbiotic bacterial cells from their intestines into the hemocoel, then bacterial cells divide in the insect hemolymph and the infected host usually dies within 24 to 48 hours. Nematodes continue to feed on the host tissue, mature and reproduce after the death of the host. A variety of commercialised products are used worldwide which target specific pest species such as Capsanem, Entonem, Larvanem, NemaTrident-CT, Slugtech-SP and so forth. They target borer beetles, *Bradysia, Otiorhynchus, O. sulcatus*, molluscs and so forth (Kenis et al. 2017; Sankaranarayanan and Askary 2017; Heriberto et al. 2017; Sulistyanto et al. 2018).

17.4 The Role of Microbes in Biofertilisers

In modern agricultural science, biofertilisers are the best tool for sustainable agriculture. In the last decades, the continuous use of chemical fertilisers in the agriculture field caused an imbalance in soil profiling that directly affects the sustainability which was overcome by manure. Biofertiliser is used in farming as an organic manure substitute. Organic manure is in compost form that is formed by household wastes, plant waste, farmyard waste and agriculture waste, which helps in maintaining the quality and sustainability of soil in the long run (Singh et al. 2016; Mohanty and Swain 2018).

Biofertiliser contains micro-organisms that provide an adequate supply of nutrients to the host plants and ensures the proper improvement of growth and regulation in physiology, which meets the immediate requirement of the crop. For the preparation of biofertiliser, living micro-organisms are used which have specific functions to increase plant growth, reproduction and yield. Biofertiliser plays a crucial role in organic farming by maintaining soil fertility and sustainability for a long time. Some of the commonly used biofertilisers in agricultural fields contain micro-organisms such as *Rhizobium*, *Azotobacter*, *Azospirillum*, cyanobacteria, *Azolla*, etc., which mainly help in fixing the gaseous nitrogen directly from the atmosphere into nitrogen compounds, an available form of nitrogen to the plants. *Rhizobium, Allorhizobium, Azorhizobium, Bradyrhizobium, Mesorhizobium*and *Sinorhizobium* belong to the family Rhizobiaceae with different genera; they fix the atmospheric nitrogen in ammonia form which is carried out by the complex enzyme nitrogenase consisting of dinitrogenase reductase with iron as its cofactor and dinitrogenase with molybdenum. This ammonia form of nitrogen is then used by plants for their growth (Verma and Srivastav 2017; Barman et al. 2017; Kumar et al. 2017). *Azolla* is associated with the N_2-fixing blue-green algae, *Anabaena azollae*, which can fix the atmospheric nitrogen. *A. azollae* is the most potent biofertiliser of nitrogen to rice. It can fix about 40–60 kg N/ha in a rice crop, which is an alternative source of nitrogen commonly used in Southeast Asia. *A. filiculoides* and *A. caroliniana* were effective as biofertilisers on cabbage production in Daloa (Kumar et al. 2017; Kour et al. 2020). Another free-living bacterium, *Azotobacter*, present in neutral and alkaline soils fixes the nitrogen in non-leguminous plants, i.e., cotton, rice and vegetables, without any specific host. Bioinoculants of *Azotobacter* may increase crop productivity by 10–12% and vigor in young plants and grain yield especially in wheat crops by producing vitamins, such as thiamine and riboflavin, and plant hormones, such as IAA, naphthalene acetic acid (NAA), GAs and CKs. It can fix 15–20 kg/ha of nitrogen per year (El-Lattief 2016).

Anabaena, Nostoc, Calothrix, and *Aulosira* are primitive forms of cyanobacteria which help in the fertility of the soil by providing humidity requirements, nutrient availability, light, temperature and water in paddy fields (Agarwal et al. 2018). *Azospirillum* is associated with different grasses as a symbiotic nitrogen fixer that promotes the growth of plants and changes in root morphology. Different strains of *Azospirillum* are sold as biofertilisers in various countries, including Africa, Argentina, Australia, Belgium, Brazil, Germany, France, India, Italy, Mexico and the United States (Kumar et al. 2017).

Phosphate solubilising microbes (PSMs) are well-known biofertilisers that promote plant growth and have the ability to convert the insoluble form of phosphorus to a soluble form by different mechanisms which include acidification, exchange reactions and chelation. Some of the PSMs studied so far include *Achromobacter xylosoxidans, Acinetobacter baumannii, Aeromonas hydrophila* (Martínez-Gallegos et al. 2018), *Arthroderma cuniculi, Aspergillus niger* (Gore and Navale 2017), *Bacillus aerius* (Singh et al. 2019), *B. altitudinis, B. amyloliquefaciens, B. licheniformis, B. megaterium, B. mucilaginous, B. subtilis, Burkholderia cepacia, Paenibacillus taichungensis, Pseudomonas koreensis, P. luteola, P. simiae* and *Serratia nematodiphila* (Qiu et al. 2019; Leite et al. 2018; Zhang et al. 2017), which are regularly used in agriculture field.

The group of bacteria that colonise roots or rhizosphere soil and are beneficial to crops are referred to as plant growth-promoting rhizobacteria (PGPR). The PGPR inoculants promote growth through suppression of plant disease (termed bioprotectants), improved nutrient acquisition (termed biofertilisers) or phytohormone production (termed biostimulants). Some of the symbiotic microbes of soil (bacteria and fungi) inhabiting the rhizosphere of many plants have diverse beneficial effects on the host plant by fixing nitrogen. Rhizospheric bacteria that colonise with roots and exert a beneficial effect on crops (PGPRs) include *Actinoplanes, Agrobacterium, Alcaligenes, Amorphosporangium, Arthrobacter, Azotobacter, Bacillus, Cellulomonas, Enterobacter, Erwinia, Flavobacterium, Pseudomonas, Rhizobium, Bradyrhizobium, Streptomyces* and *Xanthomonas*. Inoculants of PGPRs promote growth by the production of phytohormones such as IAA, CK, GA, siderophores, ACC deaminase and hydrogen cyanate, which help the plant to tolerate plant diseases and enhance nutrient acquisition. PGPRs act as biostimulants and encourage growth by increasing the absorptive surface for water uptake and nutrients (Liu et al. 2016). IAA produced by PGPRs induces the production of nitric oxide (NO), which acts as a second messenger to activate a complex signalling network, leading to enhanced root growth and developmental processes (Dhir 2017). Inoculants of PGPRs promote growth by degradation of pollutants, antibiotic or lytic enzyme production, phytohormone production, detoxification of heavy metal, salt tolerance and enhanced nutrient acquisition (Xie et al. 2016). Strains of PGPR are also effective against the spotted wilt viruses and cucumber mosaic virus in tomato/pepper and bunchy top virus in banana.

17.5 Microbes in Plant Growth for Sustainable Production

In general, micro-organisms associated with plants belong to three major groups: endophytic, phyllospheric and rhizopheric. Endophytes settle in the interior plant parts, the phyllosphere consists of the plant aerial region that is colonised by microbes and the rhizosphere consists of plant root parts that are a storehouse of microbes. These microbe associations are specific to a particular host plant and can promote plant growth by different mechanisms. Microbes help plants by fixing atmospheric N_2; producing various plant growth hormones; solubilising fixed phosphorus, potassium and zinc for plant use and producing substances that provide resistance against other pathogens (Yadav et al. 2017). The use of microbes along with agricultural practices proves to be a better alternative when compared with conventional technologies for maintaining sustainable crop production.

Microbes associated with plants produce various types of phytohormones such as auxins, cytokinins and GAs. These plant growth hormones stimulate numerous benefits to the host plant including stimulation of cell division and differentiation that leads to sufficient growth of the root system for enhanced absorption of water and nutrients and a shoot system for increased foliage growth. Apart from growth, auxin (mostly IAA) also helps in suppressing weeds, reducing pathogen attacks, encourages phytostimulation and is used by microbes to interact with the host plant. GAs mostly promote cell division, elongation and differentiation and help in alleviating seed dormancy during plant stress conditions. Cytokinins are also important for plant development as they promote seed germination, activate dormant buds and play an important role in the biosynthesis of nucleic acid and chlorophyll (Egamberdieva et al. 2017).

The insolubility of many nutrients in the soil is also a major cause of nutrient deficiency in plants which affects the growth and development of plants. However, some classes of microbes are capable of releasing the fixed phosphorus into the soil in an accessible form, which is freely available to the plants. The rhizospheric phosphate utilising bacteria can convert inorganic phosphorus into a plant-available form of orthophosphate. In the majority of crops, phosphate solubilisation is reported to be formed by a class of microbes including *Enterobacter, Halolamina, Citrobacter* and *Azotobacter* (Gaba et al. 2017). These plant growth-promoting (PGP) microbes possibly solubilise the available bound phosphate through two prominent methods: through the use of enzymes (phytases, C-P lyase and various phosphatases) or through the release of various organic acids such as gluconate, acetate, oxalate, citrate and glycolate (Ingle and Padole 2017). However, the organic acid released by the microbes depends on the substrate and glucose; sucrose supports the maximum phosphate solubilisation. Apart from phosphorus, various microbes are also able to solubilise the fixed potassium in soluble forms that are readily available for plant uptake. The potassium solubilisation process is through the production of organic acids, chelation and various exchange reactions. Various potassium solubilisation bacterial strains (*Bacillus, Paenibacillus*) were also able to solubilise silicon and aluminium from the fixed minerals (Verma et al. 2016).

Atmospheric N_2 fixation is another attribute contributed by microbes to plant development. Biological nitrogen fixation serves as the support to the nitrogen consumption of the crop plants that mostly depends on the N-containing fertilisers. This biological fixation also improves the soil condition and leads to more production and productivity over a longer period that is needed in the case of sustainable agriculture. Moreover, various microbial inoculants that fulfil about 30–50% of total crop nitrogen requirements have been reported to be applied in various non-leguminous crops (Yadav et al. 2017). Microbes are also able to mitigate the various stresses in plants by modulating the concentration of ethylene through cleaving the ACC, the precursor for plant-produced ethylene. Ethylene is commonly known as the plant stress hormone, which is produced during various abiotic and biotic stress conditions and negatively regulates plant growth. Some classes of microbes have the ability to secrete ACC deaminase, cleave the precursor of ethylene and facilitate normal growth and development in the plant.

17.6 Conclusion

The overburdened population puts a heavy load and great expectations on the production and productivity of crops. The target is to feed the population through the heavy use of fertilisers, pesticides and weedicides, which results in the deterioration of soil health over a long period. The microbial flora present in the vicinity of the plant roots or in the soil interacts with the plants and affects many physical, physiological and biochemical processes, which have a direct effect on the growth and development of plants. These microbe species also modulate the microenvironment near the root zone and help in the absorption of many minerals. These microbes may provide an alternative source for increasing crop production and productivity. Advanced research focused on the development of transgenic microbes that have specialised functions will be welcomed. The combined use of microbes with these chemicals may pave the way for future crop production focusing on sustainable agriculture.

REFERENCES

Agarwal, P., Gupta, R. and Gill, I.K. 2018. Importance of biofertilizers in agriculture biotechnology. *Annals of Biological Research* 9(3):1–3.

Barman, M., Paul, S., Choudhury, A.G., Roy, P. and Sen, J. 2017. Biofertilizer as prospective input for sustainable agriculture in India. *International Journal of Current Microbiology and Applied Sciences* 6:1177–1186.

Brader, G., Compant, S., Vescio, K., Mitter, B., Trognitz, F., Ma, L.-J., et al. 2017. Ecology and genomic insights into plant-pathogenic and plant-nonpathogenic endophytes. *Annual Review of Phytopathology* 55:61–83.

Dhir, B. 2017. Biofertilizers and biopesticides: eco-friendly biological agents. In *Advances in Environmental Biotechnology* (pp. 167–188). Springer, Singapore.

Egamberdieva, D., Wirth, S.J., Alqarawi, A.A., Abd_Allah, E.F. and Hashem, A. 2017. Phytohormones and beneficial microbes: essential components for plants to balance stress and fitness. *Frontiers in microbiology* 8:2104.

El-Lattief, E.A. 2016. Use of *Azospirillum* and *Azobacter* bacteria as biofertilizers in cereal crops: a review. *International Journal of Research in Engineering and Applied Sciences (IJREAS)* 6(7):36–44.

Gaba, S., Singh, R.N., Abrol, S., Yadav, A.N., Saxena, A.K. and Kaushik, R. 2017. Draft genome sequence of *Halolamina pelagic* CDK2 isolated from natural salterns from Rann of Kutch, Gujarat, India. *Genome Announcements* 5(6):e01593–01516.

Gore, N.S. and Navale, A.M. 2017. *In vitro* screening of rhizospheric *Aspergillus* strains for potassium solubilization from Maharashtra, India. *South Asian Journal of Experimental Biology* 6:228–233.

Gouda, S., Nayak, S., Bishwakarma, S., Kerry, R.G., Das, G. and Patra, J.K. 2017. Role of microbial technology in agricultural sustainability. In *Microbial Biotechnology* (pp. 181–202). Springer, Singapore.

Hamilton, C.E., Bever, J.D., Labbé, J., Yang, X. and Yin, H. 2016. Mitigating climate change through managing constructed-microbial communities in agriculture. *Agriculture, Ecosystems & Environment* 216:304–308.

Heriberto, C.M., Jaime, R.V., Carlos, C.M. and Jesusita, R.D. 2017. Formulation of entomopathogenic nematodes for crop pest control–a review. *Plant Protection Science* 53(1):15–24.

Ingle, K.P. and Padole, D.A. 2017. Phosphate solubilizing microbes: an overview. *International Journal of Current Microbiology and Applied Sciences* 6(1):844–852.

Kachhawa, D. 2017. Microorganisms as a biopesticides. *Journal of Entomology and Zoology Studies* 5(3):468–473.

Kenis, M., Hurley, B.P., Hajek, A.E. and Cock, M.J.W. 2017. Classical biological control of insect pests of trees: facts and figures. *Biological Invasions* 19:3401–3417.

Kour, D., Rana, K.L., Yadav, A.N., Yadav, N., Kumar, M., Kumar, V., Vyas, P., Dhaliwal, H.S. and Saxena, A.K. 2020. Microbial biofertilizers: bioresources and eco-friendly technologies for agricultural and environmental sustainability. *Biocatalysis and Agricultural Biotechnology* 23:101487.

Kumar, R., Kumawat, N. and Sahu, Y.K.2017. Role of biofertilizers in agriculture. *Pop Kheti* 5(4):63–66.

Kumari, S., Vaishnav, A., Jain, S., Choudhary, D.K. and Sharma, K.P. 2016. *In vitro* screening for salinity and drought stress tolerance in plant growth promoting bacterial strains. *International Journal of Agriculture and Life Sciences* 2:60–66.

Lata, R., Chowdhury, S., Gond, S.K. and White Jr, J.F. 2018. Induction of abiotic stress tolerance in plants by endophytic microbes. *Letters in Applied Microbiology* 66(4):268–276.

Leite, M.C.D.B.S., Pereira, A.P.D.A., Souza, A.J.D., Andreote, F.D., Freire, F.J. and Sobral, J.K. 2018. Bioprospection and genetic diversity of endophytic bacteria associated with cassava plant. *Revista Caatinga* 31(2):315–325.

Lengai, G.M. and Muthomi, J.W. 2018. Biopesticides and their role in sustainable agricultural production. *Journal of Biosciences and Medicines* 6(6):7.

Liu, W., Wang, Q., Hou, J., Tu, C., Luo, Y. and Christie, P. 2016. Whole genome analysis of halotolerant and alkalotolerant plant growth-promoting rhizobacterium *Klebsiella sp.* D5A. *Scientific Reports* 6(1):1–10.

Mahanty, T., Bhattacharjee, S., Goswami, M., Bhattacharyya, P., Das, B., Ghosh, A. and Tribedi, P. 2017. Biofertilizers: a potential approach for sustainable agriculture development. *Environmental Science and Pollution Research* 24(4):3315–3335.

Martínez-Gallegos, V., Bautista-Cruz, A., Martínez-Martínez, L. and Sánchez-Medina, P.S. 2018. First report of phosphate-solubilizing bacteria associated with *Agave angustifolia*. *International Journal of Agricultura and Biology* 20(6):1298–1302.

Mishra, J., Dutta, V. and Arora, N.K. 2020. Biopesticides in India: technology and sustainability linkages. *3 Biotech* 10:1–12.

Mohanty, S. and Swain, C.K. 2018. Role of microbes in climate smart agriculture. In *Microorganisms for Green Revolution* (pp. 129–140). Springer, Singapore.

Nawaz, M., Mabubu, J.I. and Hua, H. 2016. Current status and advancement of biopesticides: microbial and botanical pesticides. *Journal of Entomological and Zoological Studies* 4(2):241–246.

Qiu, Z., Egidi, E., Liu, H., Kaur, S. and Singh, B.K. 2019. New frontiers in agriculture productivity: Optimised microbial inoculants and *in situ* microbiome engineering. *Biotechnology Advances* 37(6):107371.

Rashid, M.I., Mujawar, L.H., Shahzad, T., Almeelbi, T., Ismail, I.M. and Oves, M. 2016. Bacteria and fungi can contribute to nutrients bioavailability and aggregate formation in degraded soils. *Microbiological Research* 183:26–41.

Ruiu, L. 2018. Microbial biopesticides in agroecosystems. *Agronomy* 8(11):235.

Sankaranarayanan, C. and Askary, T.H. 2017. Status of entomopathogenic nematodes in integrated pest management strategies in India. In *Biocontrol Agents: Entomopathogenic and Slug Parasitic Nematodes* (pp. 362–382). CAB International, Wallingford UK.

Shade, A., Jacques, M.-A., and Barret, M. 2017. Ecological patterns of seed microbiome diversity, transmission, and assembly. *Current Opinion in Microbiology* 37:15–22.

Shahzad, R., Khan, A.L., Bilal, S., Asaf, S. and Lee, I.J. 2018. What is there in seeds? Vertically transmitted endophytic resources for sustainable improvement in plant growth. *Frontiers in Plant Science* 9:24.

Shahzad, R., Waqas, M., Khan, A. L., Asaf, S., Khan, M. A., Kang, S.-M., et al. 2016. Seed-borne endophytic *Bacillus amyloliquefaciens* RWL-1 produces gibberellins and regulates endogenous phytohormones of *Oryza sativa*. *Plant Physiology and Biochemistry* 106:236–243.

Singh, M., Dotaniya, M.L., Mishra, A., Dotaniya, C.K., Regar, K.L. and Lata, M. 2016. Role of biofertilizers in conservation agriculture. In *Conservation Agriculture* (pp. 113–134). Springer, Singapore.

Singh, R. 2019. Microbial biotechnology: a promising implement for sustainable agriculture. In *New and Future Developments in Microbial Biotechnology and Bioengineering* (pp. 107–114). Elsevier, Amsterdam.

Singh, R., Kumar, A., Singh, M. and Pandey, K.D. 2019. Isolation and characterization of plant growth promoting rhizobacteria from *Momordica charantia* L. In: Singh, A.K., Kumar, A., Singh, P.K. (Eds.), *PGPR Amelioration in Sustainable Agriculture* (pp. 217–238). Woodhead Publishing, Cambridge, UK.

Sulistyanto, D., Ehlers, R.U. and Simamora, B.H. 2018. Bioinsecticide entomopathogenic nematodes as biological control agent for sustainable agriculture. *Pertanika Journal of Tropical Agricultural Science* 41(2):304.

Umesha, S., Singh, P.K. and Singh, R.P. 2018. Microbial biotechnology and sustainable agriculture. In *Biotechnology for Sustainable Agriculture* (pp. 185–205). Woodhead Publishing, Cambridge, UK.

Verma, D.K. and Srivastav, P.P. eds., 2017. Microorganisms in Sustainable Agriculture., In *Food and the Environment*. CRC Press, Boca Raton, FL.

Verma, P., Yadav, A.N., Khannam, K.S., Kumar, S., Saxena, A.K. and Suman, A. 2016. Molecular diversity and multifarious plant growth promoting attributes of *Bacilli* associated with wheat (*Triticum aestivum* L.) rhizosphere from six diverse agro-ecological zones of India. *Journal of Basic Microbiology* 56(1):44–58.

Vimal, S.R., Singh, J.S., Arora, N.K. and Singh, S. 2017. Soil-plant-microbe interactions in stressed agriculture management: a review. *Pedosphere* 27(2):177–192.

Xie, J., Shi, H., Du, Z., Wang, T., Liu, X. and Chen, S. 2016. Comparative genomic and functional analysis reveals conservation of plant growth promoting traits in *Paenibacillus polymyxa* and its closely related species. *Scientific Reports* 6(1):1–12.

Xu, Z., Jiang, Y., Jia, B. and Zhou, G. 2016. Elevated-CO_2 response of stomata and its dependence on environmental factors. *Frontiers in Plant Science* 7:657.

Yadav, A.N., Verma, P., Kumar, R., Kumar, V., Kumar, K. 2017. Current applications and future prospects of ecofriendly microbes. *EU Voice* 3(1):21–22.

Zhang, J., Wang, P., Fang, L., Zhang, Q.A., Yan, C. and Chen, J. 2017. Isolation and characterization of phosphate-solubilizing bacteria from mushroom residues and their effect on tomato plant growth promotion. *Pol Journal of Microbiology* 66(1):57–65.

Zhipeng, W.U., Weidong, W.U., Shenglu, Z.H.O.U. and Shaohua, W.U. 2016. Mycorrhizal inoculation affects Pb and Cd accumulation and translocation in Pakchoi (*Brassica chinensis* L.). *Pedosphere* 26(1): 13–26.

Zimmer, S., Messmer, M., Haase, T., Piepho, H.P., Mindermann, A., Schulz, H., Habekuß, A., Ordon, F., Wilbois, K.P. and Heb, J. 2016. Effects of soybean variety and *Bradyrhizobium* strains on yield, protein content and biological nitrogen fixation under cool growing conditions in Germany. *European Journal of Agronomy* 72:38–46.

Index

C

F

G

H

I

J

K

L

M

N

O

P

Q

R

U

V

W

X

Y

Z